Electron Paramagnetic Resonance in Biochemistry and Medicine

TOPICS IN APPLIED CHEMISTRY

Series Editor: Alan R. Katritzky, FPS
University of Florida
Gainesville, Florida

Gebran J. Sabongi
3M Company
St. Paul, Minnesota

Current volumes in the series:

ANALYSIS AND DEFORMATION OF POLYMERIC MATERIALS
Paints, Plastics, Adhesives, and Inks
Jan W. Gooch

ELECTRON PARAMAGNETIC RESONANCE IN BIOCHEMISTRY AND MEDICINE
Rafik Galimzyanovich Saifutdinov, Lyudmila Ivanovna Larina, Tamara Il'inichna Vakul'skaya, and Mikhail Grigor'evich Voronkov

FLUOROPOLYMERS 1: Synthesis
FLUOROPOLYMERS 2: Properties
Edited by Gareth Hougham, Patrick E. Cassidy, Ken Johns, and Theodore Davidson

FROM CHEMICAL TOPOLOGY TO THREE-DIMENSIONAL GEOMETRY
Edited by Alexandru T. Balaban

ORGANIC PHOTOCHROMIC AND THERMOCHROMIC COMPOUNDS
Volume 1: Main Photochromic Families
Volume 2: Physicochemical Studies, Biological Applications, and Thermochromism
Edited by John C. Crano and Robert J. Guglielmetti

ORGANOFLUORINE CHEMISTRY
Principles and Commercial Applications
Edited by R. E. Banks, B. E. Smart, and J. C. Tatlow

PHOSPHATE FIBERS
Edward J. Griffith

RESORCINOL
Its Uses and Derivatives
Hans Dressler

Electron Paramagnetic Resonance in Biochemistry and Medicine

Rafik Galimzyanovich Saifutdinov
Irkutsk Medical University, Institute of Surgery (RAMS)
Irkutsk, Russia

Lyudmila Ivanovna Larina
Institute of Chemistry, Russian Academy of Science (RAS)
Irkutsk, Russia

Tamara Il'inichna Vakul'skaya
Mikhail Grigor'evich Voronkov
Institute of Chemistry, Russian Academy of Science (RAS)
Irkutsk, Russia

Springer Science+Business Media, LLC

Library of Congress Cataloging-in-Publication Data

Electron paramagnetic resonance in biochemistry and medicine/edited by Rafik G. Saifutdinov ... [et al.].
p. cm. — (Topics in applied biochemistry)
Includes bibliographical references and index.
ISBN 978-1-4757-8689-7 ISBN 978-0-306-46925-1 (eBook)
DOI 10.1007/978-0-306-46925-1
1. Electron paramagnetic resonance spectroscopy. 2. Electron paramagnetic resonance spectroscopy—Diagnostic use. I. Saifutdinov, Rafik G. II. Series.

QP519.9.E433 E436 2000
616.07'56—dc21

00-052738

ISBN 978-1-4757-8689-7

Originally published by Kluwer Academic/Plenum Publishers, New York in 2001
Softcover reprint of the hardcover 1st edition 2001

http://www.wkap.nl/

10 9 8 7 6 5 4 3 2 1

A C.I.P. record of this book is available from the Library of Congress.

Preface

Scientific and technical progress contributes a great deal to the development of medical science. During the last years the essence of many pathogenic mechanisms of various diseases has been discovered and new effect methods of their diagnosis and treatment have been developed.

Pathologic states are caused by not only disorders of lipid, hydrocarbon, protein and mineral metabolism. Alterations in redox processes are also of importance. Paramagnetic centers (PMC) presented in many biological tissues and fluids can serve as indicators of these changes owing to the presence of free radicals and cations of transition microelements. These centers presented in different organic and inorganic biomolecules are cofactors of some enzymes, which provides a valuable, and sometimes unique, information concerning qualitative and quantitative changes in biological objects.

This book demonstrates the unlimited possibilities of the application of Electron Paramagnetic Resonance (Electron Spin Resonance) spectroscopy in biochemistry and medicine. We use the term Electron Paramagnetic Resonance (EPR) spectroscopy in the book.

EPR spectroscopy is an objective and accurate method of recording paramagnetic centers widely accepted in chemistry, biochemistry, biology and medicine.

The EPR method was discovered in 1944 by Eugene Konstantinovich Zavoisky in Kazan. In the middle fifties, nearly simultaneously in the United States and the Soviet Union, EPR spectroscopy was first used for the investigation of tissues and fluids in human and animals. In that time, owing to low sensitivity of EPR spectrometers, the method of tissue lyophilic drying was used. The development of low-temperature registration of biological tissues in the early seventies extended the scope of application of EPR spectroscopy in biochemistry and medicine. By the present time a high sensitivity and precision of this method has been achieved, which allows investigation of even native water-containing tissues.

At first, the use of EPR spectroscopy in biochemistry and medicine was aimed at studying malignant tumors. The synthesis and introduction of spin labels, probes and traps made it possible to discover many mechanisms of pathologic states. During the last years the possibility of using EPR spectroscopy *in vivo* has been intensively examined.

The literature data devoted to the investigation of paramagnetic centers in human substrates and to the possible practical use of the results, are scarce, dissipated and non-summarized. This monograph is an attempt to fill this gap and to give an outline of pathways and prospects for the use of EPR spectroscopy in medicine and biochemistry.

Practically all publications dealing with the use of EPR spectroscopy in medicine and biochemistry and the authors' own results on EPR investigation of biological fluids from the organism of normal individuals and patients with different pathologies have been summarized and analyzed in the present monograph.

The EPR spectra of blood and blood components, saliva, nasal secretion, synovial fluid, gastric and duodenal contents, feces, teeth, bones and some other tissues and biological fluids have been considered in this book. The quantitative and qualitative composition of paramagnetic centers in normal individuals and patients suffering from different diseases is presented.

Special attention is given to the EPR examination of bio-molecules (enzymes, polypeptides, vitamins, lipids, hydrocarbons, etc.) playing an essential role in the vital activities of man. Also the possible use of this method for the investigation of drugs, toxicants and metabolites as well as food products is discussed. There is a section deals with the use of EPR spectroscopy in dosimetry and diagnosis of irradiation injury in the book.

Fundamental problems related to the study of mechanisms of the development of oncologic, cardia-vascular and other diseases including AIDS have been elucidated. Much attention is given to the use of the EPR method diagnostics, investigation of metabolism, free-radical reactions in the development of pathologies, as well as in EPR-dosimetry and archaeology. EPR technique extends our concepts concerning the role of transition microelements in human activities and in fine mechanisms of pathogenic processes occurring in various diseases. This method also opens up fresh approaches to new therapeutic and prophylactic measures.

Special attention in the monograph is given to the application of EPR spectroscopy for a better understanding of pathogenesis of diseases, diagnosing and choosing the proper methods of treatment.

The monograph contains a large volume of EPR-derived information on biological processes occurring in the human body, such as generation of oxygen radicals ("respiratory oxygen burst"), nitric oxide metabolism, *etc*. It also considers the possibility of using spin labels and spin probes for the diagnosis of diseases and scavenging (trapping) of free radicals of various origin. The potential uses of the EPR method in radiation medicine, stomatology, pharmacology, archaeology, and food industry are discussed.

These above studies are updated with the literature data on EPR investigations in the field of biochemistry and medicine, published by 1999.

The monograph is intended for physicians, surgeons, otorhino-laringologists, and doctors of other specialities interested in recent problems in medicine. It will undoubtedly be of help for biologists, biochemists, biophysicists and chemists engaged in ESR spectroscopy. The book can be useful for students at the senior undergraduate or graduate level either as a textbook in a course or as a self-study guide. To everyone who is interested in modern problems in medicine, biochemistry, biophysics, ecology, and cares about his health and food.

All additions, recommendations and comments of the readers will be appreciated.

Irkutsk, Siberian Branch,

Russian Academy of Medical Sciences, Russia *R.G. Saifutdinov*

Irkutsk, Siberian Branch,

Russian Academy of Sciences, Russia *L.I. Larina*

T.I. Vakul'skaya

M.G. Voronkov

Acknowledgments

We have received invaluable support from many persons in preparing this book. We acknowledge helpful comments received from Dr. Gennadii Dolgushin and Elena Petrova and also for her translation of book on English.

Our best wishes and thanks for substantial contribution in preparing the manuscript and tolerance contributed over many years of Dr. Larina's daughter Polina Larina, student of Medical University and Dr. Saifutdinov's wife Tatiana Safutdinova, physician.

Contents

Chapter 1

ELECTRON PARAMAGNETIC RESONANCE (EPR): THEORY AND METHOD

1.1 FUNDAMENTALS OF THE EPR THEORY

Paramagnetic centers (PMC) are the sites in molecules and atoms having free unpaired electrons (UE). The method of electron paramagnetic resonance (EPR) is based on the possibility of recording UEs in atomic, crystalline ionic and molecular structures; in compounds of transition metals (copper, vanadium, chromium, molybdenum, manganese, iron, nickel, *etc.*); upon cleavage of covalent chemical bonds. In the latter case UEs appear either at unoccupied atoms or at molecular and macromolecular fragments, *i.e.* at free radicals (FR) of higher activity. Many of the above electronic states have been revealed in biological objects.

The theory of EPR and its specific use in physics, chemistry, biology and medicine have been considered in many monographs and reviews [1-19]. During the last years the EPR technique has been constantly improved [20-22] and EPR *in vivo* investigations have been extensively developed [23-27]. EPR spectroscopy enables UEs to be detected and characterized. This is possible due to the fact that odd electrons have a magnetic moment caused by spin (S) equal to 1/2 (intrinsic moment of momentum or angular moment).

In order to characterize several UEs the general coordinate axis is needed. However, there is no axis of this kind in the absence of external magnetic field. UE spins are oriented in a random way. If the external magnetic field is applied to the object to be examined a general coordinate axis would arouse around which all the electrons begin precessing in such a way that M_s, i.e., their spin projection on the axis, would be equal to either +1/2 or to -1/2 (Figure 1-1).

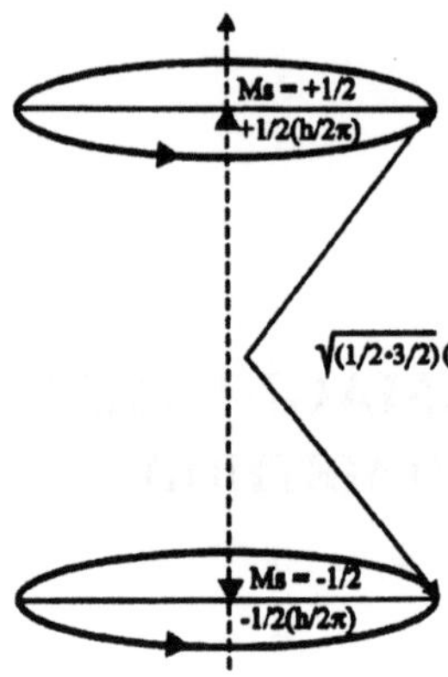

Figure 1-1. The projection of full magnetic moment on the quantum axis (the direction of the field).

In the external magnetic field UEs are subdivided into two groups: in one group the spins are oriented parallel and in the other antiparallel to the field direction. In the former the electron energy decreases and in the latter increases; this means that splitting of energy level occurs (Figure 1-2).

It is this phenomenon the EPR method is based on. When EPR spectroscopy is used the sample to be examined is placed into a strong homogeneous magnetic field and simultaneously exposed to electromagnetic radiation of a frequency with $h\nu$ (energy quantum) equal to the difference between the electronic levels of the two groups (the so called resonance frequency):

$$h\nu = g\beta H$$

Where

β is the Bohr magneton characterizing the ratio between the angular and magnetic moment,

g denotes a constant (g factor),

h stands for the Plank constant,

ν is the electromagnetic radiation frequency and

H is the magnetic field strength.

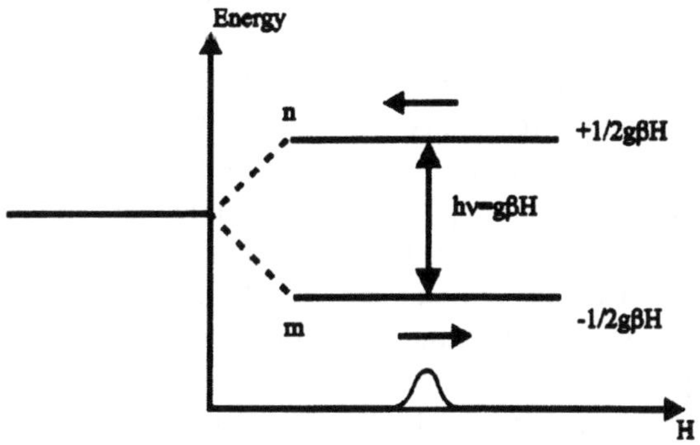

Figure 1-2. The main condition of electron paramagnetic resonance origin (under of the influence of the external magnetic field the energy levels of UEs are splitted).

Owing to radiation energy the odd electrons locating on the lower energy level pass onto the higher level with simultaneous change of the spin direction and energy absorption. The upper level electrons pass onto the lower level radiating a quantum of electromagnetic energy. This process is called "induced emission". The absorption and induced emission coefficients are equal. With the same number of electrons on the two energy levels there would be no absorption. In the reality, however, there is. This can be explained by the fact that the number of UEs on the lower energy level is always smaller than on the upper level. This difference is responsible for the observed EPR signal intensity. The latter presents one of the most important parameters characterizing the EPR spectrometer sensitivity.

Main parameters of EPR spectra are the following:

1.1.1 Integral Intensity

Integral intensity which is the difference in the energy level population and, consequently, the absorption value in the microwave region proportional to the total number of electrons in the sample. The concentration of electrons can be judged upon by the absorption line integral intensity.

1.1.2 Line Width

Two lines showing the same integral intensity may be of quite different form. The line width provides a supplementary parameter characterizing the absorption conditions and is determined by the scatter of energy levels occupied with UEs. There are two major types of interactions giving rise to the resonance line broadening: spin-lattice and spin-spin interactions.

1. Spin-lattice interaction exists between the UE spin and its surrounding (crystal lattice or molecular system). In the present case the spin-absorbed energy is restored to the lattice or to the whole molecular system, which provides for continuous resonance absorption. Without this mechanism the electrons absorbing microwave energy would be passing from the lower to the upper level until their population would equalize and the absorption would cease. When the spin-lattice interaction is great the electrons can occupy the upper level for only a very short time. The strong interaction of this type causes a rapid spin-lattice relaxation thus leading to the absorption line broadening.

2. Spin-spin interaction involves energy exchange in spins instead of giving energy to the crystal lattice or to the molecular system. Unlike spin-lattice interaction this does not favour thermal equilibration, arousing instead the resonance line broadening due to both direct spin interaction and a shorter life-span of spin states. This can be illustrated by normal dipole-dipole interaction with two bar magnets serving as an analog. Each UE interacts not only with the external magnetic field, but also with magnetic fields created by other UEs. At a high electron concentration in the sample the inter-electron space is small. This supplementary field can reach a fairly high value. In concentrated solutions of transition metal salts this value attains some scores of oersteds. In diluted solutions (including biological and biochemical systems) the UE content is much lower, so in this case odd electrons influence each other to a considerably less extent.

1.1.3 g-Factor

g-Factor characterizes the resonance point position. From the equation of resonance $h\nu = g\beta H$ it follows that with a constant microwave frequency the strength of externally applied magnetic field (H) and the g value are the only variable parameters. Consequently, the g-factor is determined by the field value giving rise to resonance. The g-factor of free electron (having no orbital moment) is 2.0023. In many FRs with UEs occupying a strongly delocalized molecular orbital the g-factors are very close to the above value which indicates a weak coupling of spin and orbital moments. The odd electron bonded to an atom of the crystal lattice is influenced by sufficiently strong internal electric field caused by the molecule structure. The g-factor value provides a detailed information on the character of intramolecular bonds.

1.1.4 Hyperfine Structure and Its Origin

Hyperfine structure (HFS) arises during the interaction between UE with the nuclear magnetic moment of the atom involved in the UE orbital. Nuclei of many atoms participating in biochemical processes (hydrogen, nitrogen, oxygen, sulfur, *etc.*) possess spin and magnetic moments. Owing to this fact the Zeeman levels are even more split (Figure 1-3).

According to the selection rule ($\Delta M_I = 0$), only the electron transitions with invariable spin nuclei orientation (I) are possible. With $I = \pm 1/2$ there are only two transitions of this kind. Correspondingly, in the EPR spectrum the absorption line is split into two components. When UEs are moving along molecular orbitals involving several atoms a hyperfine splitting caused by interaction of the electrons with different atoms occurs. In this case the spectral structure is very complex. In a general case of interaction with the nucleus spin $I = 1/2$ the number of HFS components is defined by the following formula: $2nI + 1$.

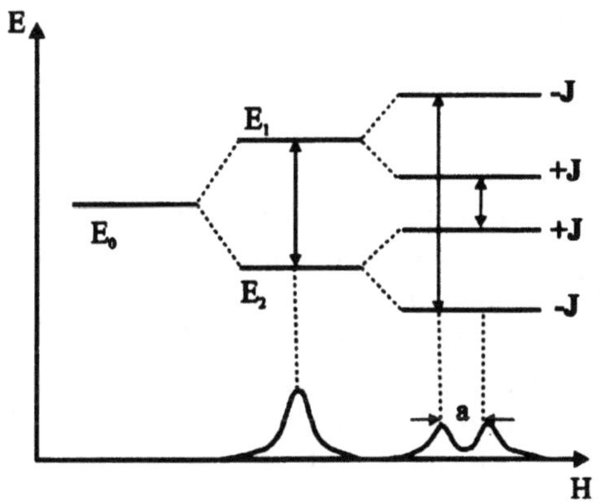

Figure 1-3. Hyperfine structure arising during the interaction between UE and nucleus with spin 1/2.

1.1.5 Saturation Effect

As mentioned above, the EPR signal intensity is determined by the difference of the Boltzmann sublevel population. This difference exists due to relaxation phenomenon. With increasing the applied SHF (Super High Frequence) power a situation when relaxation would not be able to keep pace with equilibration is quite possible. As a result the population of the upper and the lower levels would reach equilibrium and the EPR signal intensity would decrease correspondingly.

1.2 PROCEDURES FOR STUDYING PARAMAGNETIC CENTERS

In studying paramagnetic centers various EPR spectrometers are used - RE-1306, RE-1307 (Russia), "Bruker" (Germany), "Varian" (USA), "Jeol" (Japan), *etc.* In particular, the present authors made use of 3 cm range radio-spectrometers.

1.2.1 Preparation of Samples

The high potential of EPR spectroscopy in medicine became more evident with perfection of radiospectrometers, theoretical foundations of the method and ways of fixation of biological objects. In the first stages of investigation inadequate sensitivity of the instruments did not allow the EPR method to be used for studying aqueous biological systems. This made it necessary to use liophylic drying of tissues [4, 28, 29]. It turned out that the main contribution to the formation of FR signals of dried tissues is made by ascorbic acid [30-32]. Besides, a certain role is played by ubiquinone and other polyphenoles and quinones [33-35]. The PMC of lyophilically dried tissues differ from those of aqueous tissues [36, 37]. Therefore it is desirable to examine tissues containing natural amounts of water.

The present authors used deep freezing at 77°K as a method of biological tissue fixation [38]. The EPR signal intensity depends on the PMC concentration, recording conditions and the sample size. So, only the samples of standard volume should be employed. The samples of this kind are made in teflon mold slightly modified molds [39] (Figure 1-4, sizes are gave in mm).

The mold consists of the body and the plug. Another procedure of preparing a sample to be analyzed on their spectrometer is proposed recently [40]. However, the drawback of this method is that cutter basis is made of silicone rubber containing vulcanizers, potential metal chelating agents. The reaction between them can greatly distort the EPR spectrum.

Blood for analysis was taken from the ulnar vein of patients at the same time in the morning (on empty stomach) and placed into an ampoule with heparin. Then the blood was centrifuged at 1500 rpm for 10 min. To precipitate the leukocytes the plasma was separated from the erythrocytes and centrifuged at 3000 rpm for 5 min. The erythrocytes were washed thrice with a cooled physiological salt solution (1 : 10 ratio).

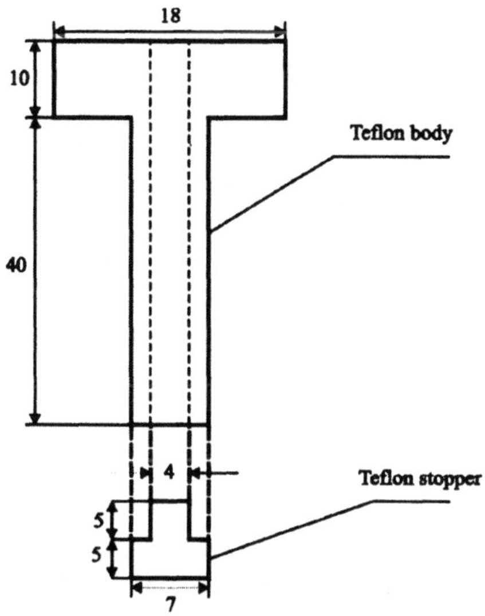

Figure 1-4. The teflon mold for the preparation of the samples at 77°K.

The synovial fluid was obtained by puncturing the joints for a diagnostic or therapeutic purpose.

The saliva was taken in the morning on the empty stomach for 20 minutes after a careful toilet of the oral cavity (without stimulation).

The nasal secretion was isolated just prior to sample preparation. For this purpose the use was made of a device the operation of which is based on mechanical stimulation of the nasal mucous membrane secretion and simultaneous NS aspiration from the surface of the superior and inferior turbinated bones and the nasal membrane into a sterile small flask. This precluded possible influence of blood, saliva and lachrymal fluid admixture to NS on the EPR spectroscopy results [15].

The device designed by the authors (Figure 1-5) consists of a metallic canula 125 mm long, ∅ 0.7 mm, flexed in the distal and proximal sections at an angle of 120° with smooth end edges, a 15 ml flask with a ground stopper, an electro-sucker and connective tubes 3 mm in diameter.

The tube connecting the canula with the receiving flask is 150 mm long; the total length of the flask with the suction hose is 1170 mm. The above mentioned ratio are most appropriate for the NS aspiration from the nasal mucous membrane (NMM) under 0.2-0.4 kg/cm^2 vacuum produced by the sucker.

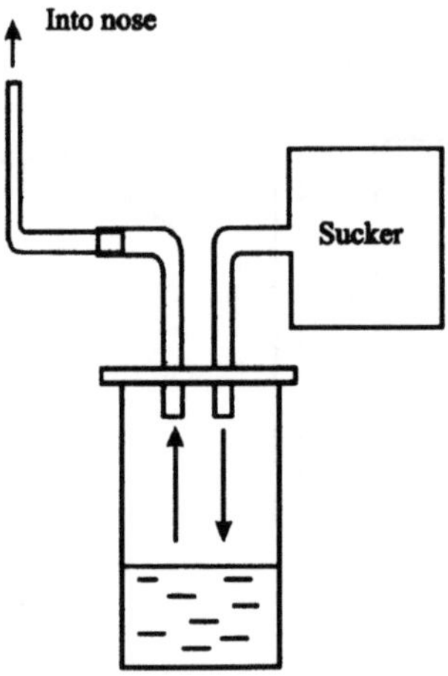

Figure 1-5. The device for isolating the nasal secretion that investigate it by EPR.

This device allows an atraumatic sighting NS aspiration from strictly defined NMM areas (visually controlled). Mechanical irritation of NMM in response to a careful touch of the canula distant end makes it possible to take necessary NS amounts even from patients with minimum secretory activity. The stomach contents were taken in the morning on an empty stomach by continuous vacuum aspiration.

Duodenal intubation was performed by a routine method in three batches (A, B and C).

The feces collected into small glass bottles were weighed in the mold before and after placement of the sample. Calculation was based on 100 mg of the substrate.

The samples prepared were frozen in liquid nitrogen, extruded from the mold, placed into quartz Dewar vessels and subjected to thermostatic control.

1.2.2 Thermostatic Control of Frozen Samples

The use of frozen samples provides much advantage. Freezing does not lead to any significant disruption of metabolic structures, and the EPR spectra of frozen samples are identical with those run at room temperature [3, 36]. When temperature drops the relaxation time of paramagnetic metal complexes increases whereas the line width narrows thus enhancing the instrument sensitivity [9, 16-19].

However this method has certain shortcomings. At 77°K accumulation of oxygen (b.p. 92°K) takes place in the sample. Since oxygen is of paramagnetic nature a broad absorption signal displacing the zero line of the instrument appears in the EPR spectrum. Besides, constant boiling of liquid nitrogen in the Dewar vessel brings about the appearance of additional noises in the spectrum. The above drawbacks can be remedied by means of thermostatic control. For this purpose the liquid nitrogen boiling level was removed outside the resonator (cavity) (Figure 1-6) in the following way. The Dewar vessel containing the sample was cooled with liquid nitrogen which was poured out 30-60 sec later. The sample was covered with a layer of cotton 3-5 mm thick and again poured over with liquid nitrogen. As a result boiling occurred above the cotton tampon, *i.e.*, outside the resonator operating zone. Under these conditions the temperature in the lower part of the sample was only 6°C higher than that of the liquid [39]. After 1-2 min no temperature fluctuations are observed.

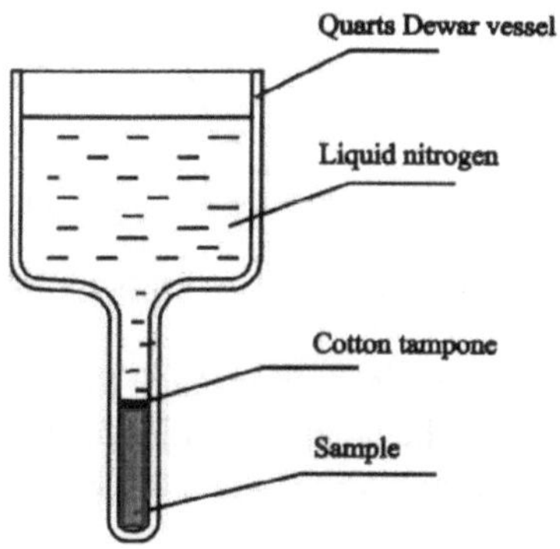

Figure 1-6. Scheme of the sample thermostatic control in the Dewar vessel at 77°K.

Thus, thermostatic control brings the liquid nitrogen boiling level outside the resonator zone and prevents accumulation of nitrogen in that part of the vessel where the sample is kept. Apart from this, the cotton tampon socked with liquid nitrogen does not allow defrosting of the sample.

1.2.3 EPR Spectra Recording

The EPR signal is recorded as first derivative. The line half-width was measured in mT between the points of maximum slope (Figure 1-7).

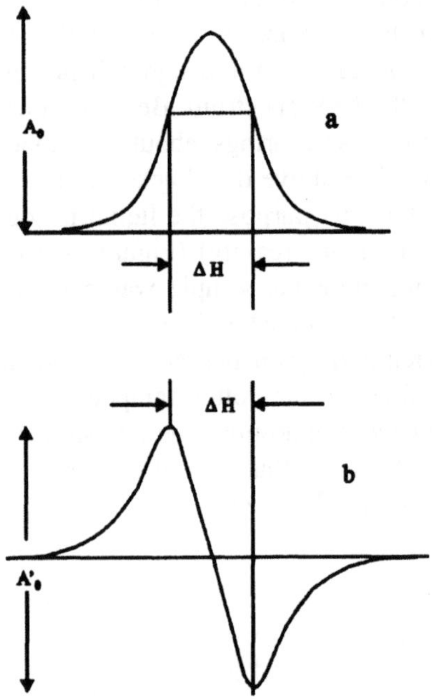

Figure 1-7. The curve of EPR-absorption (a) and its first derivative (b).

The instrument sensitivity is proportional to the high frequency (HF) modulation amplitude when the latter is smaller than the absorption line half-width, and shows the maximum value when the HF modulation amplitude is equal to the line half-width (Figure 1-8).

Measurements of the UE concentrations in free radicals in the plasma were performed with HF modulation amplitude of 0.3 mT. Other signals were recorded at 0.5 mT.

As known, saturation effect is observed in many biological objects [2, 7, 12, 18] and can lead to significant errors in intensity determinations. We examined this effect in the 0.05-50 mW range [41]. With a SHF power of above 0.6-0.7 mW the free radical signal amplitude drops (Figures 1-9, 1-10).

To get EPR spectra with the most intense absorption lines in all measurements the signal with g = 2.00 was recorded at 0.5 mW, *i.e.*, where the signal is unsaturated. In the above measurement Cu^{2+} was used as an unsaturating substance.

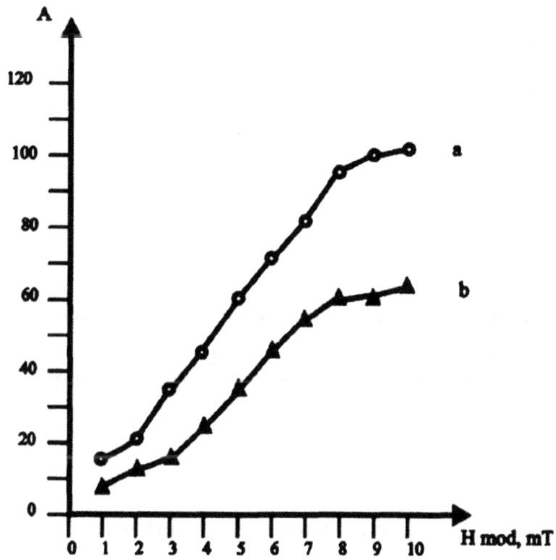

Figure 1-8. Changing of the amplitude of EPR signal free radicals in erythrocytes (g = 2.0036) (a) and in plasma (g = 2.0025) (b).

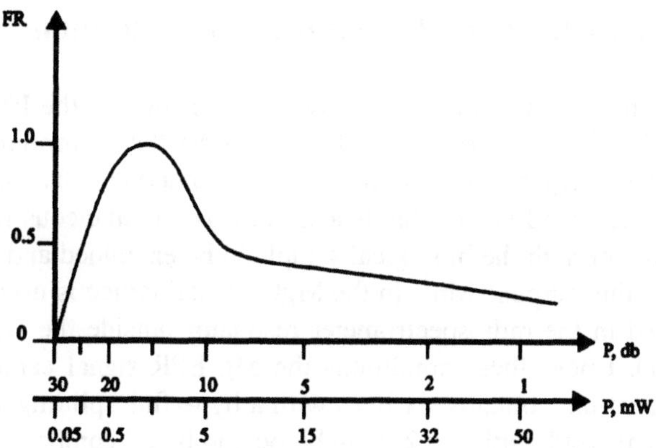

Figure 1-9. The influence of the microwave power value on the EPR signals intensity of free radicals of the blood plasma.

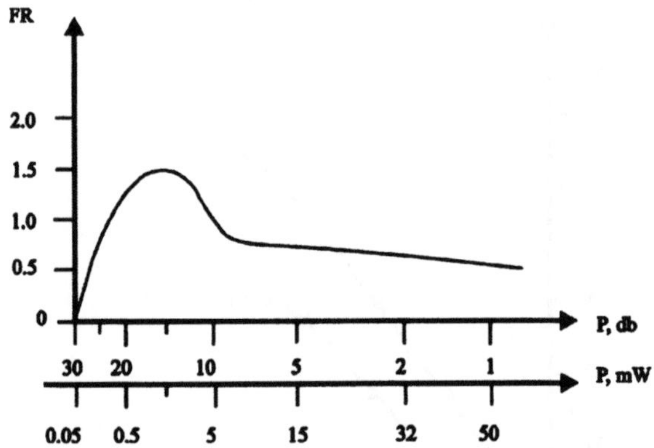

Figure 1-10. The influence of the microwave power value on the EPR signals intensity of erythrocytes free radicals.

The control and test samples were prepared under identical conditions; the same is true for EPR spectra recording. Each sample was checked in the magnetic field range from 40 to 4000 mT with magnetic field scanning of 320 mT.

1.2.4 Standardization of Measurement Conditions

The EPR method allows quantitative determination of the PMC concentration. The latter is proportional to the area under the absorption curve. On the EPR spectra registration it is necessary to undertake standardization of the measurement conditions. This is achieved by simultaneous recording of the EPR signal of both the biological sample to be examined and the internal standard. For this purpose Mn^{2+} in the MgO crystal lattice is used. The standard is placed in the radiospectrometer resonator outside the Dewar vessel (Figure 1-11). Under these conditions the Mg EPR signal cannot be saturated. The spectrum contains six lines with a hyperfine splitting value of 8.0 mT. The signal width with g = 2.00 in blood and blood components was less than 2.0 mT. Therefore free radicals were fixed between the third and the fourth SFS components of manganese (Figure 1-12).

The line half-width (ΔH, mT) was evaluated between the maximum slope points.

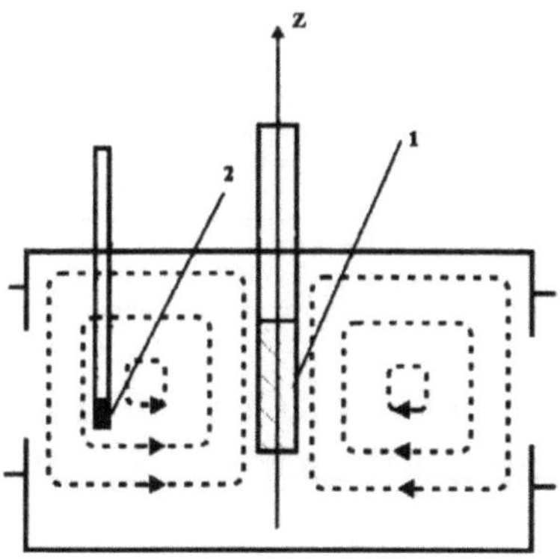

Figure 1-11. The sample position in the radiospectrometer resonator: 1 - sample, 2 - internal standard (dotted lines are force magnetic lines of the SFS field).

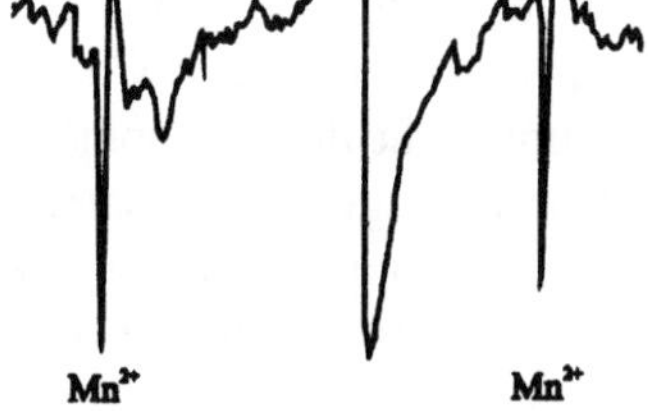

Figure 1-12. EPR signal of the free radicals recorded between 3-th and 4-th HFS components of the internal standard Mn^{2+} [distance between SFS components Mn^{2+} is (8.676 ± 0.005) mT].

To the blood to be analyzed heparin was added as an anticoagulant. It has been established that heparin does not affect the PMC concentration in erythrocytes and plasma (table 1-1).

Table 1-1. The effect of heparin on the FR level in erythrocytes and plasma (relative units)

Material to be analyzed	FR level without heparin, M ± m	FR level with heparin, M ± m
Erythrocytes	58.4 ± 4.6	56.7 ± 5.0
Plasma	31.2 ± 2.8	4 ± 4.0

1.2.5 Determination of PMC Concentration in Biological Fluids Isolated from the Human Organism

The number of PMC in biological fluids was determined using a method of EPR signal area double integration by calibration curves. The curves were drawn basing on analysis of different concentrations of the corresponding standard solutions [18, 38, 42].

The content of methemoglobin (MetHb) in the erythrocytes was evaluated as follows: in a 1/15 M phosphate buffer, pH = 7.4 hemoglobin solutions (REANAL) with concentrations from 0.1 to 6.0 g/l were prepared. Their optical density values were recorded in a 1.0 cm cuvette at wavelengths of 540, 560, 575, 630 nm, *i.e.*, in visible spectral regions with the largest differences in the light absorption of oxyhemoglobin (HbO_2) and MetHb. Then some tiny $K_3Fe(CN)_6$ crystals were added to the same cuvette in order to transform all the blood pigment to MetHb. The reaction monitoring was performed at the same wave-lengths. Samples from the obtained MetHb solutions were prepared in molds and analyzed on an EPR spectrometer, then calibration curves were constructed.

The Cu^{2+} concentration in superoxide dismutase (SOD) of erythrocytes and ceruloplasmin (CP) was determined from $CuSO_4 \cdot 5H_2O$ in human serum albumin (HSA). In the serum of healthy persons $4.65 \cdot 10^{-3}$ M of copper is biochemically determined on average. At the same time, according to EPR data the content of paramagnetic copper (Cu^{2+}) is lower. Therefore, solutions with Cu^{2+} concentrations from $1.0 \cdot 10^{-5}$ M to $1.0 \cdot 10^{-3}$ M were prepared.

In accessible literature there are no data on the amount of copper (Cu^{2+}) bonded with HSA in the human organism. In this connection we used albumin concentrations from 5 to 60 g/l in solutions to which one and the same amount of Cu^{2+} ($1.0 \cdot 10^{-3}$ M) was added. Analysis of the EPR spectra shows that the curve reaches the plateau at 20 g/l (Figure 1-13). This indicates the albumin concentration of 30 g/l to be enough for complete Cu^{2+} bonding which means that 1 g of HSA can bind approximately 3 mg of Cu^{2+}.

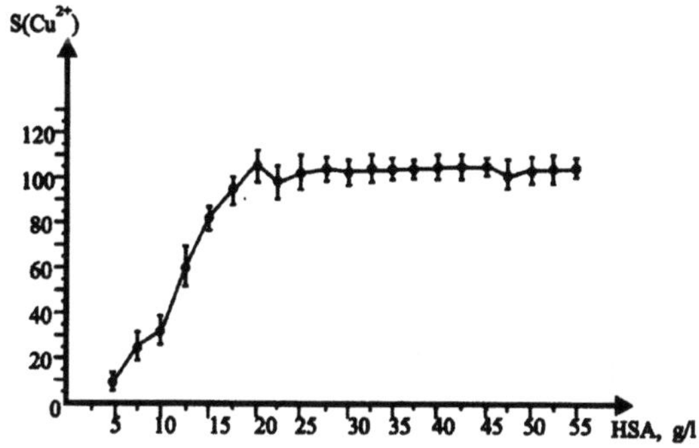

Figure 1-13. The picture of the bonding Cu^{2+} ions by HSA (S is the square under the curve of EPR-absorption).

The FR level was determined from the stable radical of 2,2,6,6-tetramethyl-4-oxypiperidin-1-oxyl (TEMPO). The concentration was from $1.0 \cdot 10^{-6}$ to $1.0 \cdot 10^{-4}$ M.

The Fe^{3+} content was established from the calibration curve drawn during the analysis when divalent iron was added to the plasma. As known, FRs oxidize Fe^{2+} to Fe^{3+} [43]. 0.01 ml of $FeSO_4$ solution in concentrations from 5 to 50 μmol/l was added to 0.4 ml of plasma. The samples were frozen in liquid nitrogen. A calibration curve was drawn up on the basis of their EPR spectra (Figure 1-14). Plasma into which certain quantify of twice distilled water was introduced served as control.

Mn^{2+} was determined with $MnCl_2$ (from $1.0 \cdot 10^{-3}$ M to $7.0 \cdot 10^{-3}$ M) in HSA. We failed to find any data concerning the Mn^{2+} amount which can bind HSA. Therefore albumin solutions in concentrations from 1.5 to 30 g/l were prepared and one and the same Mn^{2+} amount ($1.0 \cdot 10^{-3}$ M) was added. According to the EPR data the curve reaches the plateau at 7.5 g/l (Figure 1-15). Thus, the albumin concentration, sufficient for complete Mn^{2+} bonding, is 30 g/l, which means that 1 g of HSA can bind approximately 7 mg of Mn^{2+} [44].

The nitric oxide (NO) was determinate on method [38].

In the duodenal contents normally obtained by means of rubber probe it is not possible to determine exactly and quantitatively the Cu^{2+} content by the calibration curve drawn up basing on analysis of solutions of various $CuSO_4$ concentrations in HSA (as described in determining the $CuSO_4$ content in plasma and erythrocytes). This is due to the fact that being in contact with the probe the contents react with the sulfur-containing rubber vulcanizer (of Altax type). Therefore the EPR signal arising from Cu^{2+} changes its shape and its traditional curve disappears.

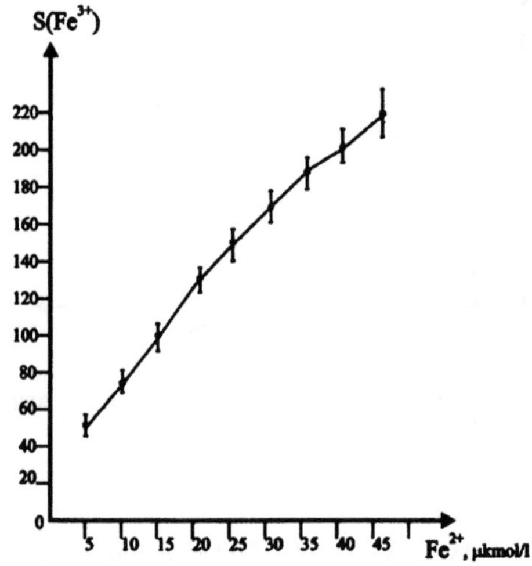

Figure 1-14. The calibration curve for the determination Fe^{3+} ions by EPR (g = 4.3) in the biological substrates (S is the square under the curve of EPR-absorption).

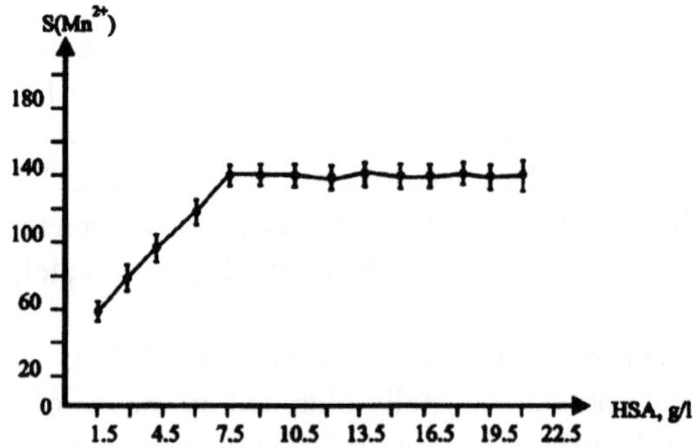

Figure 1-15. The picture of the bonding Mn^{2+} ions by HSA (S is the square under the curve of EPR-absorption).

To check this suggestion a sodium diethyldithiocarbamate (DEDTC) solution which bonds all the paramagnetic copper was added to the duodenal contents [45]. In fact, the EPR spectrum of Cu^{2+} chemically bonded by vulcanizer of Altax type proved to be identical in its form to that of sodium DEDTC of copper.

Owing to this fact the Cu^{2+} concentration was determined using the calibration curve plotted in the analysis of various $CuSO_4$ concentrations (from $1.0 \cdot 10^{-6}$ to $1.0 \cdot 10^{-3}$ M) in a DEDTC solution.

In order to examine the PMC concentration dynamics in storage the samples of erythrocytes, plasma, saliva, stomach and duodenal contents, feces in liquid nitrogen were analyzed daily for 10 days. Under these conditions the PMC level in the biological liquids studied does not change.

REFERENCES

1. L. A. Blumenfel'd, W. W. Wojewodski and A. G. Semenov, Die Anwendung der Paramagnetischen Electronen Resonanz in der Chemie. Akademische Verlagsgesellschaft, Leipzig (1966).
2. D. J. E. Ingram, Biological and Biochemical Applications of Electron Spin Resonance. Adam Hilder LTD, London (1969).
3. J. R. Mallard and M. Kent, ESR in biological tissues, Phys. Med. Biol. **14**:373 (1969).
4. M. M. Petyaev, Biophysical Methods in Diagnostic of Cancer Tumors. Meditsina, Moskva (1972) (Russ.).
5. Yu. P. Kozlov, Free Radicals and its Role in Normal and Pathological Processes. MGU, Moskva (1973) (Russ.).
6. S. A. Al'tshuler, B. M. Kozyrev, Electron Paramagnetic Resonance in Compounds of Transition Elements. 2nd ed., Halsted-Wiley, New York (1974).
7. D. Borg, Application of Electron Paramagnetic Resonance in Biology, in: Free Radicals in Biology, William A. Pryor, ed., Acad. Press, New York-San Francisco-London (1976).
8. M. C. R. Symons, Chemical and Biochemical Aspects of Electron Spin Resonance Spectroscopy, Wiley, London (1978).
9. Ya. I. Azhipa, Application of Electron Paramagnetic Resonance in Medicine. Nauka, Moskva (1979) (Russ.).
10. L. A. Blumenfel'd, Problems of Biological Physics. Springer-Verlag, Berlin (1981).
11. N. J. Blackburn, Metalloproteins, in: Electron Spin Resonance, Vol. **8**, London (1983).
12. Ya. I. Azhipa, Medical-Biological Aspects of Electron Paramagnetic Resonance Applications. Nauka, Moskva (1983) (Russ.).
13. M. A. Foster, Magnetic Resonance in Medicine and Biology, Pergamon, U.K., Oxford (1984).
14 K. M. Salikhov, Yu. N. Molin, R. Z Sagdeev and A. L. Buchachenko, Spin Polarization and Magnetic Effects in Radical Reactions, Elsevier, Amsterdam (1984).
15. A. G. Shanturov, R. G. Saifutdinov and E. V. Nosulya, EPR Spectroscopy in Clinical Rhinology, Vostochno-Sibirskoe Knizhnoe Izdatelstvo, Irkutsk (1992) (Russ.).
16. K. R. Sedov and R. G. Saifutdinov, Electron Paramagnetic Resonance in Internal Diseases Clinics, Irkutskii Dom Pechati, Irkutsk (1993) (Russ.).

17. L. A. Blumenfel'd and A. N. Tikhonov, Biophysical Thermodynamics of Intracellular Processes. Molecular Machinery of the Living Cell, Spring-Verlag, Berlin (1994).
18. J. A. Weil, J. R. Bolton and J. E. Wertz, Electron Paramagnetic Resonance. Elementary Theory and Practical Applications. John Wiley & Sons, INC, New-York (1994).
19. A. L. Buchachenko and E. L. Frankevich, Chemical Generation and Reception of Radio- and Microwaves, VCH, Weinheim (1997).
20 C. P. Keijzers, E. J. Reijerse and J. Schmidt, eds., Pulsed EPR – A New Field of Applications, North-Holland, Amsterdam (1989).
21 L. Kevan and M. K. Bowman, eds., Modern Pulsed and Continuous-Wave Electron Spin Resonance, Wiley, New York (1990).
22. A. V. Koptioug, E. J. Reijerse and A. K. Klaassen, New transmission-line resonator for pulsed EPR, J. Magn. Reson. **125**:369 (1997).
23. G. R. Eaton, S. S. Eaton and K. Ohno, eds., EPR Imaging and In-vitro EPR, CRC Press, FL, Boca Raton (1991).
24. H. M. Swartz and T. Walczak, Developing in vivo EPR oximetry for clinical use, Adv. Exp. Med. Biol. **454**:243 (1998).
25. H. M. Swartz and R. B. Clarkson, The measurement of oxygen in vivo using EPR techniques, Phys. Med. Biol. **43**:1957 (1998).
26. K. J. Liu, M. Miyake, P. E. James and H. M. Swartz, Separation and enrichment of the active component of carbon based paramagnetic materials for use in EPR oximetry, J. Magn. Reson. **133**:291 (1998).
27. I. Nicholson, M. A. Foster, F. J. L. Robb, J. M. S. Hutchison and D. J. Lurie, In vivo imaging of nitroxide-free radical clearance in the rat, using radiofrequency longitudinally detected ESR imaging, J. Magn. Reson. (B), **113**:256 (1996).
28. Yu. K. Danis, I. I. Naktinis and L. Ch. Chernyauskene, EPR spectroscopy in internal disease clinics, Terapevticheskii arkhiv. N11:70 (1976) (Russ.).
29. N. J. F. Dodd and H. M. Swartz, The nature of the ESR signal in lyophilized tissue and its relevance to malignancy, Brit. J. Cancer. **49**:65 (1984).
30. E. K. Ruuge and L. A. Blumenfel'd, Ascorbic acid free radicals forming in interaction with proteins, Biofizika. **10**:689 (1965) (Russ.).
31. M. Yu.-Yu. Tsimbolaite, The Role of Ascorbic Acid in Free Radical Formation in Animal Dry Tissues, Thesis Ph.D., Kaunas (1982) (Russ.).
32. H. Neubacher, ESR investigations of lyophilized blood: mixtures with ascorbic acid, Z. Naturforsch. **39**:174 (1984).
33. A. E. Kalmanson, On the Nature and Role of Free Radicals in Biological Processes, Thesis Ph.D., Institute of Chem. Phys., Moskva (1977) (Russ.).
34. A. V. Kozlov, D. Yu. Egorov, Yu. A. Vladimirov and S. A. Azizova, The formation of iron-ascorbic acid complexes in tissues, Biofizika. **35**:513 (1990) (Russ.).
35. V. M. Chumakov and V. M. Kalmanson, EPR study of coenzyme Q_{10} in mitochondria and tissues of rats, Dokl. Akad. Nauk SSSR. **170**:714 (1966) (Russ.).
37. H. M. Swartz, J. R. Bolton and D. C. Borg, eds., Biological Applications of Electron Spin Resonance, Wiley-Interscience, NY, New York (1972).
38. R. G. Saifutdinov, Investigation of Blood and Blood Component Free Radicals in IHD Patients, Thesis Ph.D., Cardiology Center, Moskva (1984) (Russ.).
39. O. A. Kovalenko, T. V Anfalova, V. S. Sokolov and V. M. Chibrikin, Quantitative study of the EPR spectra of frozen samples, Biofizika. **16**:663 (1971) (Russ.).
40. S. Stolc, L. Valko, M. Valko and V. Lombardi, A technique for the fast sampling of biological tissues for EPR spectroscopy, Free Radic. Biol. Med. **20**:89 (1996).

41. R. G. Saifutdinov and K. R. Sedov, An EPR study of human blood erythrocytes and plasma by at 77°K, Biofizika. **28**:87 (1983) (Russ.).
42. Yu. N. Molin, V. M. Chibrikin, V. A. Shabalkin and V. F. Shuvalov, EPR method accuracy in measuring PMC concentration, Zavodskaya laboratoriya. N 88:933 (1966) (Russ.).
43. T. Pocklington, M. A. Foster and J. R. Mallard, Plasma total iron-orbinding capacity measured by ESR, Phys. Med. Biol. **22:**98 (1977).
44. R. G. Saifutdinov, Mn(II) quantitative determination in biological fluids, Pat SSSR N 1323934 (1985), Bull. Izobr. N26:196 (1987) (Russ.).
45. R. D. Borodulin and A. F. Vanin. EPR determination of copper in cells and tissues, Biofizika. **17**:1083 (1972) (Russ.).

Chapter 2

PARAMAGNETIC CENTERS IN THE HUMAN BIOLOGICAL MEDIA

2.1 ERYTHROCYTES

Erythrocytes, responsible for oxygen transport to tissues, are cells devoid of nucleus and ribosomal apparatus. Therefore they cannot synthesize proteins, nucleic acids and lipids. During maturation the erythrocytes lost mitochondria and therefore they have got no citric acid cycle. In spite of the absence of a number of biochemical systems they are not inert protoplasmatic particles, since active metabolism takes place in them. Glucose is the only energy source of erythrocytes. With glucose deficiency the erythrocytes lose their ability of supporting the sodium and potassium gradient and inhibit the accumulation of calcium in the membrane. Glucose deficiency causes the formation of oxidized glutathione and methemoglobin. Devoid of energy, the erythrocytes are transformed to echinocytes, then to spherocytes and, finally, are subjected to osmotic lysis.

Erythrocyte glucose metabolism occurs in two main directions: Embdene-Meyerhoff and hexosomonophosphate shunt (HMP-S). In the former glucose is catabolyzed to pyruvate or lactate. The energy derived from glucose is accumulated in the form of adenosinetriphosphate (ATP). The principal function of the HMP-S is to support a sufficient amount of reduced nicatinamide dinucleotidephosphate (NADP.H). The latter is needed for reduction of hemoglobin (Hb) and deactivation of lipid peroxidation (LPO). The glutathione-reductase system (GRS) generating the reduced form of glutathione works actively in erythrocytes. Inactivation of oxygen radicals is effected by enzymes (superoxide dismutase, catalase and peroxidase). Thus, erythrocytes are metabolically very active. They contain essential microelements such as iron, copper, manganese, calcium, sodium, potassium *etc.*

During their relatively short life-span the erythrocytes are strongly injured by free radicals in contact with tissues. Superoxide anion radicals ($O_2^{-\bullet}$) form in tissues in the process of hemoglobin oxidation and metabolism of many medicines [1]. These radical anions also generate along with singlet oxygen when erythrocyte protoporphyrin is exposed to visible light. Therefore, the interaction of erythrocytes with these reactive oxygen species is suggested to be light-dependent. The effect of light on the generation of hydroxyl radicals ($HO^{\bullet}$) in human erythrocytes was demonstrated using spin trapping EPR spectroscopy [2] The mechanism of erythrocyte protection from oxidation decomposition by free radicals is very powerful in both cytosole and membrane. The cytosole antioxidant system involves super-oxide dismutase, catalase and glutathione peroxidase. In the membrane vitamin E and ubiquinone Q_{10} are used for this purpose. The vitamin E level in the erythrocyte membrane is much lower compared with that in the plasma. Therefore there is a mechanism of generation of the initial membrane vitamin E level after its consumption as an antioxidant.

Free radicals produced in human erythrocyte membranes using an enzymatic oxidation system composed of lipoxygenase and arachidonic acid were examined by EPR. The erythrocyte membranes contain NADP.H-cytochrome-C-reductase and cytochrome b_5. Owing to this fact the protec-tion against the loss of vitamin E can be provided either by NADH.H-cytochrome b_5-dependent enzymatic recycling or by a non-enzymatic pathway involving ascorbate and dihydrolipoic acid [3]. In the human organism the erythrocyte form of cytochrome-b_5 (97 residues in length) is structurally and functionally different from liver microsomal cytochrome-b_5 (93 residues in length) [4].

The erythrocyte membrane consists of a bimolecular lipid layer sandwiched between two protein monolayers. Lipid molecules are arranged reciprocally parallel and perpendicular to the membrane plane, the polar phospholipid heads being oriented outwards whereas long hydrocarbon chains towards the membrane center. Protein chains are adsorbed on the polar heads. The interaction between protein and phospholipids is suggested to be ensured by electrostatic attraction and Van-der-Waals forces. The length of lipid molecules is approximately 3 nm; the thickness of protein monolayer and cell membrane does not exceed 1 and 8 nm, respectively. In the human erythrocyte membrane there are cholesterol, phosphatidylserine, sphingomyelin, cerebrosides and other lipids. However, many structural and functional aspects of erythrocyte membranes are not well understood yet.

Phospholipid outside-inside translocation rates and equilibrium distribution in human erythrocytes of three age groups at 37°C and 4°C were determined by EPR using spin labeled phosphatidylserine, phosphatidylethanolamine, and phosphatidylcholine. In the two cases the initial rate of aminolipid translocation in age cells turned out to be reduced. Equilibrium distribution for phosphatidylserine was not changed much. The phosphatidylethanolamine fraction remained on the outer monolayer of older cells. The phosphatidylserine transmembrane diffusion was identical in the three fractions. Cytosolic ATP required for aminophosphinolipid translocation was not responsible for the above processes. Thus, aminophospholipid translocation seems to be directly altered in aged cells, possibly due to the presence of lipid peroxidation products. This is confirmed by experiments with malonic dialdehyde (MDA) or hydrogen peroxide (H_2O_2)-treated cells. Consequently, the defects in endogenous lipids asymmetry observed in aged human erythrocytes may be due to altered activity of the translocase [5].

The existing model of transverse lipid movement in the human erythrocyte membrane is based on a system of differential equations describing the time-dependence of phospholipid redistribution and the steady-state distribution between the inner and outer membrane monolayer. This has been carried out by EPR using spin labeled lipid analogs. Several mechanisms of translocation should be taken into account: ATP-dependent transport, protein facilitated and transbilayer diffusion. A reasonable modeling lipid asymmetry could only be achieved by introducing the latter mechanism. This pathway was called the compensatory flux, which is proportional to the gradient of phospholipids between both membrane leaflets. This model makes it possible to trace the transbilayer motion and distribution of endogenous phospholipids of the human erythrocyte membrane under various biological conditions. Moreover, the model can also be applied to experiments performed to assess phospholipid re-distribution in biological membranes. Thus, simulation of the transbilayer motion of exogenically added phospholipid analogs in erythrocyte membranes is quite possible [6].

EPR was used to investigate the effect of benzyl alcohol on the transverse mobility and repartition of phospholipids in membrane and morphological changes of human erythrocytes. The transmembrane rate and equilibrium distribution in erythrocytes of short-chain spin labeled phosphatidylcholine, phosphatidylethanolamine and phosphatidylserine were strongly modified by treatment with 10-70 mM benzyl alcohol solutions. In this case two effects were observed: 1) at 4 and 37°C there was an N-ethylmaleimide-sensitive long-lasting and fully reversible increase in the internalization rate of spin labeled phosphatidylserine and phospatidylethanolamine, 2) at 37°C an enhancement of N-ethylmaleimide-sensitive fluxes of all the labeled phospholipids through the membrane occurred. Both effects were dose-dependent. Erythrocytes submitted to benzyl alcohol incubation also showed dose-dependent shape changes: immediate one from discocytes to echinocytes followed by a slower N-ethylmaleimide- and ATP-dependent change to stomatocytes. Benzyl alcohol intensifies the hydrolysis of intracellular ATP. All the effects of benzyl alcohol can be considered as an accumulation of labeled phosphatidylethanolamine (and labeled phosphati-dylcholine at 37°C) in the inner leaflet of the erythrocyte membrane [7].

Normally in the EPR spectra of erythrocytes the following types of signals are observed (Figure 2-1): g = 6.00 arisen from methemoglobin Fe^{3+} and g = 2.063 corresponding to superoxide dismutase Cu^{2+}. Apart from the signals those at g = 2.00 and 4.3 were fixed (Figure 2-1) [8]. The latter was also observed in the spectra of erythrocytes of mammals [9]. The EPR signal at g = 4.3 caused by transferrin Fe^{3+} (TF) is observed not in all cases. It is met in 47% and 52% of male and female donors, respectively. This index increases when younger erythroid cells appear in the blood channel.

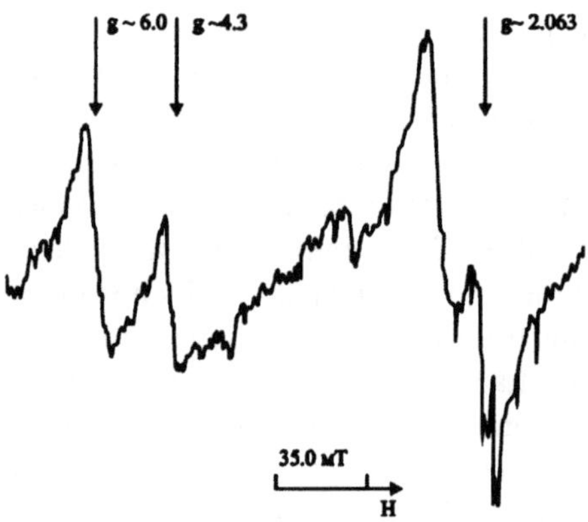

Figure 2-1. The EPR spectrum of human erythrocyte free radicals. The registration conditions: temperature – 77°K, SHF power – 5 mW, magnetic field scanning velocity - 50 mT/min, HF modulation – 0.5 mT, time constant – 0.3 sec.

An asymmetrical signal in the g = 2.00 region with a half-width 1.2 - 1.5 mT is readily saturated in the SHF field (Figure 2-2). A detailed analysis of the signal suggests the presence of two types of radicals in erythrocytes. One of these corresponds to the EPR signal with g = 2.0033 and ΔH 1.2-1.5 mT and the other gives rise to a narrow singlet at g = 2.0032 and ΔH 0.4-0.6 mT which appears at low modulation values [8]. The intensity of the above signals was quantitatively evaluated by double-integration method: the ratio of broad to narrow singlet areas is 6 : 1 (Figure 2-3). Unfortunately, the narrow signal defies interpreting and will not be considered further.

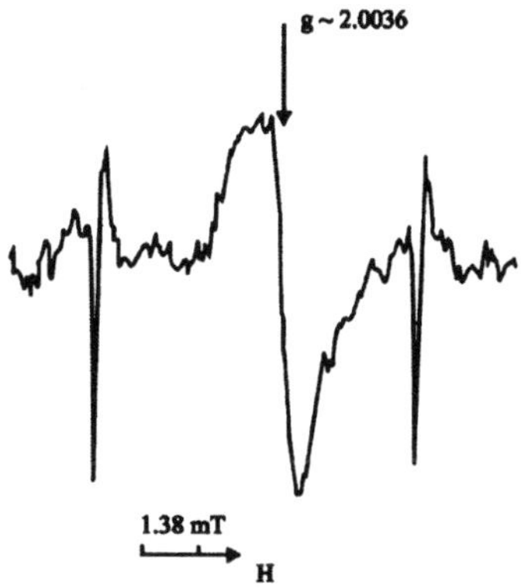

Figure 2-2. The EPR spectrum of human erythrocyte free radicals. The registration conditions: the temperature – 77°K, SHF power – 5 mW, magnetic field scanning velocity - 3 mT/min, HF modulation – 0.5 mT, time constant – 0.3 sec.

The results of parallel determination of free radical concentration and glutathione reductase activity in erythrocytes are presented in Table 2-1 [8]. Correlation between the free radical concentration responsible for the EPR signal at g = 2.0033-2.0040 and the glutathione reductase activity suggests glutathione reductase to be a possible generator of free radicals ($r = -0.64$, $P < 0.05$) [8, 10].

Table 2-1. Free radical (FR) concentration and glutathione reductase (GR) activity of donor erythrocytes

N	FR Concentration (10^{-6} M)	GR Activity (opt.density units)
1.	2.5	0.14
2.	2.0	0.16
3	1.7	0.15
4.	1.6	0.20
5.	1.3	0.19
6.	1.2	0.18
7.	1.2	0.18
8.	1.1	0.16
9.	0.9	0.20
10.	0.8	0.18

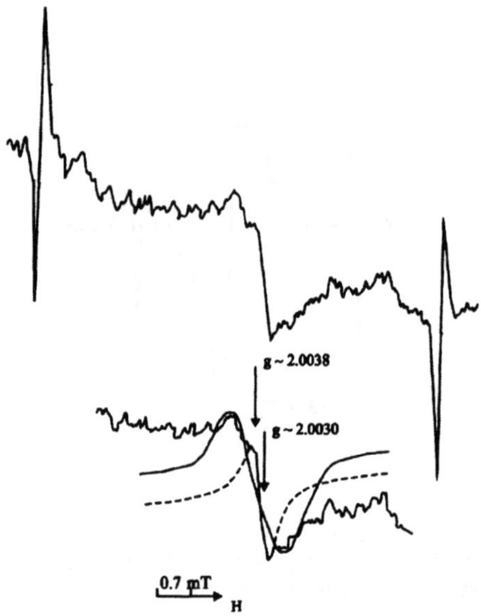

Figure 2-3. The EPR spectrum of human erythrocytes free radicals at 77°K (above) and graphic decomposition of the spectral signal (below). The registration conditions: SHF power – 2.5 mW, H_0 – 350 mT, ΔH – 72 mT, magnetic field scanning velocity - 1.5 mT/min, HF modulation – 0.4 mT, the time constant – 0.1 sec.

The GR system facilitates the activation of hexosomonophosphate shunt [11]. The rate of this system is limited by accumulation of reduced nicatinamidedinucleotidephosphate whose oxidation is mainly related with reduction of the oxidized glutathione (GSSG) to the reduced form of glutathione (GSH) (Figure 2-4).

In metabolizing erythrocytes kept from 3 to 18 days in a medium, prepared at the Central Institute for Blood Transfusion (medium 7b) the free radical concentration and HMP-S activity were determined simultaneously. The HMP-S activity was assessed judging by an increase in the ribose concentration and a decrease in glucose in the sample after 3 h incubation [12].

At the beginning of experiment the correlation coefficient (r) between these values was +0.6. During storage a tendency towards decreasing was observed: +0.6, 0.5, 0.4, 0.3 and 0.2 on the third, sixth, nineth, fifteenth and eighteenth experimental days, respectively.

This seems to be due to reduction of the enzymatic activity and drop of the concentration of substrates by the eighteenth day. Thus, HMP-S activity on the first day was 10.2 ± 1.0 opt.dens.units and 7.5 ± 0.4 opt. density units by the eighteenth day. A similar decrease was observed in the free radical level: (1.19 ± 0.15) x 10^{-6} M and (0.75 ± 0.05) 10^{-6} M on the first and eighteenth experimental days, respectively [13].

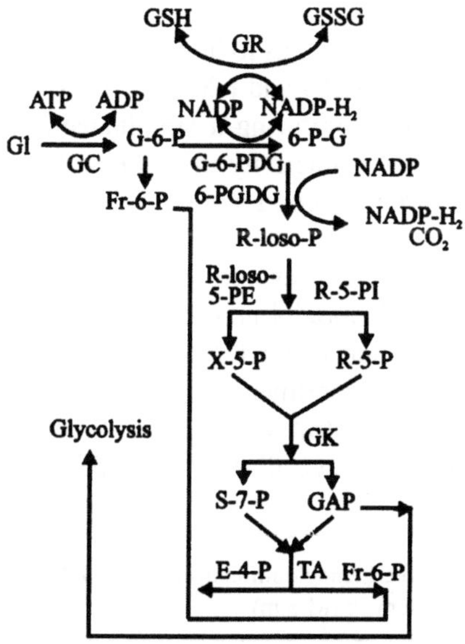

Figure 2-4. The ratio of hexosomonophosphate shunt to glutathione reductase system and glycolitic glucose metabolism.

Thus, there is a relationship between the glutathione reductase system activity determined by free radicals and erythrocyte HMP-S concentration.

The hexosomonophosphate pathway of glucose metabolism and the GRS activity are related with the red cell integrity. The hemolysis rate increases fast with a decrease in the GSH content. To support the red cell integrity the following units are involved [11]:

$$1)\ G\text{-}6P + NADP \xrightarrow{G\text{-}6\text{-}PDG} NADP.H_2 + 6\text{-}PG$$

$$2)\ GSSG + NADP.H_2 \xrightarrow{GR} NADP + 2GSH$$

Disturbance of one of these leads to hemolysis.

Taking into account that GRS and HMP-S are closely related with erythrocyte membrane strength monitoring a simultaneous determination of the blood hemolysis level, HMP-S activity in erythrocytes and free radical content depending on the time of blood storage in medium 7b was carried out [11].

The hemolysis degree was measured by the Polishchuk method [14]. On keeping, the blood free hemoglobin content was increasing, *i.e.*, latent hemolysis went to the explicit form: 1.7 ± 0.1 mg/% and 66.7 ± 14.1 mg/% on the first and eighteenth experimental days, respectively. A simultaneous drop in the free radical concentration and HMP-S activity took place by the end of conservation period. Thus, all the three indexes are interrelated (table 2-2).

Table 2-2. Free radical and free hemoglobin concentration, HMP-S activity in erythrocytes during blood storage in a medium 7b (M ± m)

Time of storage (day)	FR (10^{-6} M)	Free Hb (mg/%)	HMP-S (optical density unit)
1	1.19 ± 0.15	1.7 ± 0.1	10.2 ± 1.0
3	1.18 ± 0.15	3.3 ± 0.3	9.8 ± 0.9
6	1.12 ± 0.12	7.6 ± 0.9	9.9 ± 1.0
9	1.05 ± 0.10	10.6 ± 2.0	9.4 ± 0.9
12	0.91 ± 0.07	16.8 ± 2.6	9.2 ± 0.8
15	0.84 ± 0.06	47.1 ± 9.9	8.4 ± 0.6
18	0.75 ± 0.05	66.7 ± 14.2	7.5 ± 0.4

The above data were clinically confirmed by examination of erythrocyte and plasma the free radical level prior to and after surgery with assisted circulation [15]. A group of patients with congenital or acquired heart disease, operated under identical assisted circulation conditions, was examined. Perfusion was carried out using a PEMCO apparatus with a disc oxygenator. Blood samples were taken in the start and after the completion of perfusion. During perfusion the free radical concentration in erythrocytes increases, the free radical level before and after assisted circulation being (1.26 ± 0.15) x

10^{-6} M and (1.97 ± 0.22) x 10^{-6} M, respectively (P < 0.05). Variation in the free radical concentration in plasma in the course of perfusion are doubtful (0.47 ± 0.03) x 10^{-6} M and (0.51 ± 0.04) x 10^{-6} M, respectively (P > 0.05) [16]. The relation between human erythrocyte hemolysis and free radicals has also been studied by [17].

Mechanical traumatism of erythrocytes during assisted circulation leads to hemolysis, which can be compensated for by an increased activity of GRS unaffected cells. This seems to be a protective mechanism strengthening the erythrocyte stroma and improving the oxygen transport to tissues.

The above data allow the following conclusions to be drawn:

In erythrocytes the EPR signal at g = 2.0033-2.0040 and line half-width 1.2-1.5 mT correspond to flavosemiquinone, glutathione reductase co-factor.

The signal intensity depends on the ratio of reduced to oxidized glutathione acting as a redox pair setting optimal redox-potential for the highest content of flavine radicals. The redox-pair functions due to reversibility of oxidation reaction GSH←→GSSG catalyzed by glutathione reductase [8].

The signal of erythrocyte free radicals does not correspond to a symmetrical narrow line occurred closer to the g-factor of the Fe^{3+} complex free spin electron in the samples of animal and plant tissues as reported by Naughton and Symons [18]. Nevertheless, the above said suggests erythrocyte free radicals to be responsible for this signal.

Human erythrocyte hemolysis can proceed in a normal way as well. This is the so-called physiological hemolysis related to natural aging of erythrocytes. Erythrocyte hemolysis as a pathological process can arise owing to many factors such as incompatible blood transfusion, infusion of hypotonic salt solutions, effect of hemolytical poisons and some medicines, inherited insufficiency of enzymatic systems in erythrocytes, presence of anomalous hemoglobins in erythrocytes, appearance of erythrocyte antibodies.

Hemolysis as a biophysical process is studied *in vitro* as the destruction of erythrocytes cannot be traced in detail *in vivo*. Understanding of erythrocyte hemolysis is needed for the determination of erythrocyte life span. This makes it possible to avoid hemolysis in storage of preserved blood and erythrocytar mass, in carrying out the reaction of compliment binding, and other tests with blood and erythrocytes.

Osmotic hemolysis is caused by water penetrating the erythrocyte, which leads to volume increase and membrane extension. This widens the membrane pores the hemoglobin molecules get out through. The erythrocyte volume wherein hemolysis starts is called critical. In human erythrocytes it is 146% of the initial volume.

Erythrocyte water volumes were measured using EPR spectroscopy under equilibrium and under perturbed osmotic conditions. Cell water contents were determined using a spin label (TEMPO) and potassium chromium oxalate as a membrane impermeable quencher [19].

In clinical practice the erythrocyte osmotic hemolysis of different origin is examined by test for erythrocyte resistance to hypotonic sodium chloride solutions. The sodium chloride concentration whereby osmotic hemolysis starts is taken as an index of minimum erythrocyte osmotic resistance. The concentration at which complete hemolysis occurs is an index of maximum erythrocyte resistance. Erythrocyte hemolysis in volunteers starts upon administration of a 0.44-0.48% NaCl solution and is over in a 0.28-0.32% solution.

The effect of surfactant (polyoxyethylenenonylphenol) on erythrocyte hemolysis has been investigated using the EPR method. At low concentrations the surfactant inhibits hypotonic hemolysis, whereas at high concentrations it induced hemolysis in both isotonic and hypotonic buffer [20]. Erythrocyte hemolysis by free radicals produced from a lipid-soluble initiator 2,2-azobis-(2,4-dimethylvaleronitrile) [21].

Responsibility of band 3 (anion exchange protein in human erythrocytes) for hemolysis in hypotonic solutions was examined using ESR spectroscopy. The ESR spectra of spin labeled erythrocytes show the change of the osmotic conditions to have no effect on the membrane fluidity at pH 7.4. At pH 6.0, however, the fluidity in the central part of the lipid bilayer is dependent on osmotic stress, the fluidity change as a function of KCl concentration being similar to the hemolysis curve. Judging by the circular dichroism spectra of erythrocyte ghosts the conformational changes of the membrane proteins correlate with the hypotonic hemolysis index. In contrast to the osmotic stress, sodium dodecyl sulfate (SDS) changes the fluidity of membranes rather than the conformation of membrane proteins independently of the medium pH. In this case the changes of fluidity correspond to the hemolysis curves in SDS solutions. The use of anion transport inhibitors, 4.4'-diisothiocyanostilbene-2,2'-disulfonate, eosin-5-isothiocyanate and eosin-5-maleimide, suggested a correlation between the effect of hypotonic hemolysis and the conformational change of band 3. These results demonstrate band 3 to be responsible for hypotonic hemolysis [22, 23].

The proteolysis of protein (band 3) *in situ* was studied by EPR and fluorescent/phosphorescent methods using eosinyl-5-maleimide that reacts with membrane-spanning Mr 17K peptide [24].

The effects of diethylether on the ordering of membrane lipids and on the rotation dynamics of anion-exchange protein (band-3) in intact human erythrocytes were studied by EPR using 5-doxylstearic and 16-doxylstearic acids as spin labels. With increasing the dose, diethyl ether monotonically decreased the ordering of membrane lipids near the polar head-group. This proved by lipid-soluble spin probe 5-doxylstearic acid. Diethyl ether produces comparatively little change in the ordering of lipids in the hydrophobic midzone as supported by 16-doxylstearic acid. The rotational mobility of band-3 also increases monotonically with increasing the ether concentration as shown by the affinity of spin label bis(sulfo-N-succinimidyl)doxyl-2-spiro-5'-azelate. At concentrations of up to 2 vol% ether, hemolysis of erythrocytes was negligible and the spectroscopic changes were completely reversed following its removal. Thus, the ether-induced decrease in the lipid membrane ordering near the polar head group region significantly increase the rotational mobility of band 3 and change only negligibly its state of oligomerization [25].

Among 56 patients examined, 6 were heterozigous with respect to the elongated version of the band-3 protein of erythrocyte membranes. For these 6 persons the EPR spectra of maleimide-spin labeled membranes of erythrocyte containing the variant band 3 protein exhibited no morphological abnormalities. However they did show an increased membrane binding of hemoglobin and aldolase [26].

EPR spectroscopy was used for the assessment of transverse distribution and mobility of labeled phospholipid analogs in biological erythrocyte membranes by selective chemical destruction of the fluorescent label in the outer monolayer with dithionite. Their distribution was determined directly in the cell suspension provided that dithionite permeation through the membrane is suppresses. It is shown that the passage of dithionite can be mediated by anion-transport systems such as band 3 [27].

Heating of erythrocyte suspension to 49°C leads to erythrocyte swelling and at 62-63°C to destruction of erythrocytes with release of hemoglobin. Erythrocyte hemolysis upon repeated freezing and thawing is caused by mechanical injury with ice particles and by increased concentration of the intracellular substances. Cooling of various biological samples from 40°C to 0°C leads to destruction. Any reason for this phenomenon (called "temperature shock") has not been clearly defined until the present time. As shown by EPR using spin probe, in hypertonic salt solutions non-permeating the membranes of low molecular substances dehydration of erythrocytes occurs at a temperature decreasing from 25°C to 0°C. At the same time nothing of this kind is observed in solutions of high molecular compounds. Thus, the erythrocyte dehydration to a critical value (70 ± 5)% is one of the main reasons for erythrocyte destruction by temperature shock. From EPR data the temperature shock can also be brought about at a temperature above 20°C [28]. The EPR spin-probe method was used to examine the dependence of erythrocyte cytosole microviscosity on the outer medium composition and temperature. In this case the critical erythrocyte cytosole microviscosity values after reaching which erythrocytes are destroyed were determined. These values depend not only on NaCl concentration, but on temperature too [29].

Human erythrocyte hemolysis on exposure to heavy ions irradiation has been studied. The intensity of irradiation-induced erythrocyte hemolysis depends on the wave length. Human erythrocytes were irradiated with heavy ions of energies from 4 to 18 MeV/u having linear energy transfer values between 92 and 14.000 keV/μ. In this study hemolysis served as a macroscopic parameter of membrane destruction whereas fluidity was used as a microscopic parameter. The fluidity was determined by EPR spectroscopy using 12-doxylstearic methyl ester as a spin label. The fluidity changed depending on the irradiation dose. On exposure to heavy ions hemolysis proceeds more efficiently compared with X-rays [30].

Ether, sodium oleate, nitroglycerol and sodium salts of cholic and desoxycholic acids represent active chemicals responsible for erythrocyte hemolysis. Most of these hemolytical agents injure the erythrocyte membrane structure inducing disorder of lipid molecules and pore formation.

During erythrocyte hemolysis denatured hemoglobin particles (Heinz bodies) are attracted on the membrane inner surface. The formation of Heinz bodies is observed when erythrocytes are incubated with certain medicines having side hemolytic effect. In persons sensitive and insensitive to primaquine the Heinz bodies are formed *in vitro* in acetylphenylhydrazine-treated cells.

In phenylhydrazine-treated cells a slight deformation of human erythrocytes occurs prior to the formation of Heinz bodies and starts to decrease as soon as Heinz bodies reach the inner surface of the membrane. In methylhydrazine-treated cells in which Heinz bodies are formed far from the membrane surface the erythrocyte deformability remains at a low level. Dithioerythritol reduces in part the methylhydrazine-induced deformability of erythrocytes. However, it has no effect in phenylhydrazine-treated cells. The increased mobility of maleimide-labeled membrane proteins indicates that phenylhydrazine has a greater immobilizing effect than methylhydrazine [31].

The EPR method using spin label has been proposed for the investigation of erythrocyte permeability for cryoprotectors. Human erythrocyte permeability for 1,2-propanedioles, glycerol and dimethylsulfoxides (DMSO) at different temperatures has been studied. A clearly defined dehydration of the cell is observed when erythrocytes are exposed to concentrated cryoprotector solutions. Cell rehydration occurs as the solutions penetrate the cell its a water-soluble spin probe easily penetrates the cell membrane with a flow of water and cryoprotector solution. In this case, broadening of the extracellular spin probe spectrum using a paramagnetic broadening agent (potassium ferrocyanide) makes it possible to record changes of the cell volume and thereby kinetics of the cryoprotector penetration. TEMPO is the most appropriate spin label. In the presence of potassium ferrocyanide the spin label is not reduced and the cryoprotector permeability is not blocked [32].

In order to understand how calcium in human erythrocyte cytosole influences the transmembrane redistribution of phospholipids, the effect of this cation on the transmembrane movements of spin labeled phospholipids (phosphatidylserine and phosphatidylcholine) incorporated into inside-out vesicles was studied. The degree of the Ca^{2+}-induced lipid scrambling was dependent on the level of phosphatidylinositol 4,5-bisphosphate outer leaf of inside-out vesicles. The level of phosphatidylinositol 4,5-bisphosphate normally was 80% of the total membrane phosphatidylinositol 4,5-bisphosphate. Loading of the outer monolayer of the intact erythrocyte membrane with exogenous phosphatidylinositol 4,5-bisphosphate leads to a dose-dependent scrambling of spin labeled phosphatidylethanolamine, sphingomyelin, phosphatidylserine and phosphatidylcholine and in parallel the stomatocytic conversion of the cells. The Mg^{2+} ion could replace the Ca^{2+} ion but requires a 10 times higher concentration. The effect of these ions was specific only for phosphatidylinositol 4,5-bisphosphate. Other phospholipids fail to induce lipid redistribution. A normal ATP level is needed for the shape change but not for the lipid scrambling. This suggests that the Ca^{2+} and Mg^{2+} ions serve as triggers for the redistribution of lipids from the inner or the external sides of the erythrocyte membrane on condition that enough phosphatidylinositol 4,5-bisphosphate is present on that side. Thus, this process requires no specific protein. The ATP-dependent shape of erythrocytes after incubation with phosphatidylinositol 4,5-bisphosphate and Ca^{2+} ion is caused by bilayer disbalance due to activation of amino-phospholipid translocase which transposes phosphatidylcholine and phosphatidylethanolamine toward the internal monolayer without simul-taneous outward diffusion of phosphatidylserine and sphingomyelin [33].

Water and small ions diffuse the erythrocyte membrane at a high rate through pores 0.3-0.4 nm in diameter. The pores are impermeable for calcium and magnesium ions, for saccharides and large-size molecular colloids. The half time of exchange through membrane is 0.004 sec for water, 0.2 sec. for Cl anion, about 30 and 20 h for K^+ and Na^+ cations, respectively.

Erythrocytes in isotonic solutions of low ionic strength lose their cations. This is due to a change in the erythrocyte permeability for K^+, which depends on the transmembrane potential. The passive Rb^+ (K^+) efflux from erythrocytes of seven mammals was investigated by EPR in physiologic solutions of low ionic strength from erythrocytes of 7 mammals species. The rate constant of Rb^+ (K^+) efflux in solution of high ionic strength correlates with the mean number of double bonds in the membrane phospholipids of fatty acids [34].

Stimulation of human erythrocytes with Ca^{2+} and Rb^{+} ions causes potassium efflux (the so-called Gardos effect). Monitoring of time dependent changes in the volume and shape of erythrocytes was performed on the basis of mean cell volume and light scattering. Changes in the erythrocyte cytoplasma were traced by EPR using cytoplasmic glutathione labeled with TEMPO [35].

Nitroxide free radicals are under active investigation for their potential use as metabolically responsive contrast agents in EPR imaging. The metabolism of human erythrocytes of lipid-soluble nitroxides (doxyl stearic acids) was investigated by EPR spectroscopy. It turned out that under normoxia this free radical remains intact in erythrocytes for at least 2 h, but under hypoxia it is fast reduced. Complete signal recovery after air or potassium ferricyanide oxidation suggested the formation of hydroxylamine during hypoxia. The reduction of doxyl stearic acids depends on the nitroxide ring position in fatty acid. The reduction kinetics of doxyl stearic acids with the dioxyl ring nearest to the COOH group (5-doxyl stearic acids) is a di-phase one. A rapid reduction of half of 5-doxyl stearic acids was observed during the first hour. Then the process became much slower. The process of slow reduction abruptly stopped at a concentration below 5 mcM. This suggests a concentration-dependent membrane-cytoplasmatic translocation of 5-doxyl stearic acids. Under hypoxia purified hemoglobin and myoglobin reduce 5-docyl stearic acids and completely recover the signal resulted from air reoxidation. Globin does not reduce 5-doxyl stearic acids whereas methemoglobin does, but to only a small extent. All this suggests ferrous-heme to be involved in the hypoxic reduction of doxyl stearic acids. The high stability of doxyl stearic acids in erythrocytes is due to both doxyl stearic acids localization and characteristics of intracellular reductant (hemoglobin) [36].

2.2 BLOOD PLASMA

The EPR spectrum of human blood plasma shows the signals at g = 4.3 corresponding to a high-spin Fe^{3+} form and at g = 2.05 arising from Cu^{2+} atoms incorporated into ceruloplasmin (Figure 2-5) [37].

Ceruloplasmin, a copper-containing glycoproteid of the plasma α_2-globulin fraction (KF 1.16.3.1.), exhibits ferroxidase, catecholaminoxidase, ascorbatoxidase and superoxide dismutase activity, provides copper transport and takes part in the synthesis of connective tissue [38, 39]. As with other "blue" copper-proteides (laccase, ascorbatoxidase), the ceruloplasmin copper-containing centers are divided into three types which differ in parameters of EPR spectra (types I and II) and optical spectra (types I and III)

[40-42]. The data on the properties and relative contents of paramagnetic centers in the ceruloplasmin molecules are rather contradictory. This is mainly due to a high lability of ceruloplasmin preparations during isolation and storage. According to EPR spectroscopy and spectrophotometry, Cu^{2+} ions in human ceruloplasmin "non-blue" centers are present in two stable forms different in EPR parameters and optical absorption. The two forms were detected by the EPR method in the blood serum of healthy and sick individuals. In the latter the relative content of these forms depends on the origin of the disease. This finding opens new perspectives in the diagnostic application of the EPR method [43, 44].

Ninety-nine per cent of the blood plasma iron is believed to be bound to transferrin and completely determinable by the bathophenanthroline method [45]. However, it is likely that not all the blood plasma iron is present in transferrin [37, 46].

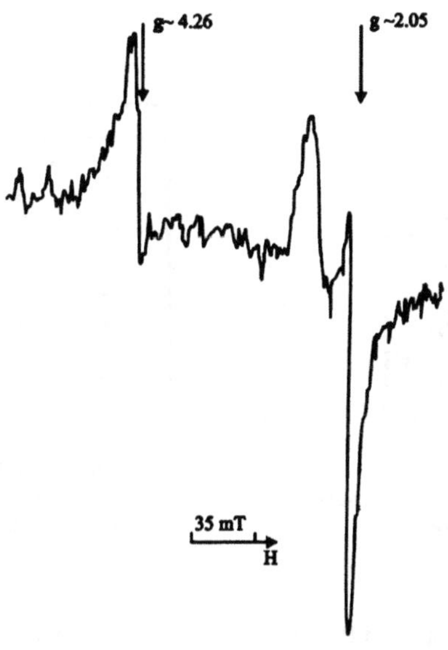

Figure 2-5. The EPR spectrum of human serum blood. The registration conditions: temperature – 77°K, SHF power – 5 mW, magnetic field scanning velocity - 50 mT/min, HF modulation – 0.5 mT, the time constant – 0.3 sec.

Some its portion seems to be bound to other plasma protein such as albumin, globulins, *etc.* This is supported by a parallel study of the iron content carried out by the BP method and EPR spectroscopy (table 2-3) [8].

Table 2-3. Fe Concentration (mcmol/l) in donor plasma (M ± m)

Sex	BP method	EPR method	Difference in Fe content determined by both methods (%)
W	17.5 ± 0.5	15.9 ± 0.3	8.97 ± 1.35
M	23.7 ± 0.5	21.6 ± 0.3	9.52 ± 1.06

BP - bathophenanthroline method

Thus, the Fe concentration in blood plasma determined by EPR spectroscopy is 10% lower than that assessed by BP method. Theoretically, this seems to be true, since EPR spectroscopy records only Fe^{3+} in the g = 4.3 region, whereas using the BP method both Fe^{2+} and Fe^{3+} can be fixed (Figure 2-6).

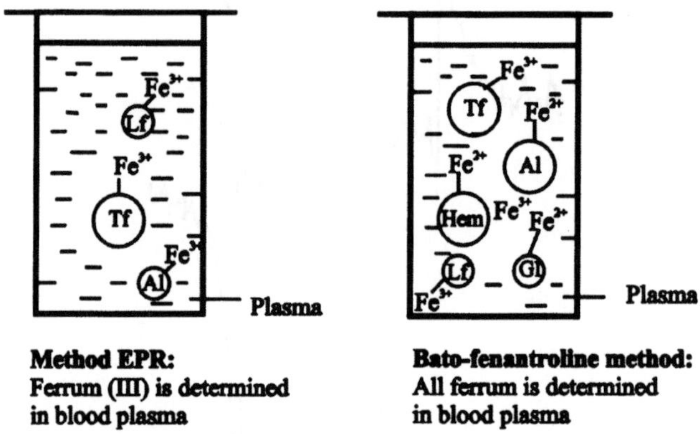

Figure 2-6. EPR spectroscopy and BP method potentialities in blood serum Fe determination.

Albumin is a versatile transporting blood protein. A 30 g/l albumin solution (REANAL) in twice distilled water gives in the g = 4.3 region a Fe^{3+} EPR signal not corresponding to transferrin (Figure 2-7). This suggests that human serum albumin also serves as Fe^{3+} transporter.

The presence of transferrin in human serum albumin is impossible since nearly all globulins [47] including transferring (β-globulin) are removed during HSA preparation [48]. Moreover, the shape of HSA Fe^{3+} EPR spectrum is quite different from that of the transferrin complex (there is neither splitting in the upper maximum, no jags on the descending line) (Figure 2-8).

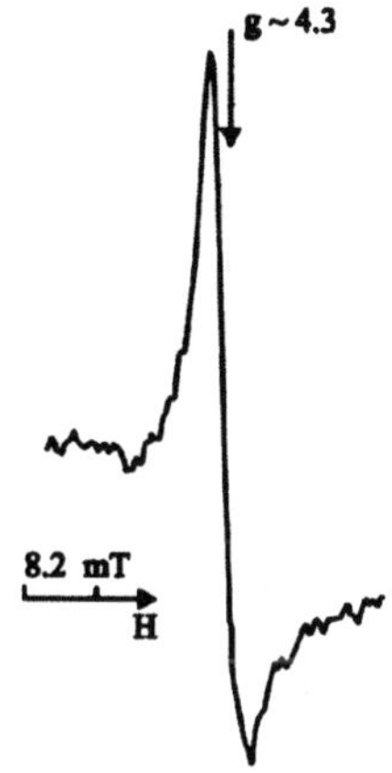

Figure 2-7. The EPR spectrum of human serum albumin. The registration conditions: temperature – 77°K, SHF power – 5 mW, magnetic field scanning velocity - 10 mT/min, HF modulation – 0.5 mT, the time constant – 0.3 sec.

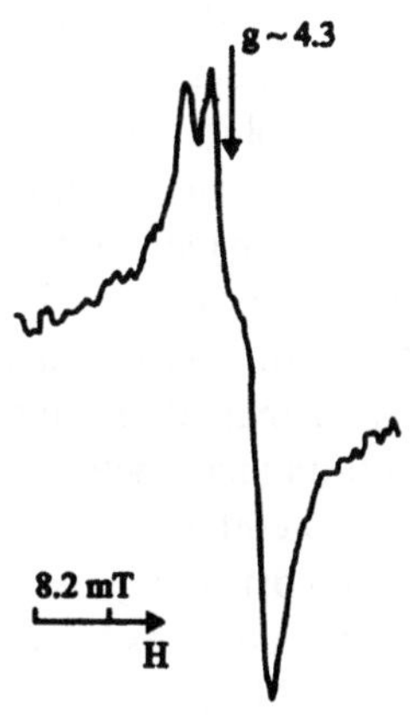

Figure 2-8. The EPR spectrum of human blood plasma. The registration conditions: temperature – 77°K, SHF power – 5 mW, magnetic field scanning velocity - 10 mT/min, HF modulation – 0.5 mT, the time constant – 0.3 sec.

In most tissues of guinea pig and rat the EPR signal at g = 4.3 fixed at 77°K arises from both Fe^{2+} and Fe^{3+} complexes [49]. An analogous signal at g = 4.3 is caused by readily dissociated Fe^{3+} - ascorbic acid paramagnetic complexes at pH = 7.4 [50]. The formation of these complexes reduces the Fe^{2+} concentration in the solution. However, the addition of phenanthroline or pH decrease to 6.1 induce breakdown of Fe^{3+} complexes and the reduction of iron to Fe^{2+}. Since in tissues there are reducers stronger than ascorbic acid the presence of Fe^{3+}-ascorbic acid complexes is unlikely. In particular, they have not been found in liver perfusate and hemogenate [50]. In plasma Fe^{3+} EPR spectrum the signal corresponding to albumin seems to be imposed on the transferrin complex signal. Therefore the total plasma Fe^{3+} content of transferrin, albumin and other proteins is recorded by EPR spectroscopy.

Transferrins, *i.e,* Fe^{3+}-transporting proteins (serum transferrin, ovo-transferrin, lactoferrin) consist of a polypeptide chain folded into two compact domains. These domains, called N- and C-terminal sites, can reversibly bind not only Fe^{3+} but also ions of other metals (such as Cu^{2+}, for example) as well.

In human blood there are four serum transferrin forms: apo-, mono(N)-Fe^{3+}, mono(C)-Fe^{3+} and di-Fe^{3+} transferrins. The apo-form contains no iron. The other three forms transport one and two Fe^{3+} ions, respectively. HCO_3^- and CO_3^{2-} are transferrin-assisted Fe^{3+}-binding synergistic anions *in vivo. In vitro* bicarbonate can be replaced by other ions, such as oxalate, for example. The spectral and magnetic properties of human serum Fe^{3+} transferrin were studied by EPR and Mössbauer spectroscopy.

The Mössbauer spectra of Fe^{3+} in diferric transferrin were interpreted using Hamiltonian spin with the values of the zero field splitting parameter, $D = 0.55 \pm 0.05$ cm^{-1} and the rhombicity of crystalline field, $E/D = 0.045 \pm 0.005$. The EPR spectrum was calculated with $D = 0.58$ cm^{-1} and E/D 0.057 using a g-strain model for the line shape based on a Gaussian distribution of D and E/D with Gaussian line widths $\sigma(D) = 0.35$ cm^{-1} and $\sigma(E/D) = 0.013$, respectively. The rhombicity of iron surrounding for the transferrin-oxalate complex is almost one order of magnitude lower than for the bicarbonate complex and the zero field splitting parameter is twice as large in the oxalate as in the bicarbonate complex. Therefore it is concluded that the crystal field symmetry in the iron site is almost tetragonal in the oxalate complex, but rhombohedral in the bicarbonate complex, thus reflecting different coordination geometry of the oxalate and bicarbonate [51].

The periodate-modified human serum transferrin possesses species that bind weakly Fe^{3+} at pH 7.4. These data contradict previous reports that the Fe^{3+} binding ability is completely lost in this case. The EPR spectra of ternary complexes of periodate-modified transferrin and oxalic or citric or ethylenediamine tetraacetic acid are similar to the spectra of ternary complexes of Fe^{3+} nitrilotriacetate formed with native transferrin in the absence of bicarbonate. At higher concentrations of bicarbonate anions the formation of Fe^{3+}-nitrilotriacetate ternary complex is hindered, whereas other complexes readily remove the transferrin-bound Fe^{3+}. The HCO_3^-, SO_4^{2-} and $P_2O_7^{4-}$ anions are capable of binding the transferrin Fe^{3+} binding sites in the absence of metal ions, though to a small extent. This suggests an inhibition of protein conformational change by intramolecular cross-linking preventing formation of the stable metal-carbonate ternary complex from the less stable metal chelate ternary complex [52].

The selectivity of Cu^{2+} binding by C-and N-sites of serum human- and ovo-transferrin was investigated by EPR spectroscopy and by the use of copper binding constants and copper dissociation constants. At pH 7.4 the copper binding constants of serum human transferrin were $K_1 = (1.8 \pm 0.6) \times 10^{12}$ M^{-1} and $K_2 = (1.2 \pm 0.5) \times 10^{11}$ M^{-1}, whereas those of serum ovo-transferrin were $K_1 = (1.9 \pm 0.5) \times 10^{11}$ M^{-1} and $K_2 = (2.1 \pm 0.4) \times 10^{11}$ M^{-1}. The Cu^{2+}-apo-transferrin complexes were separated by the kinetic method. Serum (human)-TF C-sites bind copper better than N-sites. At the same time, the copper-binding affinity of C- and N-sites of ovo-TF is nearly the same. In the presence of Cl^- and ClO_4^- anions (0.1M KCl and $NaClO_4$ solutions) the copper-binding constant of serum (human)-transferrin was almost 10 times smaller than that in the absence of the anions. The selectivity of Cu^{2+} binding of both sites of serum (human)-transferrin in the presence of 0.1M $NaClO_4$ solution is much smaller than in the presence of 0.1M KCl solution or in the absence of the anions. The EPR spectrum of copper ions in the serum transferrin N-residue containing two coppers ions is dramatically changed on addition of 0.1 M KCl, 0.1 M NaCl and 0.1 M $NaClO_4$, respectively. This suggests a change of the coordination geometry of copper ions in the N-site [53].

The EPR method was used to study the effect of site-directed mutation and NaCl on the iron-binding site of recombined half-molecule of the terminal lobe of human transferrin. In the EPR spectra of all variants studied (D63S, D63C, G65R, K206Q, H207E, H249E, H249Q, K296E and K296Q) there were changes comparable with those observed with wild type protein. The changes in the metal-binding sites were caused by replacement of the coordinating Asp-63 and His-249 residues and the non-coordinating residue, Lys-296, located in the hinge region of the iron-binding cleft. Some differences in the EPR spectra indicated small changes in the Fe^{3+} coordination to the protein. The coordination environment of Fe^{3+} is sensi-tive to structural changes [54].

Apart from the above mentioned signals, the EPR spectrum of human blood plasma shows a slightly asymmetric singlet with a line width of 0.6-

0.8 mT and g = 2.0024-2.0027, which is readily saturated with SHF power (Figure 2-9) [8].

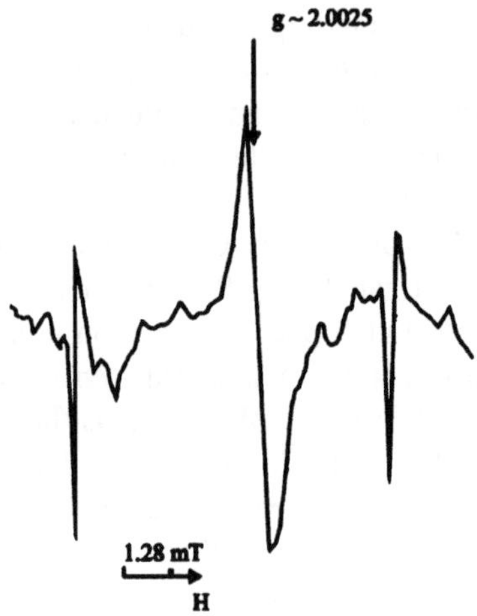

Figure 2-9. The EPR spectrum of human blood plasma free radical. The registration conditions: temperature – 77°K, SHF power – 0.5 mW, magnetic field scanning velocity - 3 mT/min, HF modulation – 0.3 mT, the time constant – 0.3 sec.

Recently compounds responsible for EPR signals of lipids have been identified [55, 56]. It turned out, however, that the EPR signals of blood plasma free radicals and lipids were of different nature [8]. Lipids were isolated by extraction of 1 ml of blood plasma with a (1 : 1) $CH_3OH/HCCl_3$ mixture followed by evaporation in a flow of nitrogen under vacuum to a constant volume in a device specially designed for this purpose (Figure 2-10). Later the lipids were solved in ethyl alcohol.

Taking into account that in the process of blood withdrawal and extraction of lipids under aerobic conditions the semiquinones present in lipids may be oxidized to quinones, alcohol-alkaline solutions were used to reduce quinones to semiquinones [55]. For this purpose several drops of 1% KOH were added to the solution of lipids in a flow of nitrogen.

In plasma lipids of donors and patients FR EPR signals of the same type were registered [8, 57]. Under the conditions of high resolution it was possible to find a clearly defined hyperfine structure consisting of at least 17 lines (Figure 2-11).

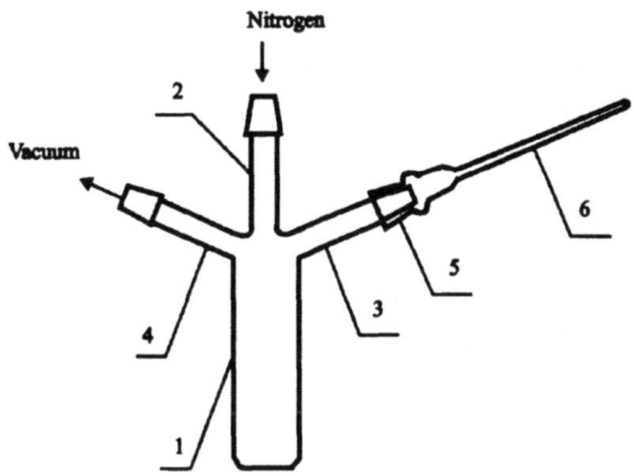

Figure 2-10. A device for the evaporation of lipids in a nitrogen flow: 1 – glass framework, 2 – nitrogen inflow elbow, 3 – elbow to ampoule connection, 4 – elbow to vacuum pump, 5 – glass grounding, 6 – ampoule.

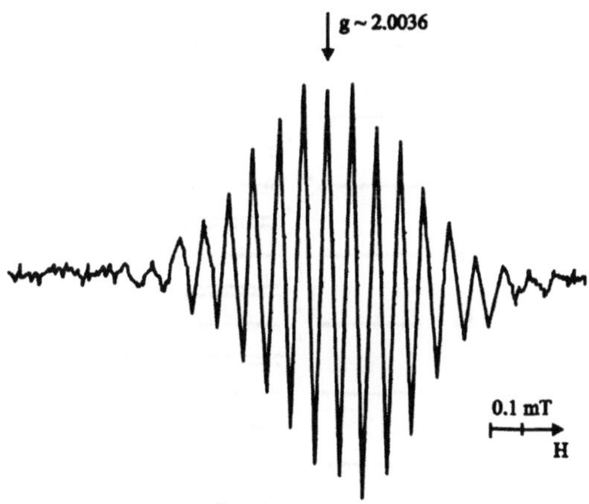

Figure 2-11. The EPR signal of α-tocopherylquinone (α-Tq) semiquinone in human blood lipids. The registration conditions: temperature – 298°K, SHF power – 0.5 mW, magnetic field scanning velocity - 0.1 mT/min, HF modulation – 0.03 mT, the time constant – 0.3 sec.

This can be interpreted as a result of unpaired electron spin coupling of α-tocopherylquinone (α-Tq) (Figure 2-12) with 9 protons of three methyl groups (the distance between spectral lines is about 0.19 mT) and a supplementary interaction with two different protons of the methylene group, which leads to an extra splitting of each of 10 components into three lines 0.09 mT apart (Figure 2-13).

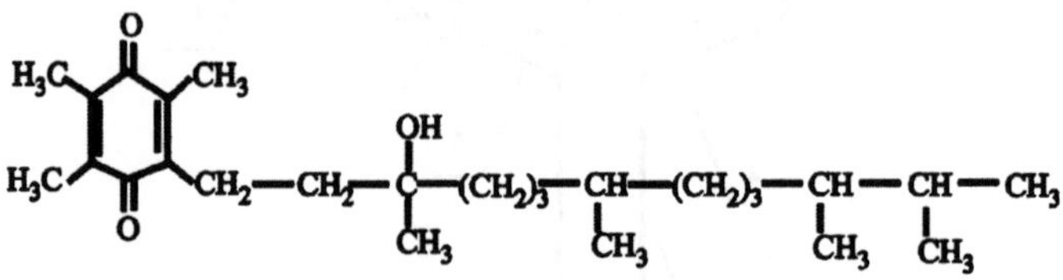

Figure 2-12. The chemical formula of α-tocopherylquinone.

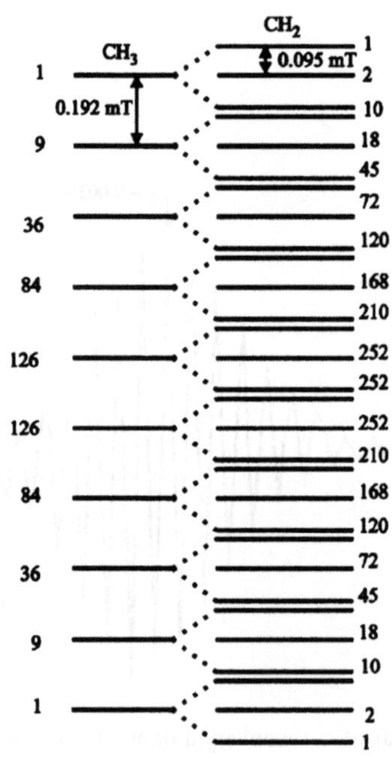

Figure 2-13. The theoretical scheme of the origin of energy level splitting.

The coincidence of the HFS row led to a summarized 21-component signal. According to theoretical calculation there should be 21 components, but due to the fact that two extreme absorption lines are of too low intensity and can be lost in noises, only 17 lines remain to be fixed [8, 58].

However, the EPR signal of plasma free radical at 77°K has line half-width 0.6-0.8 mT and g = 2.0024-2.0027, whereas α-Tq shows 0.6-0.8 mT and 2.0035-2.0037, respectively. The correlation between these signals is negligible (r = +0.16) [8].

Thus, in erythrocytes and plasma there are various paramagnetic centers (PMC) which can provide valuable information concerning the state of the organism and some mechanisms of pathogenic diseases.

2.3 SALIVA

Human saliva is a complex biological fluid. The investigation of its qualitative and quantitative composition not infrequently contributes a deal to our knowledge of the pathologic processes occurring in the organism.

In a mixed human saliva (parotic, sublingual, submaxillary) the following paramagnetic centers were recorded (Figure 2-14): that at g = 4.3 and a six-component Mn^{2+} EPR signal at $g_1 = 2.14$, $g_2 = 2.09$, $g_3 = 2.03$, $g_4 = 1.98$, $g_5 = 1.93$, $g_6 = 1.88$ with the line intensity decreasing with an increase in the magnetic field strain (Figure 2-15) [8].

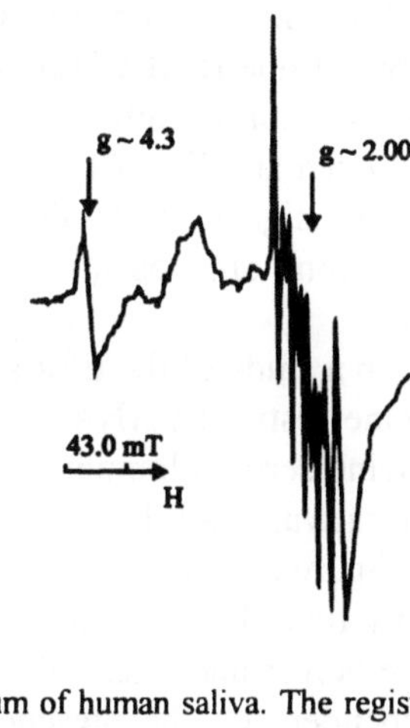

Figure 2-14. The EPR spectrum of human saliva. The registration conditions: temperature – 77°K, SHF power – 5 mW, magnetic field scanning velocity - 50 mT/min, HF modulation – 0.5 mT, the time constant – 0.3 sec.

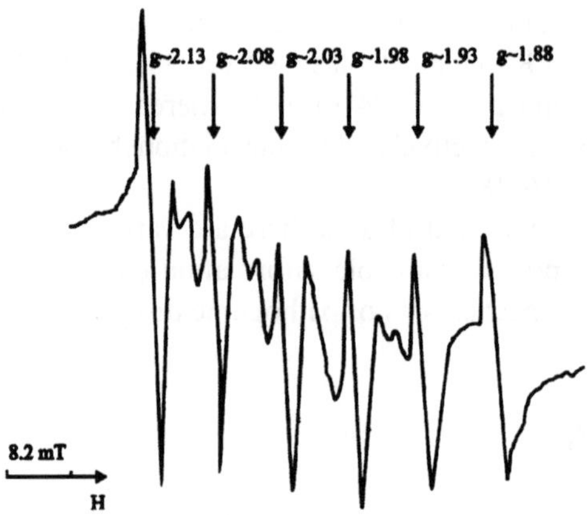

Figure 2-15. The EPR spectrum of human saliva. The registration conditions: temperature – 77°K, SHF power – 5 mW, magnetic field scanning velocity - 10 mT/min, HF modulation – 0.5 mT, the time constant – 0.3 sec.

The signal at g = 4.3 corresponding to Fe^{3+} is met only in 20-50% of healthy donors.

In order to elucidate the nature of Mn^{2+} six-component signal an EPR spectrum of a 1.0 x 10^{-4} M $MnCl_2$ solution was run at 20°C (Figure 2-16) [8]. The spectrum represents a broad line (half-width ~40 mT). However, Mn^{2+} signal is absent at 77°K. On addition of human serum albumin to this solution the Mn^{2+} signal is fixed even at 77°K. At the same time, when HSA is introduced into saliva the Mn^{2+} signal amplitude remains unchanged. This suggests that hydrated Mn^{2+} ions are absent in saliva and all the Mn^{2+} amount is bound to proteins.

To confirm this suggestion a study of the EPR spectra of 3% solutions of HSA, hyaluronidase of bovine testis (REANAL), lysozyme (3-crystallized from the egg-white) and bovine serum albumin, containing 1.0 x 10^{-4} Mn^{2+} has been carried out. Like in saliva, in all these solutions Mn^{2+} gave rise to a six-component EPR signal (Figure 2-17).

Consequently, in human saliva Mn^{2+} is bound to proteins in fact. The saliva Mn^{2+} content is the same before and after a thorough toilet of the mouth cavity (table 2-4) and independent on the presence of carious teeth, steel bridges and fillings. This shows that the presence of Mn^{2+} is not caused ether by forming elements or by microorganisms.

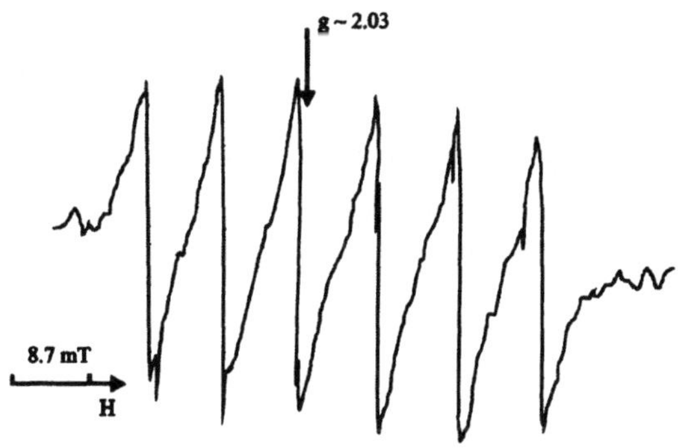

Figure 2-16. The EPR spectrum of Mn^{2+} ion aqueous solution. The registration conditions: temperature – 293°K, SHF power – 5 mW, magnetic field scanning velocity - 10 mT/min, HF modulation – 0.5 mT, the time constant – 0.3 sec.

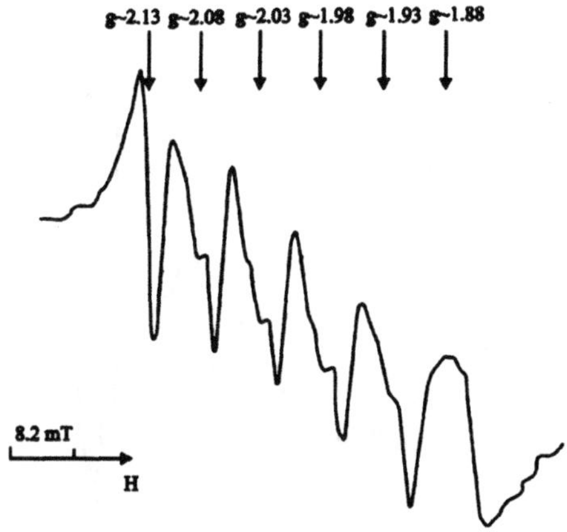

Figure 2-17. The EPR spectrum of Mn^{2+} ions in HSA. The registration conditions: temperature – 77°K, SHF power – 5 mW, magnetic field scanning velocity - 10 mT/min, HF modulation – 0.5 mT, the time constant – 0.3 sec.

Table 2-4. The Mn^{2+} concentration in the saliva of donors prior to (-) or after (-) toilet of the mouth cavity (M ± m)

Sex	Mn^{2+} (10^{-6} M), (-)	Mn^{2+} (10^{-6} M), (+)
M	11.6 ± 0.35	11.5 ± 0.31
W	7.0 ± 0.54	7.1 ± 0.42

In mixed human saliva about 50 enzymes related to oxyreductases, transferrases, hydrolases and isomerases have been found [59, 60]. Some of enzymes such as RNA-ase, DNA-ase, phosphatase [61, 62] contain manganese [38].

2.4 NASAL SECRETION

Patients without inflammatory diseases of the upper respiratory tract (URT) served as nasal secretion donors. Their occupation was not concerned with exposure to industrial gases, chemicals, sprays and dust. Paramagnetic center with g = 4.3 and 2.05 and free radicals with g = 2.0024-2.0050 have been revealed in human nasal secretion (Figure 2-18). The Mn^{2+} signal was present in the EPR spectrum of mucous nasal discharge of 7% of donors, whereas PMC signals of small intensity at g = 4.3 occurred in the spectra of 38% of donors [63].

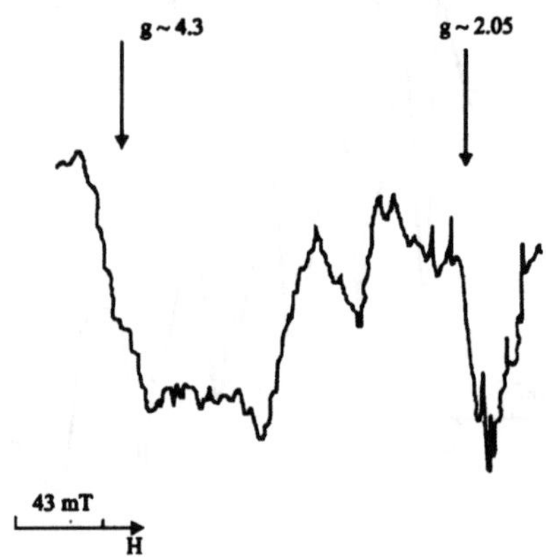

Figure 2-18. The EPR spectrum of human nasal secretion. The registration conditions: temperature – 77°K, SHF power – 5 mW, magnetic field scanning velocity - 50 mT/min, HF modulation – 0.5 mT, the time constant – 0.3 sec.

2.4.1 Free Radicals

As mentioned previously, one of the most important parameters of the EPR signal of free radicals is g factor whose value is determined by FR structure (table 2-5).

Table 2-5. g-Factor range for free radicals of various nature

Range of values	Radicals
2.0023	Free electron
2.0030 - 2.0050	Semiquinones
2.0050 - 2.0060	Nitroxides
2.01 - 2.02	Peroxides
2.02 - 2.06	Thiyl radicals

g-Factor is of great importance in FR identification by EPR spectroscopy [64-66].

Using the EPR method the g-factor and the signal line half-width (ΔH) of free radicals were determined in the discharge of nasal mucous membrane. For this purpose 23 men and women 18-57 years of age, without inflammatory changes of the upper respiratory tracts were chosen as donors. Their occupation was not connected with exposure to unfavourable environmental factors.

The FR EPR signal of nasal secretion represents a slightly asymmetrical singlet in the region of g-factor = 2.00, which is easily saturated with SHF power.

In most cases (70% of examined individuals) the FR EPR signal g-factor was within a range from 2.0024 to 2.0028. Its average value 2.0026 ± 0.0002 did not differ from the corresponding g-factor observed in the blood plasma EPR spectrum of healthy donors 2.0025 ± 0.0002. The ΔH value of EPR signal of NMM discharge 1.20 ± 0.02 mT is somewhat larger than that of blood plasma 1.10 ± 0.02 mT; $P < 0.01$) [67].

It is most likely that FR EPR signals of normal donor nasal secretion commonly correspond to cations and anions of polycyclic hydrocarbons [65] and more seldom to those of other classes of free radicals, in particular, flavosemiquinones.

The similarity of spectral characteristics of the EPR signal of NS and blood plasma free radicals suggests that, like in blood plasma, similar electron transport mechanisms responsible for final optimal level of redox processes function occur in NS. Therefore, the NS free radical EPR spectrum can be considered as a summarized EPR signal arisen from not only free radical complexes present in the secretion of glands and NMM goblet cells, but also from free radicals incoming due to transudation from the microcirculation channel. Some changes of the EPR spectral parameters can be related with fluctuations in the content of semi-oxidized ubiquinone forms in nasal secretion [68].

2.4.2 Nasal Secretion Metal-Proteides in Volunteers

For assignment of the EPR signal at $g = 4.3$ the nasal secretion (1 ml) with addition of 0.01 ml of 0.01 M $FeSO_4$ solution was cooled to 77°K and examined using EPR spectroscopy. The spectrum showed a 13-fold increase in the content of PMC at $g = 4.3$ after $FeSO_4$ introduction. The spectral shape was characteristic of Fe^{3+} with albumin Fe^{3+} (absence of splitting on the top and of jag on the descending curve typical of the transferrin signal) (Figure 2-19).

For assignment of the EPR signal at $g = 2.05$ the EPR spectrum of 0.05 ml of the nasal cavity discharge with addition of 0.01 ml of 0.01 M sodium diethyldithiocarbamate solution was studied. In this case a signal with a characteristic superfine structure caused by the Cu^{2+}-DEDTC complex was observed in the spectrum (Figure 2-20) [8, 69].

On $FeSO_4$ introduction the nasal secretion EPR spectrum shows a decrease in the Cu^{2+} concentration from $(0.89 \pm 0.09) \times 10^{-5}$ M to $(0.57 \pm 0.099) \times 10^{-5}$ M; $P < 0.05$) alongside with a decrease in the signal $g = 4.3$ (Figure 2-21).

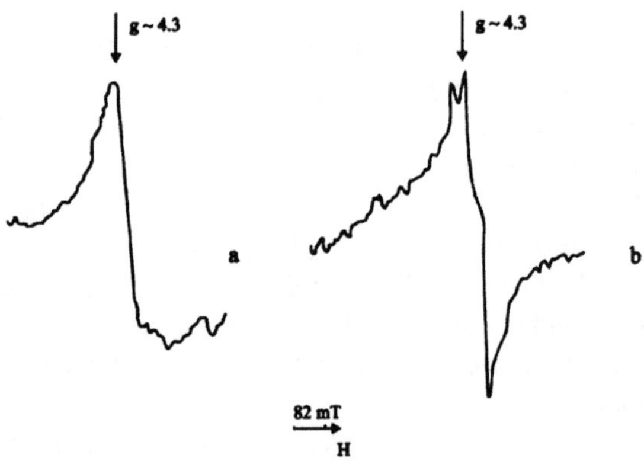

Figure 2-19. EPR spectra: A - of human nasal secretion after incubation with a $FeSO_4$ solution; B – blood plasma transferrin. The registration conditions: temperature – 77°K, SHF power – 5 mW, magnetic field scanning velocity - 10 mT/min, HF modulation – 0.5 mT, the time constant – 0.3 sec.

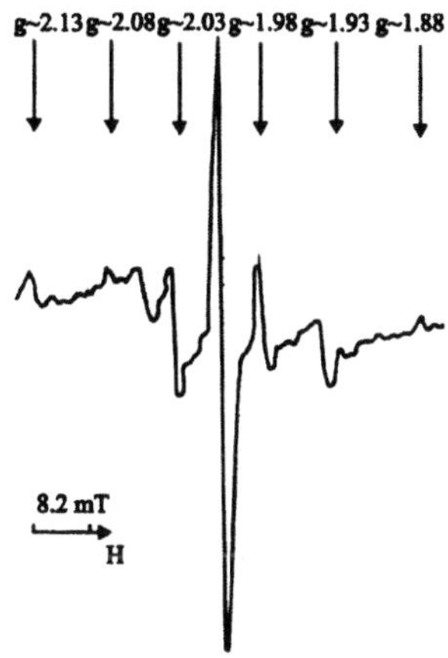

Figure 2-20. The EPR spectrum of human nasal secretion after incubation with DEDTC solution. The registration conditions: temperature – 77°K, SHF power – 5 mW, magnetic field scanning velocity - 10 mT/min, HF modulation – 0.5 mT, the time constant – 0.3 sec.

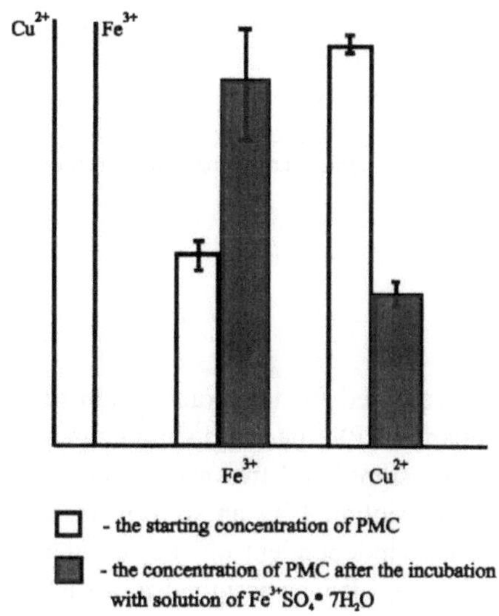

Figure 2-21. The Fe^{3+} and Cu^{2+} contents of human nasal secretion after incubation with $FeSO_4$ solution.

This points to ferroxidase activity of the nasal mucus copper-containing protein. After introduction of 0.01 ml of 5% ascorbic acid solution the intensity of the Cu^{2+} signal at g = 2.05 in the EPR spectrum of nasal secretion is reduced from (1.01 ± 0.13) x 10^{-5} M to (0.73 ± 0.08) x 10^{-5} M (P < 0.05) (Figure 2-22). This suggests copper-containing protein to possess ascorbatoxidase activity [68].

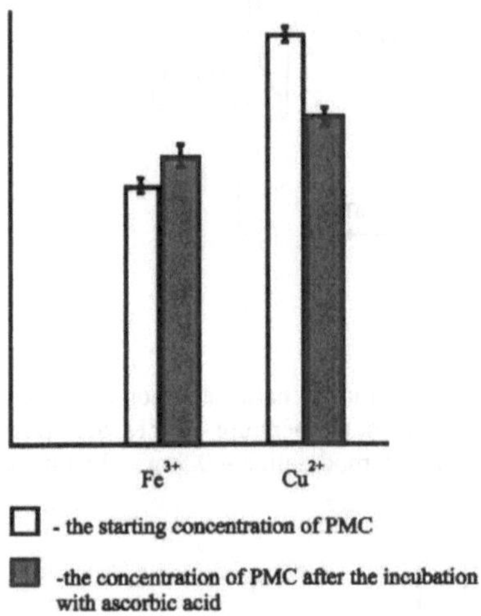

Figure 2-22. The ion Fe^{3+} and Cu^{2+} contents of human nasal secretion after incubation with an ascorbic acid solution.

2.5 TEETH

EPR spectroscopy was used for the examination of human tooth enamel dried at 400°C. The enamel EPR spectrum shows three spectral components: one corresponds to CO_3^{3-} signal and two arise from CO_2^- signals. The latter two radicals can be differentiated by anisotropy of the tooth enamel EPR spectrum. One of them is independent on the orientation of enamel blocks and disappears upon heating. It was assigned to a surface radical. The other CO_2^- component is mainly responsible for the angular orientation of enamel blocks and belongs to the bulk position. The CO_3^{3-} ion angular orientation is discussed. These results can be of help in interpreting the EPR spectra of tooth enamel [70].

2.6 CONTENTS OF STOMACH

The following paramagnetic centers have been found in the stomach contents: Mn^{2+} (a six-component EPR signal), a Hem-NO (signal with line half-width 7.5 mT and maxima at g = 2.1, 2.07, 2.007 and 1.98 with a triplet structure) and centers characterized with g = 2.007 and splitting of 1.5-1.6 mT (Figure 2-23) [8, 16, 71].

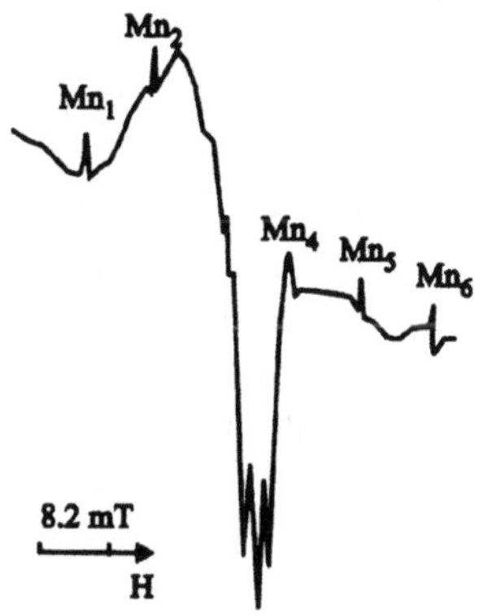

Figure 2-23. The EPR spectrum of human stomach contents. The registration conditions: temperature – 77°K, SHF power – 5 mW, magnetic field scanning velocity - 10 mT/min, HF modulation – 0.5 mT, the time constant – 0.3 sec.

The application of EPR spectroscopy Hem-NO studies has been considered in a recent review [72]. The Hem-NO signal was fixed in tumor cells and in the blood of animals exposed to nitrites and other toxic compounds [73-76]. This signal was also present in the spectra tumors of human liver, large intestine and stomach. It is most often found in necrotic area of tumor [77]. In the EPR spectrum of the hemoglobin-NO (Hb-NO) complex a signal at $g_x = 2.065$, $g_y = 2.01$ and $g_z = 2.005$ was observed. After introduction of sodium dodecylsulfate or salicylate into a Hem-NO solution a triplet superfine structure at g = 2.01 (g_x, = 2.094, g_y = 2.051 and g_z = 2.009) and splitting 1.7 mT. This suggests that the unpaired electron in Hem-NO is mainly localized on the Fe atom orbital. Antibonding of the hemoglobin structure with sodium dodecylsulfate or salicylate increases the unpaired electron density on the NO nitrogen atom [78-80]. A triplet superfine structure with a center at g = 2.01 also appears on addition of inositol-hexaphosphate, a hemoglobin allosteric effector [81]. Analogous data were obtained in EPR studied of Hb-NO and Mb-NO the Q-range [82].

The nitrosyl complexes can form cytochromes, peroxydases, catalases and other enzymes. The signal shape is specific for each type of hemoproteins which allows them to be distinguished. For instance, a Hem-NO signal was recorded in the EPR spectrum of human hemoglobin [83], mosquito monomeric hemoglobin [84], cytochrome P_{450} of animal and bacterial microsomes [85], cytochrome "C" [86].

In healthy animals hemoglobin-NO is produced only after introduction of considerable amounts of nitrite [73, 83].

Hem-NO EPR signals are recorded in animal tissues during metabolism of biologically active substances containing nitro- and nitroso groups (5-nitrofuran, dimethylnitrosourea derivatives, *etc.*) [87]. These signals appear as a result of NO production from the above compounds.

In the stomach contents, in the presence of hem-containing groups (hemoglobin, myoglobin, catalase, cytochromes, *etc.*), Hem-NO can form nitrates or nitrites, nitrate- and nitrite-reducing enzymes or microorganisms containing the above enzymes. Hem-NO also appears at pH above 3.0, *i.e.*, under reduced acidity, which promotes the development of microflora.

Maximum Hem-ON concentration is recorded in the sample taken from the empty stomach (Figure 2-24).

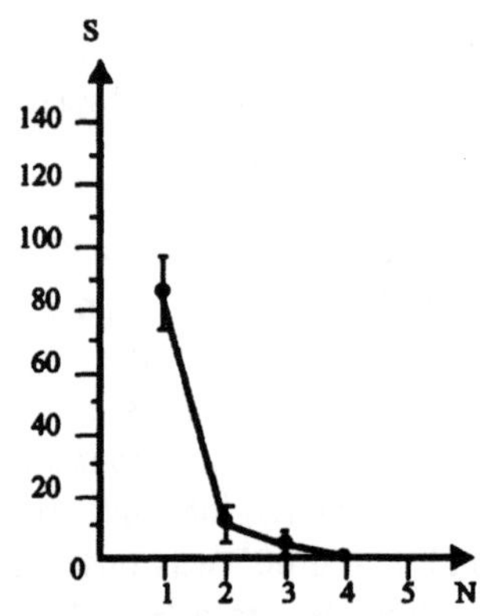

Figure 2-24. The dependence of Hem-NO concentration (S is an area below the Hem-NO absorption curve) on the gastric juice portion (N).

The probe-assisted extraction of the stomach contents causes irritation of the stomach, thus stimulating peristalsis and evacuation. This reduces the Hem-NO concentration in the basal and stimulated portions.

Thus, it is likely that nitrate- and nitrite-reducing microorganisms act as NO-transporters to the human stomach. This can be supported in the following way. The addition of 0.5 ml of hemolyzed erythrocytes to 10 ml of stomach contents increases two-fold the Hem-NO signal intensity (2 h incubation at 37°C), whereas the introduction of 0.5 ml of 1 M $NaNO_3$ results in a 1.4-fold increase (table 2-6). This suggests that in the stomach the content of NO prccursors is higher than that of Hem.

Incubation of 10 ml of gastric contents with addition of 0.5 ml of 1 M $NaNO_3$ and 0.5 of hemolyzed erythrocytes leads to a sharp increase in Hem-NO concentration at 37°C (table 2-6).

Table 2-6. The effect of hemolyzed erythrocytes and $NaNO_3$ on the Hem-NO concentration in the stomach contents (M ± m)

	Hem-NO concentration (10^{-5}M)		Concentration gain
Ingredients added	before the test	after the test	
Hemolyzed erythrocytes	8.4 ± 1.16	15.8 ± 2.11	1.9 ± 0.06
$NaNO_3$	8.4 ± 1.16	12.0 ± 1.67	1.4 ± 0.07
Hemolyzed erythrocytes and $NaNO_3$	8.4 ± 1.16	121.7 ± 40.78	11.8 ± 2.85

Maximum concentration is observed after 1.5-3 h, further it decreases and completely disappears after 24-48 h (Figure 2-25).

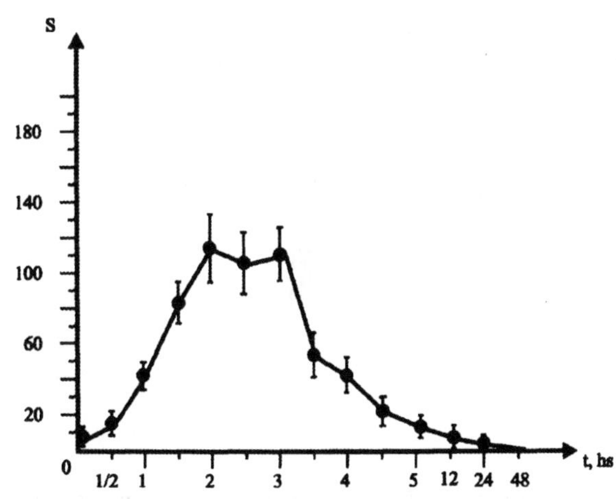

Figure 2-25. The dependence of Hem-NO concentration (S is an area below the Hem-NO absorption curve) in the donor gastric juice on the incubation time at 37°C with addition of 0.5 ml of 1 M $NaNO_3$ and 5 ml of hemolyzed erythrocytes.

There is a close relation between the Hem-NO level and the content of microorganisms in the stomach. From the empty stomach portion the following bacteria were sown out in concentrations ranging from 10^4 to 10^7 cells in 1 ml: *Bifidobacteria, Lactobacilli, Bacteroides species, Clostridium species, Yeasts, Proteus mirabilis, Proteus vulgaris, Streptococcus viridans, Streptococcus anhaemoliticus, Bacteroides fragilis, Citrobacter, Enterobacter* [88].

Thus, Hem-NO appears in the stomach contents owing to the reproduction of nitrate- and nitrite-reducing microorganisms which form NO bound to hem-containing groups [8, 89].

The Mn^{2+} signal having a characteristic six-component superfine structure can be detected in the spectrum of stomach contents only in the presence of duodenal-gastric reflux (Figures 2-26, 2-27) [90, 91].

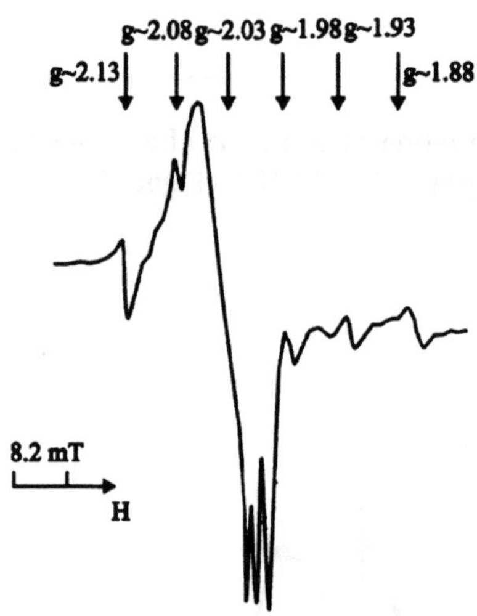

Figure 2-26. The EPR spectrum of human gastric juice at a moderate duodenum-gastric reflux. . The registration conditions: temperature – 77°K, SHF power – 5 mW, magnetic field scanning velocity - 10 mT/min, HF modulation – 0.5 mT, the time constant – 0.3 sec.

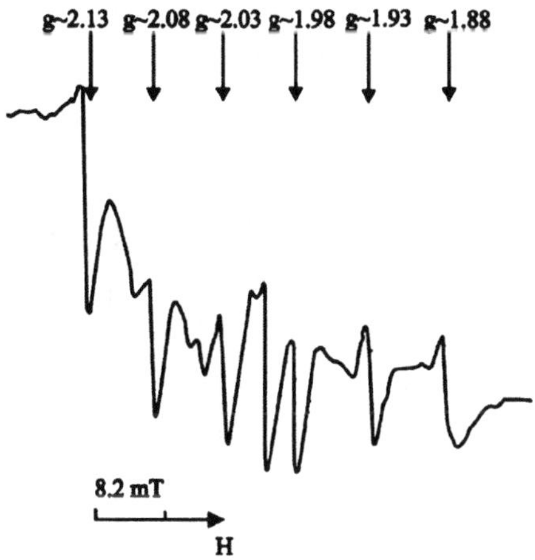

Figure 2-27. The EPR spectrum of human gastric juice at a clearly-expressed duodenum-gastric reflux The registration conditions: temperature – 77°K, SHF power – 5 mW, magnetic field scanning velocity - 10 mT/min, HF modulation – 0.5 mT, the time constant – 0.3 sec.

2.7 DUODENAL CONTENTS

Normal duodenal contents consists of three portions: A, B and C. Portion A represents a mixture of stomach and duodenum contents including pancreatic juice, some bile, intestinal juice. Portion B contains mainly bile from the gallbladder. Portion C represents a bile from the hepatic duct.

Portion A of the duodenal contents gives a complex EPR spectrum (Figure 2-28) representing imposition of PMC signals from gastric and duodenal contents, pancreatic juice and bile [8, 92].

In the spectra of portions B and C there is a six-component Mn^{2+} signal and a signal with SFS at g_{II} and $g_{\perp}$ - 2.077 and 2.018, respectively (Figures 2-29, 2-30) [8, 92].

As mentioned above, portion B is closely related to the bile cyst. Therefore, bile of 22 patients with chronic calculous cholecystitis collected during cholecystectomy was examined.

The bile was taken under strictly sterilized conditions without any contact with rubber items. The bile sample of one of the patients, a woman of 55 years of age, can serve as an example.

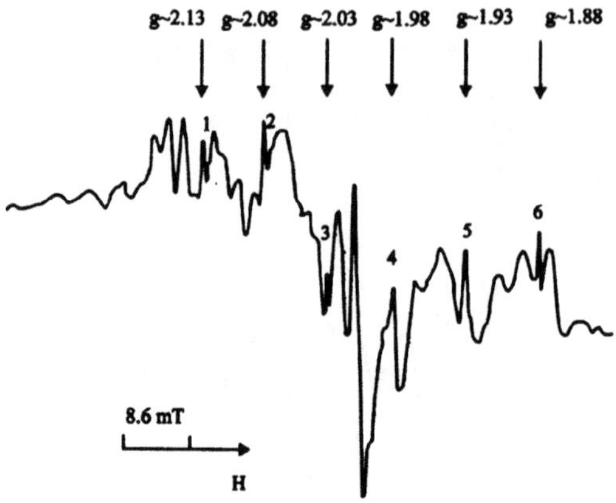

Figure 2-28. The EPR spectrum of human duodenal contents (portion A). The registration conditions: temperature – 77°K, SHF power – 5 mW, magnetic field scanning velocity - 10 mT/min, HF modulation – 0.5 mT, the time constant – 0.3 sec.

In the EPR spectrum of her bile there were Cu^{2+} and Mn^{2+} signals as well as signals at g = 6.0 and 4.3 corresponding to high-spin Fe^{3+} and low-spin Fe^{3+}, respectively (Figure 2-31) [8, 92].

The signal arising from bile Cu^{2+} (Figure 2-32) is like that in the EPR spectrum of ceruloplasmin in an alkaline medium [93].

At a 2-3 h contact of bile with rubber there appears an EPR Cu^{2+} signal (Figure 2-33) identical to that in the spectrum of duodenal contents taken with a rubber probe (figs 2-29, 2-30).

A similar signal was fixed in the EPR spectra of cells of microorganisms and animal tissues treated with xanthogenates and some organic solvents (dimethylformamide, acetone) [69]. An analogous signal was also observed in the EPR spectrum of human plasma and blood serum specimens after being in contact with rubber items (gloves, plugs, *etc.*) [94]. This was explained by the fact that ceruloplasmin Cu^{2+} was bound to sulfur-containing rubber vulcanizes of Altax type bringing about an EPR signal identical to that observed in the duodenal contents and bile after extraction of the samples with a rubber probe.

In order to support this suggestion DEDTC, an Altax analog was added to the bile. The shape of Cu^{2+} EPR spectra was identical to that of the spectrum of duodenal contents withdrawn by a rubber probe. Consequently, the EPR signal observed corresponds to Cu^{2+}.

Copper in the bile seems to be of serum origin and bound to ceruloplasmin [95].

The EPR signal at g = 6.0 is caused by MetHb Fe^{3+} and its derivatives which are removed with the bile in large amounts from the organism [96]. In the EPR spectrum of human 5-lipoxygenase, a non-hem iron-containing enzyme, a strong multicomponent signal with g = 6.0 appears upon titration with an oxidant (13-hydroperoxyoctadienoic acid). The signal approaches its maximum when approximately 1 mol of the titration reagent is added. The signal at g = 4.3 in the EPR spectrum of this enzyme is less than 3% of the total iron content. Addition of a reducing agent (hydroxyurea) cancels the g = 6.0 signal. This indicates the signal at g = 6.0 signal to arise from the Fe^{3+} form of 5-lipoxygenase [97].

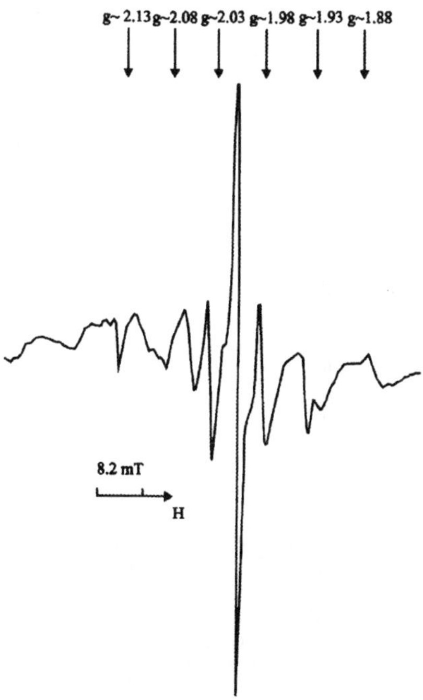

Figure 2-29. The EPR spectrum of human duodenal contents (portion B). The registration conditions: temperature – 77°K, SHF power – 5 mW, magnetic field scanning velocity - 10 mT/min, HF modulation – 0.5 mT, the time constant – 0.3 sec.

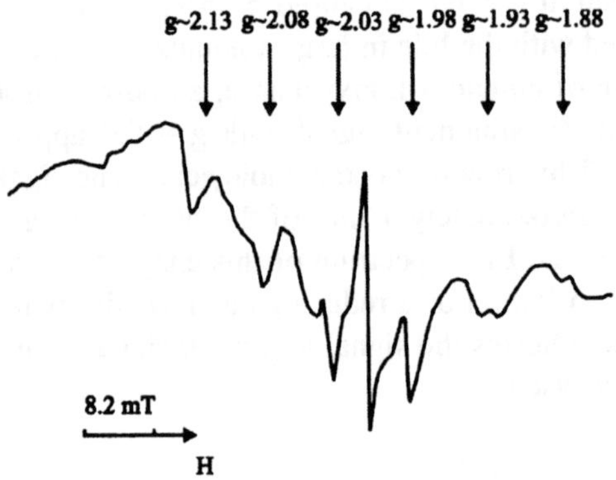

Figure 2-30. The EPR spectrum of human duodenal contents (portion C). The registration conditions: temperature – 77°K, SHF power – 5 mW, magnetic field scanning velocity - 10 mT/min, HF modulation – 0.5 mT, the time constant – 0.3 sec.

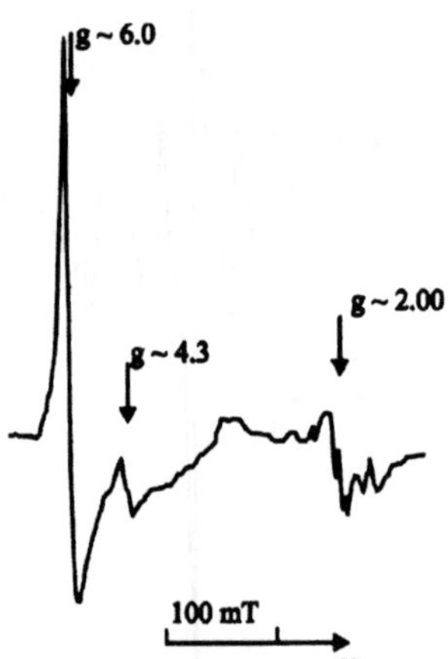

Figure 2-31. The EPR spectrum of human B-bile (Patient Ts-va, case record N 6284). The registration conditions: temperature – 77°K, SHF power – 5 mW, magnetic field scanning velocity speed - 50 mT/min, HF modulation – 0.5 mT, the time constant – 0.3 sec.

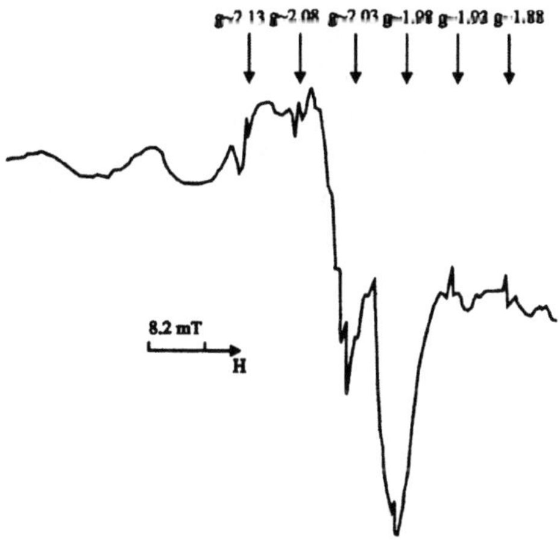

Figure 2-32. The EPR spectrum of human B-bile (prior to any contact with a rubber item). The registration conditions: the temperature – 77°K, SHF power – 5 mW, magnetic field scanning velocity - 10 mT/min, HF modulation – 0.5 mT, the time constant – 0.3 sec.

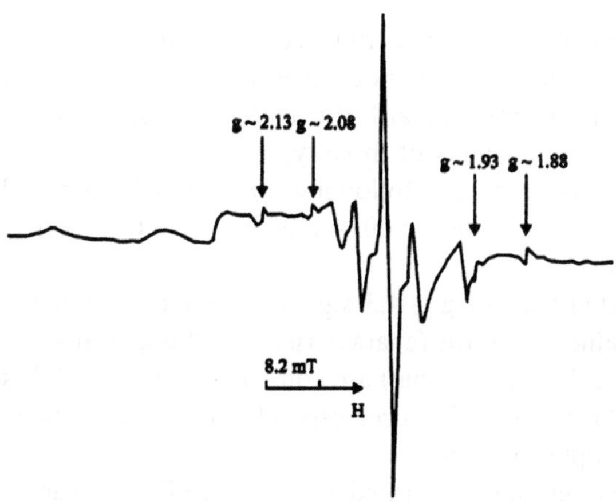

Figure 2-33. The EPR spectrum of human B-bile (after the contact with a rubber item). The registration conditions: temperature – 77°K, SHF power – 5 mW, magnetic field scanning velocity - 10 mT/min, HF modulation – 0.5 mT, the time constant – 0.3 sec.

The EPR signal at g = 4.3 is related to Fe^{3+} occurring most likely in lactoferrin [98]. Many other blood plasma proteins such as albumin, transferrin, lactoferrin are excreted with the bile [99].

Mn^{2+} in the duodenal juice seems to be involved in phosphatase, DNA-ase, RNA-ase [30]. Some portion of the juice comes from the pancreas, since its content in portion C is higher than in portion B $(6.1 \pm 0.60) \times 10^{-6}$ M and $(3.1 \pm 0.36) \times 10^{-6}$ M, respectively. The pancreas contains a considerable amount of manganese releasing into the duodenum [100].

The Cu^{2+} and Mn^{2+} contents in portions B and C differ only quantitatively. Portion B contains slightly more Cu^{2+} $(1.0 \pm 0.06) \times 10^{-5}$ M and $(0.9 \pm 0.07) \times 10^{-5}$ M and $(3.1 \pm 0.36) \times 10^{-6}$ M and $(6.1 \pm 0.60) \times 10^{-6}$ M, respectively.

2.8 SYNOVIAL FLUID

Synovial fluid (SF) resulted mainly from plasma dialysis acts as synovia. Only small and, on rare occasions, slightly larger size molecules penetrate SF due to gel-filtration through sinovial cells. The SF viscosity is ensured by hyaluronic acid. The major SF properties are dependent on various pathological conditions. This allows differentiation of dystrophic and pathologic diseases of joints [101].

Investigation of the synovial fluid composition provides an informative method in the diagnostics of various diseases of joints. In the synovial fluid of patients with rheumatoid arthritis (RA) and deforming osteoarthrosis (DOA) paramagnetic centers of three types were observed (Figures 2-34, 2-35). They correspond to the following EPR signals: g = 4.3 (Fe^{3+}), 2.05 (Cu^{2+}) and free radical with g 2.0024-2.0029, ΔH 0.6-0.8 mT of unidentified nature [102, 103].

It is suggested that the g = 4.3 signal is caused by transferrin Fe^{3+} since the g factor value and form (characteristic splitting in the upper maximum and a jag on the descending line) are coincident with the Fe^{3+} spectrum. The g = 2.05 signal is most likely to correspond to Cu^{2+} atoms incorporated into a protein of ceruloplasmin type.

This suggestion was confirmed using an EPR spectrum of rheumatoid synovial fluid (1 ml) after addition of 0.2 ml of 0.01M $FeSO_4$ solution. It turned out that the amplitude of the EPR signal at g = 4.3 having a characteristic spectral form of three valence iron on transferrin had increased more than ten-fold, whereas that of the signal at g = 2.05 decreased 4 times (Figure 2-36).

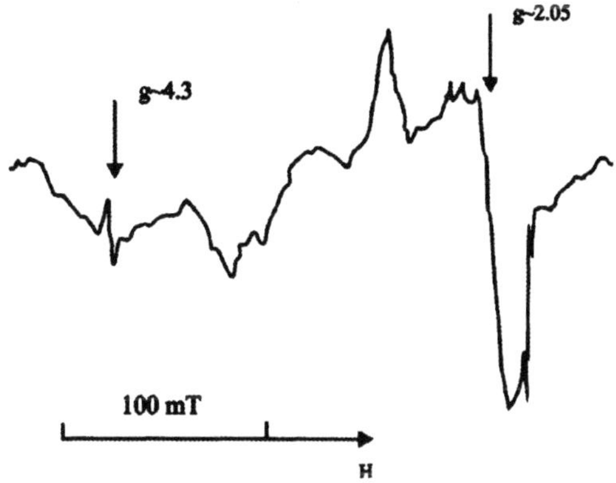

Figure 2-34. The EPR spectrum of synovial fluid of a patient with rheumatoid arthritis (Patient M-ova, case record N 967/9423). The registration conditions: temperature – 77°K, SHF power – 5 mW, magnetic field scanning velocity - 50 mT/min, HF modulation – 0.5 mT, the time constant – 0.3 sec.

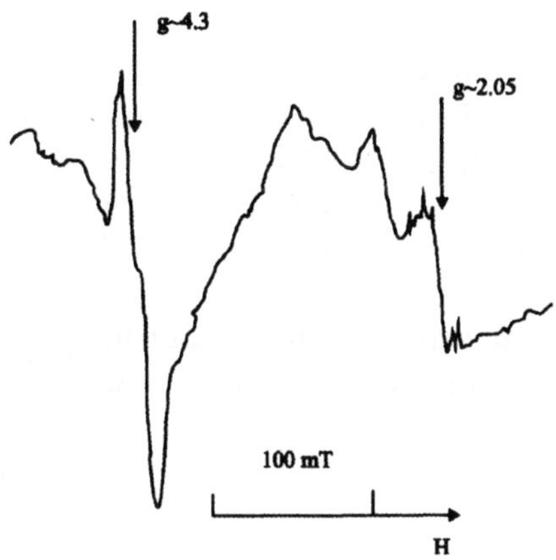

Figure 2-35. The EPR spectrum of a patient with osteoarthrosis deformans (Patient Ch-va, case record N 414/4488). The registration conditions: temperature – 77°K, SHF power – 5 mW, magnetic field scanning velocity - 50 mT/min, HF modulation – 0.5 mT, the time constant – 0.3 sec.

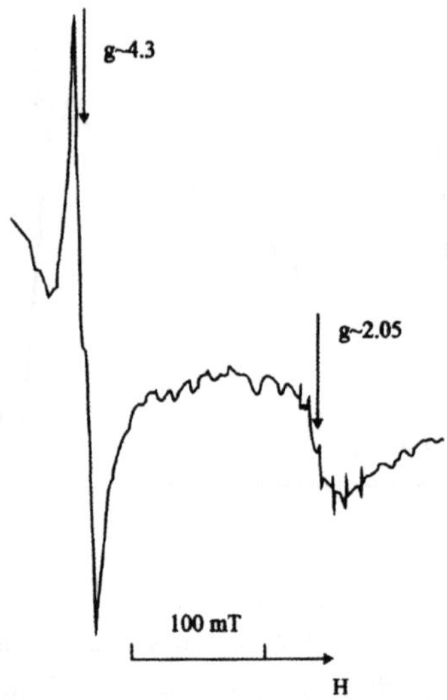

Figure 2-36. The EPR spectrum of synovial fluid of a patient with rheumatoid arthritis (after exposure to a $FeSO_4$ solution) (Patient M-ova, case record N 967/9423). The registration conditions: temperature – 77°K, SHF power – 5 mW, magnetic field scanning velocity - 50 mT/min, HF modulation – 0.5 mT, the time constant – 0.3 sec.

This shows that Fe^{2+} is oxidized to Fe^{3+} by Cu^{2+} of the rheumatoid synovial fluid. Besides, this is also indicative of the presence in the fluid of a considerable amount of protein of TF type unbound to iron.

Quite a different picture is observed in the rheumatoid synovial fluid (1 ml) drawn from DOA patients. The addition of 0.2 ml of 0.01M $FeSO_4$ solution did not cause such an intensive rise of the Fe^{3+} level. This indicates either poor ferroxidase activity of the fluid or the presence of a decreased number of transferrin molecules, or saturation with Fe^{3+} [102, 103].

2.9 FECES

The EPR spectra of normal human feces display the following signals (Figures 2-37, 2-38): at g = 6.0 corresponding to Fe^{3+} of MetHb and its derivatives, at g = 4.3 arising from low-spin Fe^{3+}, Cu^{2+} with characteristic SFS, a six-component spectrum of Mn^{2+}, Hem-NO, and free radicals [8].

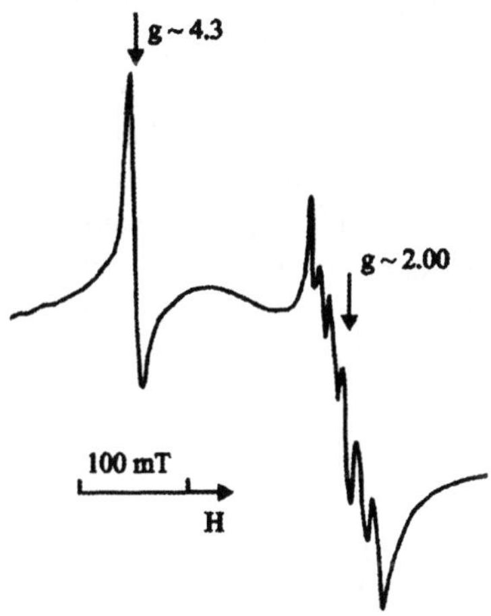

Figure 2-37. The EPR spectrum of human feces. The registration conditions: temperature – 77°K, SHF power – 5 mW, magnetic field scanning velocity - 50 mT/min, HF modulation – 0.5 mT, the time constant – 0.3 sec.

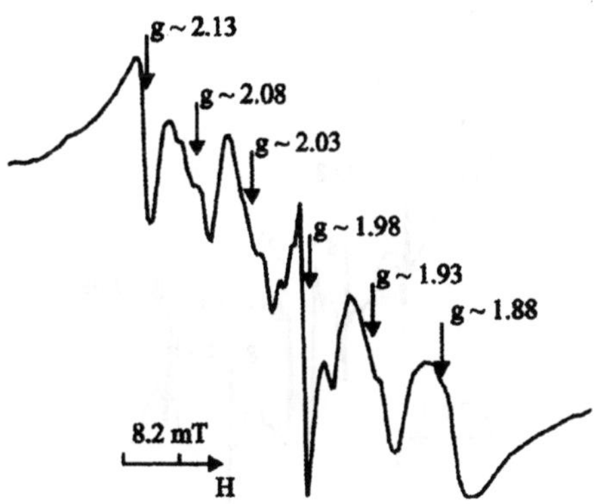

Figure 2-38. The EPR spectrum of human feces. The registration conditions: temperature – 77°K, SHF power – 5 mW, magnetic field scanning velocity - 10 mT/min, HF modulation – 0.5 mT, the time constant – 0.3 sec.

To elucidate the source of paramagnetic centers in feces the large intestine microorganisms were isolated. In feces anaerobes of the *Bacteroides* genus as well as *E.coli* and *Citobacter* aerobes were mainly found.

Figures 2-39 and 2-40 show the EPR spectra of *E.coli* aerobes, whereas in Figures 2-41 and 2-42 *Bacteroides* anaerobes are presented (their content is nearly the same in all samples: approximately 10^9 of cells in 1 ml). In the EPR spectra of anaerobes the Mn^{2+} signal is fixed whereas Fe^{3+} signal is absent. In anaerobes, on the contrary, the Mn^{2+} content is very low (Figure 2-40) (Mn^{2+} is observed only under maximum amplification of the spectrometer), the Fe^{3+} content being high.

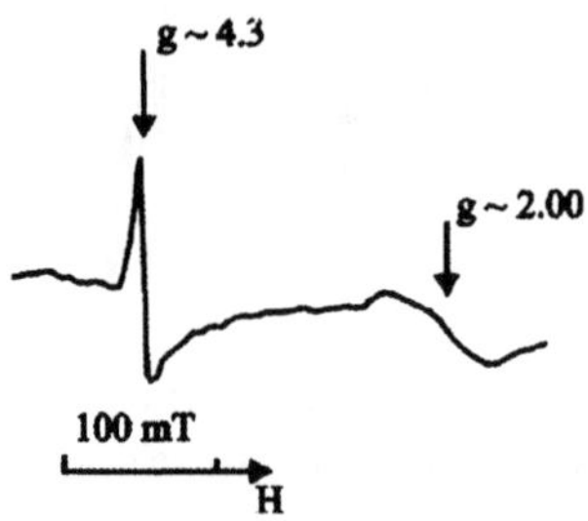

Figure 2-39. The EPR spectrum of pure culture of E.coli aerobes from human feces. The registration conditions: temperature – 77°K, SHF power – 5 mW, magnetic field scanning velocity - 50 mT/min, HF modulation – 0.5 mT, the time constant – 0.3 sec.

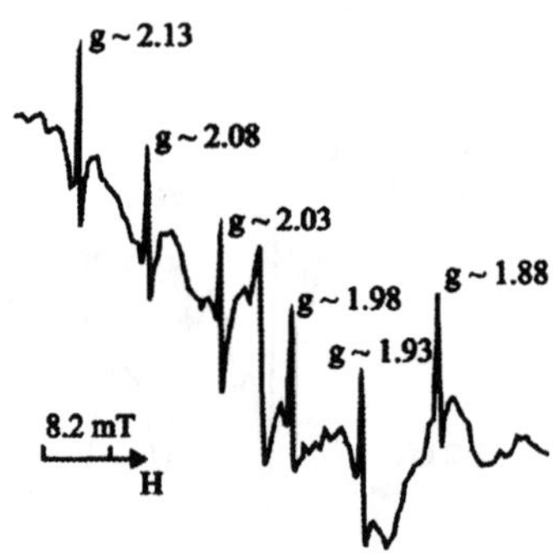

Figure 2-40. The EPR spectrum of pure culture of E.coli aerobes from human feces. The registration conditions: temperature – 77°K, SHF power – 5 mW, magnetic field scanning velocity - 10 mT/min, HF modulation – 0.5 mT, the time constant – 0.3 sec.

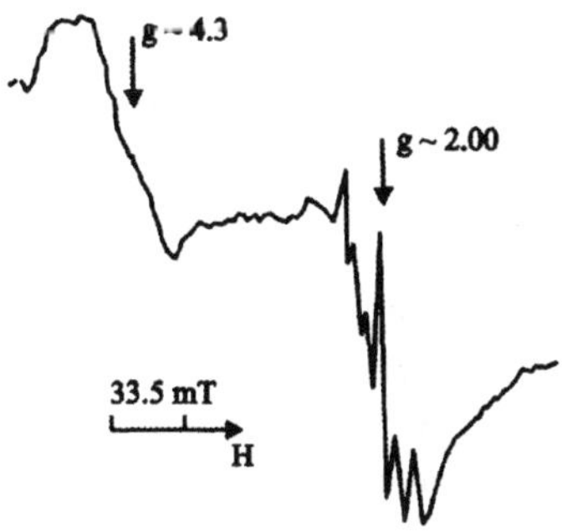

Figure 2-41. The EPR spectrum of pure culture of Bacteroides anaerobes from human feces. The registration conditions: temperature – 77°K, SHF power – 5 mW, magnetic field scanning velocity - 50 mT/min, HF modulation – 0.5 mT, the time constant – 0.3 sec.

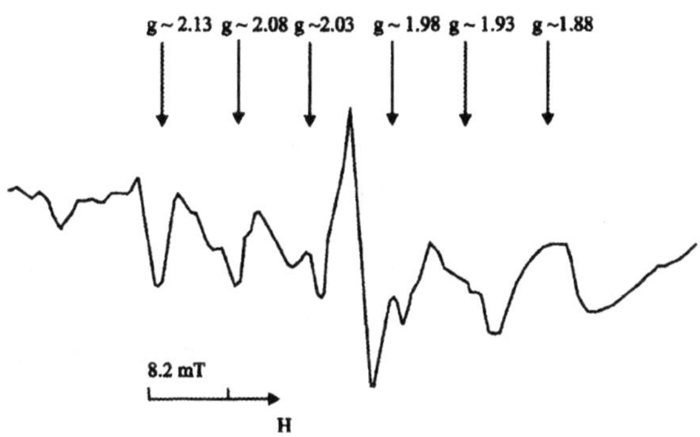

Figure 2-42. The EPR spectrum of pure culture of Bacteroides anaerobes from human feces. The registration conditions: temperature – 77°K, SHF power – 5 mW, magnetic field scanning velocity - 10 mT/min, HF modulation – 0.5 mT, the time constant – 0.3 sec.

In human ferrochelatase non-heme Fe was revealed (approximately 2 mole per 1 mole of the purified enzyme, M(r) = 40.000). The EPR spectrum of the dithione-reduced enzyme showed the presence of a $[2Fe\text{-}2S]^{+}$ cluster with 1/2 spin and a high-spin Fe^{2+} heme. The educed enzyme gives rise to an EPR signal g = 2.00, 1.94, 1.91 which readily saturates at low microwave power below 10°K. The signal is observed without significant broadening at temperatures up to 100°K. The Fe-S cluster is labile and gradually disappears with concomitant loss of enzyme activity when the enzyme is stored aerobically at 4°C over a period of 24 h [104].

Colon bacillus (*E.coli*) contains significant amount of Fe in both heme and non-heme forms. Recombinant human microsomal heme oxygenase-2 contained in *E.coli* was studied using EPR spectroscopy. At pH = 7 the ferric heme is six coordinate high-spin and six coordinate low-spin at pH > 7 (pK_α = 8.5). The reaction with hydrogen peroxide converts the heme of the heme oxygenase-2 fragment complex into a verdoheme-like product. The oxidation of 3-chloroperbenzoic acid leads to an oxoferryl derivative. As the spectroscopic properties of heme oxygenase-1 and heme oxygenase-2 are similar their catalytical mechanism seems to be identical [105].

The same authors [106] have shown that it is His-45, but not His 152, that represents the heme proximal ligand in hemoxygenase-2, whereas His-152 plays a structuring role in the enzyme protein stabilization [106].

Thus, in the EPR spectrum of feces main contribution to the Mn^{2+} signal is made by anaerobes, whereas aerobes are responsible for the Fe^{3+} signal. It should be noted that anaerobes show higher FR concentrations [107].

In feces Hem-NO is formed of nitrates, nitrites and heme brought to the intestine with food, water, bile and saliva [108]. In the intestine there are nitrate- and nitrite-reducing microorganisms [109].

In diarrhea clearly defined signals at g = 6.0 and Hem-NO appear in the EPR spectrum the Cu^{2+} and Mn^{2+} contents are decreased (Figures 2-43 and 2-44). This seems to be related to changes of the intestinal microflora and increased water content in feces [110].

Samples of feces were collected at different time of the day depending on human individual physiological features. Normally the samples were prepared not later than 18 h after evacuation. Therefore it was thought reasonable to study the effect of feces storage time at room temperature on the PMC concentration (table 2-7). The samples were stored in glass vials over a period of 24 h at room temperature. The EPR spectra were run every other 12 and 24 h.

Table 2-7. Concentration dynamics of Fe^{2+} (mcmol/l), Mn^{2+} (10^{-6} M) and Hem-NO (10^{-5} M) in donors' feces during storage (M ± m)

S*	Initial values			12 h later			24 h later		
	Fe^{3+}	Mn^{2+}	Hem-NO	Fe^{3+}	Mn^{2+}	Hem-NO	Fe^{3+}	Mn^{2+}	Hem-NO
M	66.0 ± 5.10	10.3 ± 0.83	14.4 ± 1.09	67.6 ± 4.30	10.6 ± 0.55	14.6 ± 1.19	63.4 ± 5.30	9.9 ± 0.80	14.8 ± 1.19
W	95.2 ± 7.40	7.3 ± 0.29	13.5 ± 2.06	99.8 ± 7.90	6.9 ± 0.39	14.3 ± 2.23	91.3 ± 7.30	6.9 ± 0.34	14.8 ± 2.11

Note: P > 0.05 in all cases; S - sex

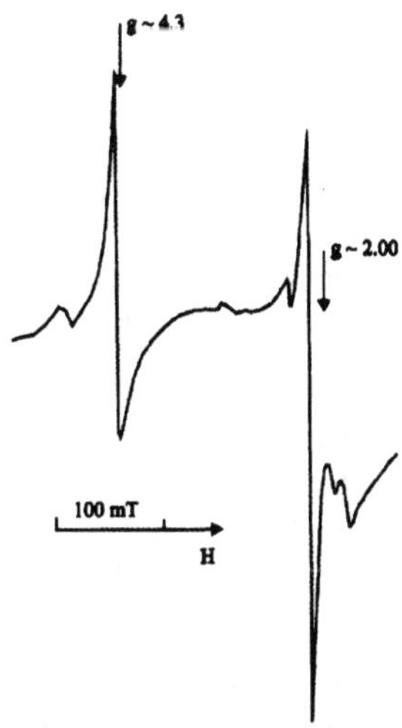

Figure 2-43. The EPR spectrum of human feces. The registration conditions: temperature – 77°K, SHF power – 5 mW, magnetic field scanning velocity - 50 mT/min, HF modulation – 0.5 mT, the time constant – 0.3 sec.

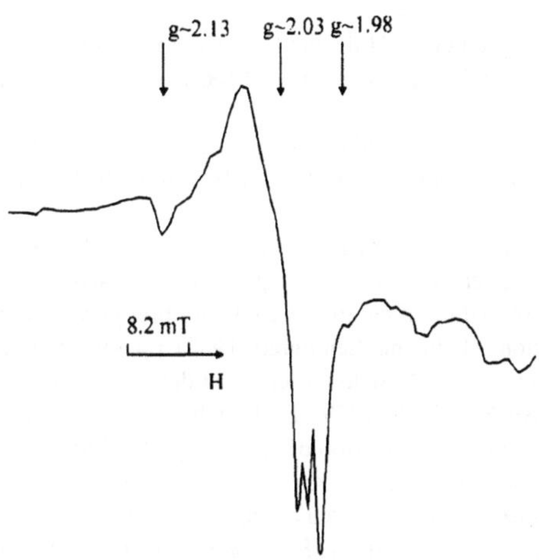

Figure 2-44. The EPR spectrum of human feces. The registration conditions: temperature – 77°K, SHF power – 5 mW, magnetic field scanning velocity - 10 mT/min, HF modulation – 0.5 mT, the time constant – 0.3 sec.

A tendency toward decreasing the Fe^{3+} and Mn^{2+} concentration and increasing the Hem-NO content after 24 h storage has been established.

REFERENCES

1. B. Halliwell and J. M. C. Gutterige, Free Radicals in Biology and Medicine, Clarendon Press, Oxford (1989), pp. 324-332.
2. L. A. Bynoe, S. Pou, J. D. Gottsch and G. M. Rosen, Light-dependent spin trapping of hydroxyl radical from human erythrocytes, Biochem. Biophys. Res. Commun. **179**:1305 (1991).
3. A. Constantinescu, D. Han and L. Packer, Vitamin E recycling in human erythrocyte membranes, J. Biol. Chem. **268**:10906 (1993).
4. E. Lloyd, J. C. Ferrer, W. D. Funk, M. R. Mauk and A. G. Mauk, Recombinant human erythrocyte cytochrome b_5, Biochemistry. **33**:11432 (1994).
5. A. Herrmann and P. F. Devaux, Alteration of the aminophospholipid translocase activity during in vivo and artificial aging of human erythrocytes, Biochim. Biophys. Acta. **1027**:41 (1990).
6. M. Brumen, R. Heinrich, A. Herrmann and P. Muller, Mathematical modelling of lipid transbilayer movement in the human erythrocyte plasma membrane, Eur. Biophys. J. **22**:213 (1993).
7. F. Basse, J. Sainte-Marie, L. Maurin and A. Bienvenue, Effect of benzyl alcohol on phospholipid transverse mobility in human erythrocyte membrane, Eur. J. Biochem. **205**:155 (1992).
8. R. G. Saifutdinov, Paramagnetic Centers in Human Biological Fluids and Their Diagnostic and Pathogenetic Role in Some Internal Diseases, Thesis Dr.Sci., Medical Institute, Tomsk, (1989) (Russ.).
9. M. Lohman, J. Schreiber, D. Holz and F. R. Hopkins, The effect of alcohol oxidase on the ascorbic acid-erythrocyte interaction: ESR investigations, Z. Naturforsch. **38**:864 (1983).
10. R. G. Saifutdinov and K. R. Sedov, Method of glutathione reductase activity determination in erythrocytes, Pat SSSR N 1057866 (1981), Bull. Izobr. N44:167 (1983) (Russ.).
11. Kh. N. Rubina, Biochemistry of Erythrocytes, in: Physiology of blood. Physiology of Erythropoiesis, Nauka, Moskva (1979), pp. 211-232 (Russ.).
12. I. Shonka, Determination of pentose cycle in erythrocytes, in: Pentose Cycle and Ionizing Radiation, Meditsina, Leningrad (1968), pp. 49-59 (Russ.).
13. R. G. Saifutdinov and K. R. Sedov, Method of determining blood safety in long storage, Pat SSSR N 912169 (1979), Bull. Izobr. N10:19 (1982).
14. R. S. Polishchuk, On determination of preserving blood hemolysis degree, Gematologiya i perelivanie krovi (Kiev), 110 (1965) (Russ.).
15. S. S. Bryukhanenko, Artificial Circulation, Nauka, Moskva (1964) (Russ.).
16. K. R. Sedov and R. G. Saifutdinov, Electron Paramagnetic Resonance in Internal Diseases Clinics, Irkutskii Dom Pechati, Irkutsk (1993) (Russ.).
17. Y. Sato, S. Kamo, T. Takahashi and Y. Suzuki, Mechanism of free radical-induced hemolysis of human erythrocytes: hemolysis by water-soluble radical initiatorm, Biochemistry. **34**:8940 (1995).

18. D. P. Naughton and M. C. Symons, When is a radical not a radical? Free Radic. Res. **25**:1 (1996).
19. M. M. Moronne, R. J. Mehlhorn, M. P. Miller, L. C. Ackerson and R. I. Macey, ESR measurement of time-dependent and equilibrium volumes in red cells, J. Membr. Biol. **115**:31 (1990).
20. E. Galembeck, A. Alonso and N. C. Meirelles, Effects of polyoxyethylene chain length on erythrocyte hemolysis induced by poly[oxyethylene (n) nonylphenol] non-ionic surfactants, Chem. Biol. Interact. **113**:91 (1998).
21. Y. Sato, S. Kanazawa, K. Sato and Y. Suzuki, Mechanism of free radical-induced hemolysis of human erythrocytes. II. Hemolysis by lipid-soluble radical initiator, Biol. Pharm. Bull. **21**:250 (1998).
22. Y. Sato, H. Yamakose and Y. Suzuki, Participation of band 3 protein in hypotonic hemolysis of human erythrocytes, Biol. Pharm. Bull. **16**:188 (1993).
23. E. J. Hustedt and A. H. Beth, Analysis of saturation transfer electron paramagnetic resonance spectra of a spin-labeled integral membrane protein, band 3, in terms of the uniaxial rotational diffusion model, Biophys. J. **69**:1409 (1995).
24. C. E. Cobb and A. H. Beth, Identification of the eosinyl-5-maleimide reaction site on the human erythrocyte anion-exchange protein: overlap with the reaction sites of other chemical probes, Biochemistry. **29**:8283 (1990).
25. C. E. Cobb, S. Juliao, K. Balasubramanian, J.V. Staros and A.H. Beth, Effects of diethyl ether on membrane lipid ordering and on rotational dynamics of the anion exchange protein in intact human erythrocytes: correlations with anion exchange function, Biochemistry. **29**:10799 (1990).
26. W. Retelewska, M. Gaczynska, G. Bartosz and L. Judkiewicz, Consequences of the presence of elongated variant of the major transmembrane protein (Band 3 protein) in the human erythrocyte, Clin. Chim. Acta. **198**:255 (1991).
27. T. Pomorski, A. Herrmann, A. Zachowski, P. F. Devaux and P. Muller, Rapid determination of the transbilayer distribution of NBD-phospholipids in erythrocyte membranes with dithionite, Mol. Membr. Biol. **11**:39 (1994).
28. S. Kh. Mezhidov and V. A. Monseev, The effect of temperature on erythrocyte hydration, Biofizika. **36**:294 (1991) (Russ.).
29. V. I. Grishchenko, S. Kh. Mezhidov, V. A. Moiseev and O. A. Nardid, The effect of temperature and concentrations of various substances on the microviscosity of erythrocyte cytosol, Biofizika. **40**:106 (1995) (Russ.).
30. W. Schon, C. Ziegler, H. Gartner and G. Kraft, Heavy ion induced membrane damage: hemolysis of erythrocytes and changes in erythrocyte membrane fluidity, Radiat. Environ. Biophys. **33**:233 (1994).
31. Y. Fukushima and H. Kon, On the mechanism of loss of deformability in human erythrocytes due to Heinz body formation: a flow EPR study, Toxicol. Appl. Pharmacol. **102**:205 (1990).
32. S. Kh. Mezhidov, O. A. Nardid and V. A. Moiseev, The ESR spin-probe method in the study of erythrocyte permeability for cryoprotectors, Biofizika. **41**:1278 (1996) (Russ.).
33. J. C. Sulpice, A. Zachowski, P. F. Devaux and F Giraud, Requirement for phosphatidylinositol 4,5-bisphosphate in the Ca(2+)-induced phospholipid redistribution in the human erythrocyte membrane, J. Biol. Chem. **269**:6347 (1994).
34. A. Erdmann, I. Bernhardt, A. Herrmann and R. Glaser, Species-dependent differences in the influence of ionic strength on potassium transport of erythrocytes, The role of membrane fluidity and $Ca2^{+}$, Gen. Physiol. Biophys. **9**:577 (1990).

35. L. E. Eriksson and H. Beving, Calcium- and lead-activated morphological changes in human erythrocytes: a spin label study of the cytoplasm, Arch. Biochem. Biophys. **303**:296 (1993).
36. M. Minetti and G. Scorza, Hypoxia-stimulated reduction of doxyl stearic acids in human red blood cells. Role of hemoglobin, Biochim. Biophys. Acta. **1074**:112 (1991).
37. T. Pocklington, M. A. Foster and J. R. Mallard, Plasma total iron-orbinding capacity measured by ESR, Phys. Med. Biol. **22**:98 (1977).
38. N. J. Blackburn, Metalloproteins, Electron Spin. Reson. **8**:413 (1984).
39. A. Mickalad, C. Deby, Dupont G. Deby and M. Lamy, An electron spin resonance (ESR) study on the mechanism of ascorbyl radical production by metal-binding proteins, Biometals. **11**:81 (1998).
40. G. Musci, M. C. Bonaccorsi di Patti and L. Calabrese, The state of the copper sites in human ceruloplasmin, Arch. Biochem. Biophys. **306**:111 (1993).
41. G. Musci, M. C. Bonaccorsi di Patti and L. Calabrese, Modulation of the redox state of the copper sites of human ceruloplasmin by chloride, J. Protein. Chem. **14**:611 (1995).
42. V. B. Vassiliev, A. M. Kachurin, M. Beltramini, G. P. Rocco, B. Salvato and V. S. Gaitskhoki, J. Inorg. Biochem. **65**:167 (1997).
43. V. V. Rylkov, M. Iu. Tarasev and K. A. Moshkov, A novel form of copper-containing centers in human ceruloplasmin, Biokhimiya. **55**:1367 (1990) (Russ.).
44. V. V. Rylkov, M. Yu. Tarasiev and K. A. Moshkov, Labile conformation of type 2 Cu2+ centres in human ceruloplasmin, Eur. J. Biochem. **197**:185 (1991).
45. L. I. Idel'son, Gipokhromnaya Anemiya, Meditsina, Moskva (1981) (Russ).
46. P. L. Indovina, ESR studies on different pathological states, Ann. Ist. Super Sanita. **14**:863 (1978).
47. E. J. Cohn, M. L. Hughes and J. R. Weare, Crystallization of serum albumins from ethanol-water mixtures, J. Am. Chem. Soc. **69**:1753 (1947).
48. V. N. Petrov, Physiology and Pathology of Iron Metabolism, Nauka, Leningrad (1982) (Russ.).
49. S. A. Kopylovskii, g – 4.5 EPR signal registration in test animal tissues, Magnitnyi Rezonans v Biologii i Meditsine, Moskva (1989), pp. 90-91 (Russ.).
50. A. V. Kozlov, D. Yu. Egorov, Yu. A. Vladimirov and S. A. Azizova, The formation of iron-ascorbic acid complexes in tissues, Biofizika **35**:513 (1990) (Russ.).
51. A. Seidel, E. Bill, L. Haggstrom, P. Nordblad and F. Kilar, Complementary Mossbauer and EPR studies of iron(III) in diferric human serum transferrin with oxalate or bicarbonate as synergistic anions, Arch. Biochem. Biophys. **308**:52 (1994).
52. D. C. Ross, T. J. Egan and L. R. Purves, Periodate modification of human serum transferrin Fe(III)-binding sites. Inhibition of carbonate insertion into Fe(III)- and Cu(II)-chelator-transferrin ternary complexes, J. Biol. Chem. **270**:12404 (1995).
53. J. Hirose, H. Fujiwara, T. Magarifuchi, Y. Iguti, H. Iwamoto, S. Kominami and K. Hiromi, Copper binding selectivity of N- and C-sites in serum (human)- and ovo-transferrin, Biochim. Biophys. Acta. **1296**:103 (1996).
54. J. K. Grady, A. B. Mason, R. C. Woodworth and N. D. Chasteen, The effect of salt and site-directed mutations on the iron(III)-binding site of human serum transferrin as probed by EPR spectroscopy, Biochem. J. **309**(Pt 2):403 (1995).
55. V. M. Chumakov, V. M. Kalmanson, EPR study of coenzyme Q_{10} in mitochondria and tissues of rats, Dokl. Akad. Nauk SSSR. **170**:714 (1966) (Russ.).
56. M. S. Blois and J. E. Maling, The quenzyme Q_{10} and Vitamin E, K semiquinone free radicals, Biochem. Biophis. Res. Commun. **3**:132 (1960).

57. K. R. Sedov and R. G. Saifutdinov, EPR study of α-tocopherilquinone in serum blood lipids in normal state, Biofizika **26**:571 (1981) (Russ.).
58. R. G. Saifutdinov and K. R Sedov, Method of α-tocopherilquinone determination in biological fluids, Pat SSSR N 679206 (1977), Bull. Izobr. N30:12 (1979) (Russ.).
59. I. B. Zbarskii and L. F. Agidamov, Nucleases of saliva and saliva gland in normal and pathological state, Vestnik Akad. Med. Nauk SSSR. N11:3 (1971) (Russ.).
60. M. Lindgren, M. Svensson, P. O. Freskgard, U. Carlsson, P. Jonasson, L. G. Martensson and B. H. Jonsson, Characterization of a folding intermediate of human carbonic anhydrase II: probing local mobility by electron paramagnetic resonance, Biophys. J. **69**:202 (1995).
61. A. N. Golovchenko and A. P. Levitskii, Saliva desoxyribonuclease and blood serum activity on pathients with acute respiratory disease and tonsillitis, Zhurnal Ushnykh, Nosovykh i Gorlovykh Boleznei. N4:61 (1978) (Russ.).
62. A. M. Solntsev, A. M. Korolenko and E. A. Kolesnikov, Acid and alkaline phosphatase activity in parotid saliva on saliva gland disease, Vrachebnoe delo. N4:50 (1981) (Russ.).
63. A. G. Shanturov, R. G. Saifutdinov and E. V. Nosulya, EPR Spectroscopy in Clinical Rhinology, Vostochno-Sibirskoe Knizhnoe Izdatelstvo, Irkutsk (1992) (Russ.).
64. J. A. Weil, J. R. Bolton and J. E. Wertz, Electron Paramagnetic Resonance. Elementary theory and Practical applications, John Wiley & Sons, INC, New-York (1994).
65. D. Borg, Application of electron paramagnetic resonance in biology, in: Free Radicals in Biology, William A. Pryor, ed., Acad. Press, New York-San Francisco-London (1976), pp. 69-147.
66. Ya. I. Azhipa, Medical-Biological Aspects of Electron Paramagnetic Resonance Applications, Nauka, Moskva (1983) (Russ.).
67. A. G. Shanturov, E. G. Nosulya and R. G. Saifutdinov, EPR study of nasal secret free radicals, VII Respublikanskii S'ezd Otolaringologov Ukrainy, Odessa (1989), p.281 (Russ.).
68. E. G. Nosulya, High Respiratory Chronic Disease: Diagnostic and Pathogenetic Role of Nasal Secret PMC, Thesis Dr.Sci., Nauchno-Issledovatel'skii Institut Ucha, Gorla, Nosa i Rechi, Sankt-Peterburg (1994) (Russ.).
69. R. D. Borodulin and A. F. Vanin, EPR determination of copper in cells and tissues Biofizika. **17**:1083 (1972) (Russ.).
70. F. Callens, P. Moens and R. Verbeeck, An EPR study of intact and powdered human tooth enamel dried at 400 degrees C, Calcif. Tissue. Int. **56**:543 (1995).
71. R. G. Saifutdinov, Nitric Oxide joining with heme in human's juice EPR, Newslette. **4**:3 (1992).
72. G. Bemski, Contribution of Electron Paramagnetic Resonance to the studies of hemoglobin: the nitrosylhemoglobin system, Mol. Biol. Rep. **24**:263 (1997).
73. Ya. I. Azhipa, Application of Electron Paramagnetic Resonance in Medicine. Nauka, Moskva (1979) (Russ.).
74. A. F. Vanin, L. V. Vakhnina and A. G. Chetverikov, On nature EPR signal in cancer tissues, Biofizika. **15**:1044 (1970) (Russ.).
75. A. N. Saprin, Nature and Role of Free Radicals and Other PMC on Cancerogenesis, Thesis Dr.Sci., Moskva (1974) (Russ.).
76. M. J. Brennan, T. Cole and J. A. Singley, A unique hyperfine ESR spectrum in mouse neoplasma analysed by computer simulatio, Proc. Soc. Exper. Biol. and Mod. **123**:715 (1966).

77. M. C. Symons, I. J. Rowland, N. Deighton, K. Shorrock and K. P. West, Electron spin resonance studies of nitrosyl haemoglobin in human liver, colon and stomach tumour tissues, Free Radic. Res. **21**:197 (1994).
78. E. Wajnberg, M. P. Linhares, L. J. el-Jaick and G. Bemski, Nitrosyl hemoglobin: EPR components at low temperatures, Eur. Biophys. J. **21**:57 (1992).
79. E. Wajnberg, G. Bemski, L. J. el-Jaick and O. C. Alves, Nitrosyl hemoglobins: EPR above 80°K, Int. J. Biol. Macromol. **18**:231 (1996).
80. L. E. Eriksson, Binding of nitric oxide to intact human erythrocytes as monitored by electron paramagnetic resonance, Biochem. Biophys. Res. Commun. **203**:176 (1994).
81. P. Ascenzi, A. Bertollini, M. Coletta, A. Desideri, B. Giardina, F. Polizio, R. Santucci, R. Scatena and G. Amiconi, Cooperative effect of inositol hexakisphosphate, bezafibrate, and clofibric acid on the spectroscopic properties of the nitric oxide derivative of ferrous human hemoglobin, J. Inorg. Biochem. **50**:263 (1993).
82. M. Flores, E. Wajnberg and G. Bemski, Temperature dependence of Q-band electron paramagnetic resonance spectra of nitrosyl heme proteins, Biophys. J. **73**:3225 (1997).
83. H. Rein, O. Ristau and W. Sheler, On the influence of allosteric effectors on the EPR spectrum on nitric oxide hemoglobin, FEBS Letts. **24**:24 (1972).
84. M. Overcamp, H. Twilfer and K. Gersonde, Conformation-controlled trans-effect of the proximal histidine in haemoglobins, An ESR study of monomeric nitrozyl-^{57} Fe-haemoglobins, Z. Naturforsch. **31**:524 (1976).
85. R. E. Ebel, D. H. O'Keeffe and J. A. Peterson, Nitric oxide complexes of cytochrome P-450, FEBS Letts. **55**:198 (1975).
86. H. Kon, Paramaganetic resonance study of nitric oxide hemoglobin, J. Biol. Chem. **243**:4350 (1968).
87. Z. V. Kuropteva, O. N. Pastushenko, Changing in animal blood and liver paramagnetic complexes by action nitroglycerin, Dokl. Akad. Nauk SSSR. **281**:189 (1985) (Russ.).
88. R. G. Saifutdinov, Method of nitrate-reducing ability determination of contents stomach, Pat SSSR N 1529115 (1986), Bull. Izobr. N46:167 (1989) (Russ.).
89. R. G. Saifutdinov, Nitric oxide joining with heme in human's juice, The World Congress of Gastroenterology, USA, California, Los Angeles, (1993), p.1365.
90. R. G. Saifutdinov, Method of duodenal-gastric reflux diagnostic, Pat SSSR N 1448889 (1986).
91. R. G. Sayfutdinov, Nitric oxide joining with heme in human's juice, YI International Symposium on Magnetic Resonance in Colloid and Interface Science, Ference (1992), p. 176.
92. R. G. Saifutdinov, EPR study of human duodenal contents and bile, Magnitnyi Rezonans v Biologii i Meditsine, Moskva, (1989), pp. 86-87 (Russ.).
93. R. M. Nalbandyan, Denaturated agents influence on ceruloplazmin EPR spectra parameters, Biofizika. **18**:821 (1973) (Russ.).
94. T. Kakuda, H. Tanaka, E. Kimoto and F. Morishige, EPR spectrum of human serum copper, Appl. Spectrosc. **34**:276 (1980).
95. B. M. Mullock and L. J. Shaw, Sources of proteins in human bile, Gut. **26**:500 (1985).
96. G. G. Ivanov, Diagnosis and bile acids determination in stomach contents, Laboratornoe delo. N5:277 (1978) (Russ.).
97. N. D. Chasteen, J. K. Grady, K. I. Skorey, K. J. Neden, D. Riendeau and M. D. Percival, Characterization of the non-heme iron center of human 5-lipoxygenase by electron paramagnetic resonance, fluorescence, and ultraviolet-visible spectroscopy: redox cycling between ferrous and ferric states, Biochemistry. **32**:9763 (1993).

98. N. N. Yasnova, Duodenal and gastric contents study for diagnosis of gastro-intestinal tract diseases, Meditsinskii Referativnyi Zhurnal (Gastroenterologiya). **XVII**(N9):1 (1983) (Russ.).
99. F. M. J. Zuyderhoudt, P. Vos, G. G. A. Jorning and J. Good, Ferritin in liver, plasma and bile of the iron-loaded rat, Biochim. Biophys. Acta **838**:381 (1985).
100. N. A. Skuya, Pancreas Diseases, Meditsina, Moskva (1986) (Russ.).
101. V. A. Nasonova and M. G. Astapenko, Clinical Rheumatology, Meditsina, Moskva (1989) (Russ.).
102. R. G. Saifutdinov, E. V. Popova and Yu. A. Goryaev, Paramagnetic centres in synovial fluid in patients with rheumatoid arthritis and osteoarthrosis deformans, EPR Newsletter. **3**:6 (1991).
103. E. V. Onuchina, Clinical and Diagnostic Value of Paramagnetic Antioxydants in Biological Fluids of Patients with Rheumatoid Arthritis and Osteoarthrosis, Thesis Ph.D., Moskva (1993) (Russ.).
104. H. A. Dailey, M. G. Finnegan and M. K. Johnson, Human ferrochelatase is an iron-sulfur protein, Biochemistry. **33**:403 (1994).
105. K. Ishikawa, N. Takeuchi, S. Takahashi, K. M. Matera, M. Sato, S. Shibahara, D. L. Rousseau, M. Ikeda-Saito and T. Yoshida, Heme oxygenase-2. Properties of the heme complex of the purified tryptic fragment of recombinant human heme oxygenase-2, J. Biol. Chem. **270**:6345 (1995).
106. K. Ishikawa, K. M. Matera, H. Zhou, H. Fujii, M. Sato, T. Yoshimura, M. Ikeda Saito and T. Yoshida, Identification of histidine 45 as the axial heme iron ligand of heme oxygenase-2, J. Biol. Chem. **273**:4317 (1998).
107. R. G. Sayfutdinov, Investigation of the Human's faeces by ESR, International Conference on Bioradicals Detected by ESR Spectroscopy, Yamagata (1994).
108. P. Fritsen, G de S. Blangnat, D. Klein, Excretion of nitrates and nitrites in saliva and bile in the dog, Food and Chemical Toxicology **23**:655 (1985).
109. B. S. Brasar and M. J. Hill, Metabolism of nitrogen compounds, in: Human Intestinal Flora, Academic Press, London (1974), pp. 72-102.
110. R. G. Sayfutdinov, Investigation of the human's faeces by ESR, The World Congress of Gastroenterology, USA, California, Los Angeles, (1993), p. 2061.

Chapter 3

PARAMAGNETIC CENTERS IN THE ORGANISM OF VOLUNTEERS AND PATIENTS IN VARIOUS PATHOLOGICAL STATES

3.1 BIOLOGICAL SUBSTRATES OF NORMAL INDIVIDUALS

Investigation of paramagnetic centers in biological substrates has opened fresh opportunities for detailed understanding of physiological state of volunteers and elucidation of in-depth pathogenic mechanisms on cellular and molecular level. Paramagnetic centers give an exact and fine picture of the state of metabolic and general functional activities of tissues and the organism as a whole [1-4]. The use of EPR spectroscopy method in medicine and biology provides valuable information on paramagnetic regularities of physiological processes.

Examination of plasma and erythrocyte PMC of volunteers was mainly performed with "regular donors" (RD) (giving blood 2-3 times a year). It has been found that independently of the length of donation, in blood plasma of men the concentration of Fe^{3+} was higher and of Cu^{2+}-ceruloplasmin was lower than in women [5]. Earlier investigation has not revealed any significant sex-dependent differences in the Fe^{3+} and Cu^{2+}-CP contents [6-8]. This appears to be due to the fact that the authors did not study differentially the plasma of donors giving blood for the first time (new donors, ND) and of regular donors. Furthermore, many physiological loads (heavy physical labor, sport, *etc.*) cause iron-deficient state. This means that selection of donors should be done very carefully.

The Fe^{3+} content in RD blood plasma is always lower than in that of new donors (regular donors giving blood for a long time exhibit latent iron-deficiency) (table 3-1) [9-11].

Table 3-1. PMC content in donor blood plasma depending on the length of donation (M ± m)

Blood plasma PMC	New donors		Regular donors	
	men	women	men	women
Iron (EPR) (mcmol/l)	21.6 ± 0.3	16.0 ± 0.3	14.4 ± 1.7	11.8 ± 1.1
Iron (BP) (mcmol/l)	23.7 ± 0.5	17.5 ± 0.5	17.0 ± 1.9	14.5 ± 1.2
Difference between the iron content determined by BP and EPR methods (%)	9.5 ± 1.1	9.0 ± 1.3	20.0 ± 1.9	26.0 ± 1.9
Cu^{2+}-CP (10^{-5} M)	1.7 ± 0.1	1.8 ± 0.02	1.5 ± 0.1	1.5 ± 0.1
FR (10^{-6} M)	0.5 ± 0.0	0.5 ± 0.01	0.5 ± 0.1	0.5 ± 0.0
α-Tq (10^{-6} M)	2.3 ± 0.1	2.4 ± 0.06	2.2 ± 0.1	2.3 ± 0.1

The Fe^{3+} content in male regular donors decreases to a greater extent than in women (34% and 26%, respectively). Evidently, in women a more efficient iron-deficiency compensation has been developed during the long evolution period.

In male and female regular donors the decrease in Cu^{2+} ceruloplasmin level is likely to be due to enhanced incidence of hepatitis and other hepatic lesions leading to liver malfunction, in particular, ceruloplasmin production. Serum hepatitis markers are known to be met not infrequently among donors giving blood for a long time [12, 13]. The MetHb concentration in erythrocytes is higher in donor giving blood for a long time compared with new donors (table 3-2). Normal MetHb content ranges from 0.5 to 6.0 g/l [14]. According to EPR data the maximum MetHb content in donors amounts to 0.9 g/l [9].

Table 3-2. PMC content in erythrocytes of healthy donors depending on the length of donation (M ± m)

Erythrocyte PMC	New donors		Regular donors	
	men	women	men	women
MetHb, (g/l)	0.7 ± 0.02	0.8 ± 0.03	0.8 ± 0.05	0.8 ± 0.05
Cu^{2+} SOD, (10^{-5} M)	0.9 ± 0.04	0.8 ± 0.04	0.7 ± 0.05	0.7 ± 0.03
FR, (10^{-6} M)	1.2 ± 0.03	1.2 ± 0.03	1.3 ± 0.06	1.4 ± 0.09
α-Tq, (10^{-6} M)	1.7 ± 0.04	1.7 ± 0.05	1.6 ± 0.05	1.6 ± 0.05

The role of erythrocyte MetHb is being discussed for a long time. Using EPR spectroscopy the mechanism of lipid reorganization in oxidized cells was examined. It is found that the products of hemoglobin oxidation absorb fatty acids from solution, thus taking part in lipid distribution in the erythrocyte membranes [15].

EPR spectroscopy was used to study the reaction of human MetHb with hydrogen peroxide. On addition of H_2O_2 the EPR signals (77°K) of high-spin and low spin MetHb rapidly decreased and two EPR signals of free radicals formed. One of these was an anisotropic signal ($g_{\parallel} = 2.035$ and $g_{\perp} = 2.0053$), the other being an isotropic singlet ($g = 2.0042$, ΔH 2 mT) [16]. The interconversion of the Fe^{3+} EPR signal in solution between the high-spin ($g = 6.0$) and low-spin forms ($g = 2.00$) was quantitatively studied as a function of hydration. Dehydration causes a loss of paramagnetism leading to the appearance of about 40% Fe^{2+} below 0.40 H_2O/gHb. The remaining 60% of Fe^{3+} EPR signals at $g = 6.0$ is distributed in both the high-spin and low-spin forms (5% and 55%, respectively). Removal of water leads to displacements and torsion of the F helix changing the iron coordination at proximal site [17].

The increased hemoglobin stability against autooxidation is very important when erythrocytes are used in clinical practice. Independently of the Hb concentration, HbO_2 shows a more stable in water-containing solvents (ethanol, butanol, ethyleneglycol, glycerol) than in water. A study of the effect of alcohol on the temperature denaturalization of MetHb has shown MetHb to be less stable in ethanol and butanol than in water. In glycerol and ethyleneglycol the opposite is observed [18].

It is possible that erythrocytes need MetHb in small amounts. Normal MetHb content in functioning erythrocytes is controlled by certain biochemical systems. Thus, a relationship between the MetHb content in erythrocytes and the glutathione reductase activity has been revealed. The correlation coefficient in men and women is -0.7 and -0.8, respectively. This means that the more active GR the lower the MetHb level. Glutathione reductase is a transporter of SH groups reducing Fe^{3+} to Fe^{2+} [19, 20].

In the hemoglobin molecule there are proximal and distal histidines (His) which control binding of ligands, thus preventing Fe^{2+} oxidation. Hemoglobin having one of the histidines substituted by tyrosine either in the α- or in β subunits, is called mutant hemoglobin (HbM). Since HbM-invaded patients are cyanotic ones, all the mutant subunits circulating in the blood are supposed to be oxidized. Nevertheless, the patients with HbM Saskatoon exhibit smaller cyanosis. Under anaerobic conditions the mutant subunit HbM Saskatoon is reduced by MetHb-reductase purified from human erythrocytes, at a rate needed for complete reduction of the HbA (MetHb) oxidized form. Some amount of mutant subunits in HbM (Boston and Saskatoon) remains reduced in the fresh blood. All the mutant subunits in HbM give an EPR signal caused by factor g anisotropy at g = 6.0 comparable with that at $g_{\perp}$ = 6.00 of the oxidized normal Hb. Nevertheless, the EPR method enables rapid identification and differentiation of the two Hbs from normal and mutant subunits in a small amount of blood. The oxidized Hb (MetHb) of normal subunits of patients has an EPR signal only at $g_{\perp}$ = 6.00. In the EPR spectrum of mutant subunits HbM Saskatoon (β-63His →Tyr) there is a triplet at g_1 = 6.7, $g_{\perp}$ = 6.00 and g_2 = 5.3, whereas in the EPR spectrum of HbM Boston (α-58His →Tyr) there is a triplet at g_1 = 6.3, $g_{\perp}$ = 6.00 and g_2 = 5.7, respectively. The extent of oxidation of mutant subunits is higher than in normal ones. Over 50% of mutant subunits remain reduces in circulating blood. Careful shaking of fresh blood at 37°C for 15 h in air leads to auto-oxidation of normal and mutant subunits at different rates. Possibly NAD.H-MetHb reductase is involved in maintenance of the reduced mutant subunits of HbM Saskatoon [21].

In regular donors the free radical level is higher compared to that of donors giving blood for the first time (new donors). This indicates glutathionereductase system (GRS) activation, seemingly related to both increasing the MetHb level and enhancing hemolysis during storage. Donors giving blood for a long time exhibit a lowered erythrocyte Cu^{2+} level. This suggests an increased activity of superoxide dismutase [22, 23].

$$O_2^{-\bullet} + Cu^{2+}\text{-SOD} \rightarrow O_2 + Cu^{+}\text{-SOD}$$

Oxygen radical anion serves as a substrate for SOD. The activation of SOD shows an increasing concentration of superoxide radical anion.

The transferrin Fe^{3+} content on the erythrocyte surface is so small that could not be detected in 53% of men and 48% of women. The α-tocopherylquinone content in erythrocyte and plasma does not change depending on sex and age.

A nasal secretion examination was carried out in 19 women and 34 men from 16 to 56 years of age admitted to the otorhinolaryngologic hospital for planned surgery of curved nasal septum. In most patients the nasal mucous membrane was of pale pink colour and pathologic discharge was absent. The Fe^{3+}, Cu^{2+} and FR PMC concentration in the nasal mucous membrane discharge was 14.1±0.2 mcmol/l, (2.0 ± 0.2) x 10^{-5} and (0.9 ± 0.04) x 10^{-6} M, respectively. The concentration was independent either on sex or on age [24-26].

The Mn^{2+} concentration in the saliva of women donors is approximately 40% lower than in male saliva (table 3-3). The presence of Fe^{3+} in male saliva is observed more frequently.

Table 3-3. The Mn^{2+} content in donor saliva (M ± m)

Examined groups	Mn^{2+}, (10^{-6} M)
Men	10.29 ± 0.38
Women	6.23 ± 0.42

In the stomach contents Hem NO has been found in the same amount in 70% of donors of both sexes (table 3-4).

Table 3-4. The Hem-NO content in stomach contents (M ± m)

Examined groups	Hem-NO, (10^{-5} M)
Women	9.8 ± 1.28
Men	6.6 ± 1.88
Total	8.4 ± 1.18

Any reliable difference in the Mn^{2+} and Cu^{2+} content in duodenal contents of male and women donors has not been revealed (table 3-5).

Table 3-5. The PMC content in duodenal contents (M ± m)

Portion	PMC	Men	Women	Total
"B"	Mn^{2+} (10^{-6} M)	3.3 ± 0.33	2.7 ± 0.79	3.1 ± 0.36
	Cu^{2+} (10^{-5} M)	0.9 ± 0.08	1.0 ± 0.11	1.0 ± 0.06
"C"	Mn^{2+} (10^{-6} M)	6.8 ± 0.75	4.9 ± 0.73	6.1 ± 0.60
	Cu^{2+} (10^{-5} M)	0.9 ± 0.13	1.5 ± 0.39	1.0 ± 0.07

In feces of women the EPR signal level at g = 4.3 is higher than that in men (table 3-6). The source of this Fe^{3+} signal is mainly located in aerobic microorganisms and, possibly, in lactoferrin [27]. As known, in normal iron metabolism lactoferrin inhibits iron absorption in the gastro-intestinal tract [28, 29]. This was also observed in the donors examined. In men the Fe^{3+} level in plasma and feces was 21.6 ± 0.3 and 57.5 ± 2.0 mcmol/l, whereas in women these values were 16.0 ± 0.3 and 86.4 ± 3.4 mcmol/l, respectively [30].

Table 3-6. The PMC content in feces (M ± m)

PMC	Men	Women
Fe^{3+} (mcmol/l)	57.5 ± 2.00	86.4 ± 3.39
Mn^{2+} (10^{-5} M)	9.8 ± 0.50	6.4 ± 0.41
Hem-NO (10^{-5} M)	13.8 ± 0.52	10.0 ± 0.79

The fecal Mn^{2+} content brought about by anaerobes of the *Bacteroides* genus is lower women than in men. Possibly Mn^{2+} is brought to feces from the duodenal or with saliva. However, there is no correlation between the Mn^{2+} concentration in the duodenal contents, feces and saliva.

The Hem-NO concentration in woman donors is reliably lower than in man donors. This seems to be caused by nitrosamines incoming from tobacco smoke [31].

Thus, in biological substrates of normal human organism there are diverse paramagnetic centers involved in biochemical processes.

3.2 ISCHEMIC HEART DISEASE

Diseases of cardiovascular system give very high morbidity and mortality rates. The ischemic heart disease (IHD) occupies the first place among these lesions. The EPR spectroscopy method is a guide to some aspects of pathogenesis of cardiac and vascular diseases. It is also of help in diagnostics of a number of diseases and assessment of the action of drugs. Lipid metabolism and blood lipoprotein structure are traditionally investigated.

EPR spectroscopy was used for measurements of the surface potential and apoprotein structure of low-density lipoproteides (LDL) and high-density lipoproteides (HDL) in the presence of Ca^{2+} and dextran sulfate, heparin and chondroitin sulfate. Addition of Ca^{2+} leads to a decrease in the absolute values of LDL and HDL surface potential. The spectral properties of spin labeled maleimide covalently attached to the apoprotein were changed under aggregation of LDL induced by dextran sulfate or heparin in the presence of Ca^{2+}. In the HDL system this effect is observed only with dextran sulfate. The influence of polyethyleneglycol (PEG) on the spin label spectral parameters depends on the molecular mass of the biopolymer. PEG 4000 decreases the mobility of the spin labeled apoprotein region of LDL. At the same time PEG of higher molecular mass only slightly increases the maleimide mobility. On the other hand, the maleimide-labeled apoprotein region of HDL is more sensitive to the mass of all PEGs. Addition of PEG causes immobilization of apoprotein A [32].

To probe the head group of human plasma low density lipoproteins and very low density lipoproteins (VLDL) the use was made of nitroxide Tempil stearate whereas the acyl chain region was probed with spin labeled cholestane (ChSl) and methyl ester of 5-doxyl palmitate MeFASL(10,3). The EPR data were compared with the model spectrum which acquires a rapid anisotropic motion of nitroxide. The results show that in the head group region of LDL and VLDL a slowing down of the rotation motion occurs at lower temperature. The acyl chain region probed with MeFASL(10,3) is homogeneous at physiological temperatures. At lower temperatures, how-ever, the fluid and immobilized components coexist. The EPR spectrum of LDL indicate a strong immobilization ChSl axial rotation already at physiological temperatures [33].

Using EPR spectroscopy it is indicated that the lipid packing with spin labeled TEMPO on the surface of LDL and VLDL may be due to different intermolecular forces between phospholipids and cholesterol temperatures [34].

The effect of enzymatic modifications of HDL(3) surface lipid composition on their physical properties was studied by EPR spectroscopy. Human HDL(3) (d: 1.125-1.21 g/ml) was treated either with phospholipase A2 from *Crotalus adamanteus* or with sphingomyelinase from *Staphylococcus aureus* in the presence of albumin. The degree of hydrolysis of the glycerophospholipids treated in this way ranges from 13% to 81%. Sphingomyelinase leads to a 31-92% degradation of sphingophospholipids. Glycerophospholiposis indicates an enhanced rigidity of HDL(3) environment, a decrease in mobility and an increase in the order of the surface lipids. Treatment of HDL(3) with sphingophospholipase induces fluidization of lipids. All this suggests the important role of the two major phospholipids in controlling the fluidity and order on the HDL(3) surface [35].

Thyroxin-binding globulin, albumin, prealbumin, immunoglobulins and lipoproteides represent components of the system of iodothyroxin-binding human blood plasma proteins. As shown by EPR spectroscopy, this multicomponent system also involves molecular versions of major proteins, the presence of which is conditioned by genetic and post-translation features of the synthesis. High-density lipoproteides (apo-1, apoA-1-HDL and HDL3 can also serve as thyroxin transporter in human blood serum [36].

EPR spectroscopy was used to evaluate the effect of lipid peroxidation on the structure of surface proteolipid layer of human serum low-density lipoprotein. The results obtained with a positively charged spin label shows that LPO decreases the mobility of phospholipid molecule in the regions of polar head and acyl chain to a depth of 1.7 nm from the water-lipid interface. At relatively high levels of oxidation (over 6 mmol MDA/g LDL phospholipid) the lipid phase polarity increases. Lipid peroxidation enhances the mobility of strongly immobilized labeled maleimide. Oxidation of LDL leads to a decreased ability to incorporate spin labeled steroid (androstane) into lipoproteides compared with native ones. It is suggested that LPO-induced structural changes of LDL surface are responsible for increased accumulation of cholesterol in human monocytes in the process of their incubation with oxidized LDL. The cholesterol content in erythrocytes is directly related to the concentration of malonic dialdehyde (MDA) in apo-B containing lipoproteides, but not in the whole serum. It is thought that free radical modification of serum LDL is an important reason for cholesterol accumulation in cells and for their transformation to foam cells in the course equilibrium of atherosclerosis [37].

One way in which oxidized LDL may be involved in atherogenesis is their ability to release interluekin-1-β (I-1-β) from monocytes. The effect of spin trap α-phenyl-*tert*-butylnitrone (PBN) on lipoprotein oxidation and subsequent release of I-1-β was examined by EPR spectroscopy. PBN attenuates LDL oxidation and decreases the ability of LPL to release I-1-β from monocytes. It is suggested that the ability of this spin trap to inhibit oxidation requires its incorporation into the LDL particle [38].

α-Tocopherol (α-tp) can promote lipid peroxidation in human LDL in the absence of co-antioxidants which eliminate α-tocopheroxyl radical (α-Tpr) [39]. The mechanism of 3-HAA antioxidation action was studied. It directly reduces (α-Tpr). Anthranilic acid lacks the phenoxyl group and is not able to do this. 3-HAA dose-dependently inhibits the peroxidation of surface phospholipids and cholesterol esters in LDL lipids treated with oxidants transforming α-Tp into α-Tpr [40-42].

LDL-transported vitamin E is suggested to protect LDL from oxidation. 1. It was shown that a chromanoxyl radical of endogenous vitamin E and of exogenously added α-tocotrienol, α-tocopherol or its synthetic homologs (6-carbon side chain), chromanol-α -C6 can be directly generated in human LDL on exposure to UV-light or by interaction with peroxyl radicals produced either by an enzymatic oxidation system (lipoxygenase + linolenic acid) or by an azo-initiator, 2,2'-azo-bis(2,4-dimethylvaleronitrile). 2. Ascorbate can recycle endogenous vitamin E and exogenously added chromanols by direct reduction of chromanoxyl radicals in LDL. 3. Dihydrolipoic acid is not efficient in direct reduction of chromanoxyl radicals but it recycles vitamin E by synergistic interaction with ascorbate (reduces dehydroascorbate thus maintaining the steady-state of the concentration of ascorbate). 4. β-Carotene is not active in recycling vitamin E, however, it can protect LDL against oxidative destruction by reductants of chromanoxyl radicals. The recycling of vitamin E and other phenolic antioxidants by plasma reductants is suggested to be an important mechanism for an enhanced antioxidant protection of LDL [43].

Phosphotidylcholine and sphingomyelin are two major high-density phospholipid lipoprotein. They are localized on the monolayer surface of the particle together with cholesterol and apolipoproteins (Apo), mainly Ano-A-1. Changes in the surface HDL composition may strongly affect the metabolism of this lipoprotein. In particular, hydrolysis of phosphotidylcholine facilitates absorption of non-esterified cholesterol. Phospholipid hydrolysis also changes the HDL activity towards lecithin-cholesterol-acetyl transferase or cholesterol ether-transporting protein. Mechanism of phospholipid degradation effect on HDL behavior is not well understood yet. This provoked an EPR study of the effect of phospholipid degradation on the physical properties of HDL3 surface. In this study spin label 16-doxylstearic, 5-doxylstearic were used. Hydrolysis of glycerophopholipids leads to rigidity of environment, a decreased mobility and an increased order of the surface lipids. Treatment of HDL(3) with sphingomyelinase, on the contrary, induces fluidization of lipids [35].

Cholesterol has always been in the focus of attentions of chemists and physicians. Along with generally accepted views on its role in the development of atherosclerosis and, finally, of stenocardia and myocardial infarction there are other viewpoints on cholesterol function in the organism. Patients with serum cholesterol low concentrations of die more frequently not from atherosclerotic lesions, but from cancer, diseases of respiratory system and gastro-intestinal tract. When concentrations of vitamins E and C and thiols in blood serum were compared in two groups of 24 patients with correspondingly low and high cholesterol level, the low cholesterol group turned out to have decreased antioxidant activity, thiols and vitamin E concentration. Thus hypocholesterolemia may be associated with low serum antioxidant reserve, possibly increasing susceptibility to oxidative stress [44].

In the treatment of many diseases in internal organs the use was made of electrochemically prepared sodium hypochlorite (NaOCl) [45, 46]. For improvement of this method in medical practice it is necessary to comprehensively study the effect of NaOCl on the blood protein-lipid complexes. For this purpose a stearic acid-based spin label was used for studying structural changes in the surface of proteolipid layer of human blood low-density lipoproteins peroxidized by sodium hypochlorite. Sodium hypochlorite decreases the mobility of polar "heads" as well as of acyl phospholipid chains of the LDL particle and increases polarity of the lipid phase at a distance of 1.7 nm from its surface. In deeper layers (2.0-2.2 nm) the structural changes in LDL are fixed only at NaOCl concentration of over 1 mM [45]. The mechanism of sodium hypochlorite action on human blood lipoproteins was refined by the use of Mn^{2+} as a positively charged spin probe. Hypochlorite (OCl^-) ion increases the LDL negative surface potential and causes its partial degradation. High-density lipoproteins seem to be more resistant to the action of (OCl^-) compared with LDL and especially with very low density protein [46].

As known, oxygen radicals can damage vascular endothelium. Therefore they are of importance in the promotion of atherosclerosis [47]. Hydrogen peroxide (H_2O_2) and Fe^{2+} generate oxidants which initiate free radical reactions. The radicals formed can cause cytolysis. In the H_2O_2 - Fe^{2+} interaction the hydroxyl radical generates first. $HO^\bullet$ is extremely active and initiates cytolysis via cellular lipid peroxidation. Besides, the reaction of H_2O_2 and

Fe^{2+} leads to the formation of Fe^{3+}, which also can facilitate lipid peroxidation. The ideal Fe^{2+}/Fe^{3+} ratio for LPO is 1 : 1. In the reaction mixture containing two parts of F^{2+} to one part of H_2O_2, $HO^{\bullet}$ will be generated. At the 1:1 ratio of hydroxyl radicals to Fe^{2+}/Fe^{3+} LPO is more strongly initiated by hydroxyl radicals formed during the interaction of Fe^{2+} and H_2O_2 in the 2:1 ratio. Exposure to Fe^{2+}/Fe^{3+} ions affords perferryl ion $Fe^{2+}OO^{\bullet} <=> Fe^{3+}O_2^-$ which also activates LPO. The Fe^{2+}/H_2O_2 ratio of 2 : 1 provides a more significant cytolysis than that observed under separate action of these ingredients. The LPO products disrupt specific cell functions without cytolysis as well. Prostaglandin-H-synthase is an enzyme sensitive to both LPO and H_2O_2. Disruption of this enzyme with LPO products reduces the synthesis of prostaglandin I_2 (PGI_2, prostacyclin).

Exposure of cells to H_2O_2 increases the generation of intracellular lipid peroxides inactivates prostaglandin-H-synthase and inhibits prostaglandin synthesis. The Fe^{2+} ion protects against peroxide cytotoxicity and prevents the rise in intracellular calcium caused by H_2O_2. Fe^{2+} preserves intracellular glutathione in H_2O_2-exposed cells. EPR using spin trapping shows that extracellular Fe^{2+} generates hydroxyl radical outside the cell. Due to this fact extracellular Fe^{2+} protects the intracellular space from H_2O_2 by initiating the Fenton reaction inside the cell. This reductive cleavage of H_2O_2 generates hydroxyl radical in the extracellular space, where $HO^{\bullet}$ will react with noncellular components, thereby protecting the cell interior [48].

Hydroperoxides cause loss of endothelial cell integrity, increase the permeability of endothelium to macromolecules, and compromise its ability to control vascular tone. Using EPR spin trapping an investigation into the metabolism of a model compound *tert*-butylhydroperoxide to free radicals in intact human endothelial cells was carried out.

The hydroperoxide is shown to reduce by a single-electron mechanism to form FRs, the mitochondrial electron-transport chain being the source of electrons for this reduction [49]. The formation of peroxide- and globin radical derivatives in the reaction of MetHb with *tert*-butylhydroperoxide has also been demonstrated [50].

The abnormal function of monocytes may accelerate the promotion of atherosclerosis. To support this suggestion the mechanism of increased superoxide generation by monocytes from patients with hypertriglyceridemia was investigated. By the EPR/spin trapping method and 2-methyl-6-[*para*-methoxyphenyl]-3,7-dihydroimidazo-[1,2-a]-pyrazin-3-one (MCLA)-dependent chemiluminescence the superoxide scavenging activity (SSA) and superoxide generation ($O_2^{-\bullet}$) by monocytes was determined. Monocytes were stimulated by 4 β-phorbol 12 β-myristate 13 α-acetate (PMA) or opsonized zymosan. As a result, a significant positive relationship was found between $O_2^{-\bullet}$ generation and plasma triglyceride level [51].

In ischemic patients the content of erythrocyte free radicals is higher than in patients with other diseases and in normal individuals, being the higher the heavier the patient's state (Figure 3-1) [52].

Thus, in the events of stable and progressive stenocardia, the FR level is $(1.3 \pm 0.05) \times 10^{-6}$ M and $(1.5 \pm 0.07) \times 10^{-6}$ M, respectively ($P < 0.01$). The same regularity is observed in the change of PMC contents of MetHb, Cu^{2+}, α-tpq (table 3-7) [53].

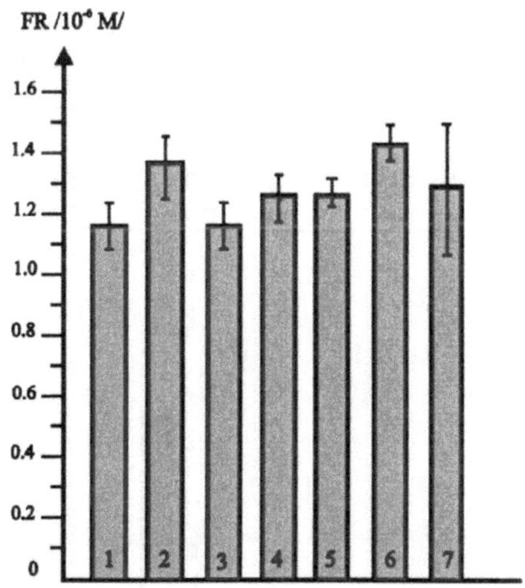

Figure 3-1. Erythrocyte free radical concentration (10-6 M) in donors (1) and in patients with: iron-deficient anemia (IDA) - 2, chromic gastritis - 3, duodenal ulcer - 4, gastric ulcer - 5, ischemic heart disease - 6, hypertensive disease - 7.

Table 3-7. PMC content in erythrocytes of IHD patients depending on the severity of disease

PMC in erythrocytes	Stable stenocardia	Progressive stenocardia
MetHb, (g/l)	0.8 ± 0.04	1.3 ± 0.05
Cu^{2+}-SOD, (10^{-5} M)	0.7 ± 0.02	0.6 ± 0.02
FR, (10^{-6} M)	1.3 ± 0.05	1.5 ± 0.07
α-Tq, (10^{-6} M)	1.5 ± 0.03	1.7 ± 0.02

The same is evidenced by FR content in patients with myocardial infarction (MI), which is enhanced in cases of macrofocal infarction. As the myocardial cicatrix forms and pain syndrome diminishes or disappears, the FR concentration decreases as well, but never reaches normal values (Figure 3-2) [54].

During pain syndrome the FR concentration in both erythrocytes and plasma in IHD patients increases significantly, whereas the plasma ceruloplasmine Cu^{2+} content decreases simultaneously. In patients with neurocirculatory dystonia (NCD) who experience pain in the region of heart the PMC content does not change (tables 3-8, 3-9) [55].

Table 3-8. Erythrocyte PMC concentration in IHD and NCD patients with (+) and without (-) pain syndrome (M ± m)

Erythrocyte PMC	IHD patients		NCD patients	
	(-)	(+)	(-)	(+)
MetHb, (g/l)	0.8 ± 0.06	0.8 ± 0.05	0.7 ± 0.03	0.7 ± 0.02
Cu^{2+} SOD, (10^{-5} M)	0.7 ± 0.03	0.6 ± 0.02	0.8 ± 0.03	0.8 ± 0.02
FR, (10^{-6} M)	1.2 ± 0.04	1.4 ± 0.05	1.1 ± 0.05	1.1 ± 0.05
α-Tq, (10^{-6} M)	1.4 ± 0.04	1.4 ± 0.04	1.8 ± 0.09	1.8 ± 0.08

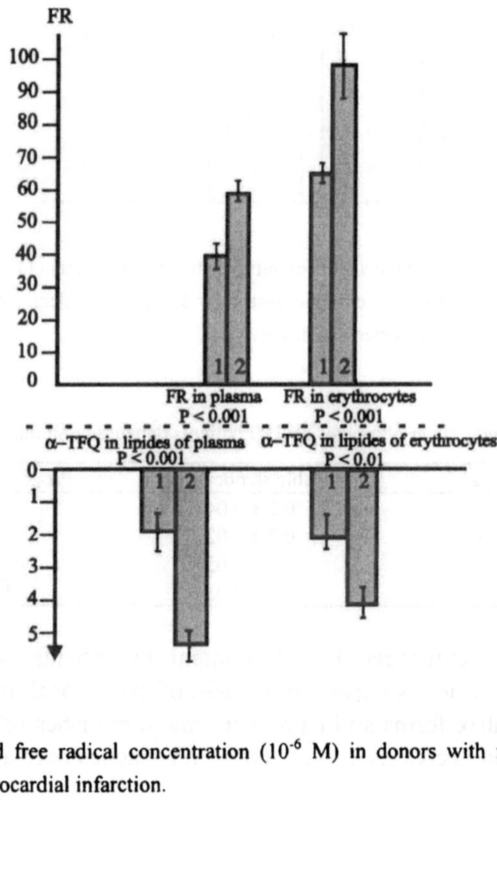

Figure 3-2. Blood free radical concentration (10^{-6} M) in donors with microfocal (1) and macrofocal (2) myocardial infarction.

Table 3-9. Blood plasma PMC concentration in IHD and NCD patients with (+) and without (-) pain syndrome (M ± m)

Blood plasma PMC	IHD patients		NCD patients	
	(-)	(+)	(-)	(+)
Fe^{3+}, (mcmol/l)	15.9 ± 0.9	15.5 ± 0.8	20.1 ± 0.7	20.2 ± 0.6
Cu^{2+}-CP, (10^{-5} M)	1.8 ± 0.03	1.4 ± 0.05	1.8 ± 0.09	1.7 ± 0.09
FR, (10^{-6} M)	0.5 ± 0.04	0.7 ± 0.04	0.5 ± 0.02	0.5 ± 0.03
α-Tq, (10^{-6} M)	1.8 ± 0.05	1.7 ± 0.03	2.2 ± 0.10	2.2 ± 0.09

Stenocardia attack is accompanied by ejection of catecholamines into blood [56], which activates the erythrocyte glutathione-reductase system [57]. This seems to be caused by an increased free radical content during pain syndrome. Evidently, free radicals form in the interaction of ceruloplasmin with adrenalin [30]. In order to confirm this suggestion two identical erythrocyte and plasma samples (0.5 ml each) were examined. One sample was frozen in liquid nitrogen immediately and the other only after addition of 0.01 ml of 0.1% adrenalin hydrochloride solution and subsequent incubation for 30 min at 25°C. It turned out that adrenalin led to a 43% decrease in the Cu^{2+} EPR signal at g = 2.05 and over a 40% increase in the FR concentration (table 3-10) [58].

Table 3-10. Changes in the plasma and erythrocyte PMC concentration upon administration of adrenalin (M ± m)

Indexes examined	Plasma	Erythrocytes
Cu^{2+} (10^{-5} M) (prior to addition of adrenalin)	1.7 ± 0.11	0.8 ± 0.13
Cu^{2+} (10^{-5} M) (after addition of adrenalin)	1.1 ± 0.09	0.8 ± 0.07
FR (10^{-6} M) (prior to addition of adrenalin)	0.3 ± 0.03	1.0 ± 0.09
FR (10^{-6} M) (after addition of adrenalin)	0.5 ± 0.05	2.6 ± 0.27

Previously it has been supposed that free radicals form in the interaction of adrenalin with ceruloplasmin due to Cu^{2+} reduction to Cu^{+} with simultaneous oxidation of adrenalin. It is quite possible that in this case adrenalin forms a complex with ceruloplasmin [59, 60]. There are grounds to assume that this mechanism also occur *in vivo* when catecholamines are ejected into the blood channel during stenocardia attack. In this case the FR concentration increases in erythrocytes as well, but without any change in the Cu^{2+} level in superoxide dismutase. This suggests that adrenalin induces direct activation of the glutathionereductase system [57].

The amplitude of plasma and serum EPR signals at g = 2.0030 and 2.0110, respectively, arisen from free radicals, increases on addition of *para*-phenylenediamine dihydrochloride (p-PDD) (ceruloplasmin substrate). A simultaneous decrease in the intensity of signal at g = 2.05 caused by Cu^{2+} incorporated into ceruloplasmin type protein [61].

Ischemia and subsequent reperfusion lead to injury of heart tissue. Free radicals are involved in this process [62].

Generation of free radicals and cholesterol hydroperoxide during reperfusion was studied. Fifteen patients undergoing subcutaneous transluminal coronary angioplasty were investi-gated. During balloon inflation ischemia was assessed by lactate level in the coronary sinus. Coronary sinus free radicals were detected by EPR spectro-scopy using spin trap PBN. The concentration of free radicals is positively correlated with changes in coronary sinus lactate [63].

Analogous data were obtained for six patients experienced open heart surgery and cardioplegia [64] and for seventeen patients with acute myocardial infarction successfully treated by primary percutaneous transluminal coronary angioplasty [65].

From these data it follows that oxygen radicals forming in endothelial cells are of importance as mediators of postischemic injury. Nevertheless the mechanism of generation of a burst of these radicals is unknown. A possible source of their production can be xanthine oxidase. However this enzyme has not been reported to be present in human endothelium. The mechanism of radical generation in human endothelial cells under the conditions of an-oxia and reoxygenation was determined by EPR measurements on cultured human aortic endothelial cells using a spin trap DMPO. The results were correlated with cellular injury, xanthine oxidase activity and changes in cel-lular nucleotides. Upon reoxygenation 60 min after anoxia an intense DMPO-HO signal ($a_N = a_H$ = 1.49 mT) and weak DMPO-R signals (a_N = 1.58 mT, a_H = 2.28 mT) were observed in the EPR spectra. Generation of these radicals is totally quenched by superoxide dismutase. Fe^{3+} Chelator desferrioxamine prevents cell death, totally quenches the DMPO-R signal and by 40% the DMPO-HO signal. Thus, the presence of superoxide dis-mutase in human endothelial cells provides the source of free radicals. The concentration of this enzyme is not changed after anoxia, whereas the con-tent of its substrate, hypoxanthine significantly increases which stimulates free radical generation upon reoxygenation. Reoxygenated vascular endo-thelial cells generate anion radicals $O_2^{-\bullet}$. Further they react with iron to form reactive $HO^{\bullet}$ which in turn cause cell death [66].

To assess the development of oxidative stress in cardiac ischemia/ reper-fusion, the resulting depletion of plasma ascorbate was monitored by EPR detection of ascorbyl radical (AR) in a homogeneous group of 12 patients undergoing aortic valve replacement. Dimethylsulfoxide was used as an en-hancer and stabilizer for ascorbyl radicals. Plasma was separated from blood samples collected 15 before incision, 10 min before aortic declamping and sequentially during the initial 15 minutes of perfusion. Plasma DMSO/AR levels of patients were significantly lower in healthy individuals (25%). Upon ischemia they further decreased (35%) and dropped to the lowest values within the first 10 min of perfusion (46%) [67].

Nitric monoxide (NO) is an important regulator of various biological functions and play a certain role in pathogenesis of cellular injury. The major mechanism of NO generation in biological tissues is conditioned solely by specific NO-synthtases (NOSs), which metabolize arginine to cerulline with the formation of NO. It is also reported that NO can be generated in the ischemic heart by direct reduction of nitrite to NO at low pH values and under highly reduced conditions. The above pathway of NO formation is not blocked by NOS inhibitors. This mechanism of NO formation is ever dominating during further ischemia progressing to necrosis. Enzyme-independent NO generation leads to myocardial injury with a loss of contractile function. This enzyme-independent mechanism of NO formation is of importance in our understanding of pathogenesis and treatment of tissue injury [68].

To protect cells and tissue from oxidation injury a variety of compounds were tested in laboratory and clinical trials. In this study a protective effect of S-2-(3-aminopropylamino) ethylphosphorothionic acid (WR-2721) on ischemic rat myocard was revealed by EPR resonance spectroscopy [69]. In clinic vitamin, superoxide dismuthase were used for this purpose. Generation of free radicals and the role of allopurinol as a cardioprotective agent were investigated in 36 patients underwent coronary artery bypass grafting. Of these 18 patients were treated with allopurinol and the other 18 patients made up a control group. Allopurinol was administrated orally twice a day for one day before the operation (300 mg) and on the morning of the day of operation a single dose of 600 mg was administrated. Lipid peroxidation products were determined by assays for thiobarbituric acid reactive substances, malonic dialdehyde in the main. During the ischemic period there were no considerable changes in the levels of MDA and creatine kinase. One minute after release of the cross-clamp the blood MDA level was raised significantly, which correlated with the severity of ischemia. In patient treated with allopurinol both the MDA level and creatine kinase activity were sufficiently lower. EPR spectroscopy using spin trapping also revealed a rise in the production of free radicals adducts during reperfusion. Thus, during aorto-coronary bypass grafting surgery there is an increase in the concentration of lipid peroxidation products and in the activity of cardiac creatine kinase on the onset of reperfusion. This indicates enhanced free radical generation. Pretreatment with allopurinol could attenuate post-ischemic reperfusion injury [70]. Free radical intermediates have an essential role in oxidation modification of low density lipoproteins. The appearance of oxidized lipoproteins in the blood vessel is directly related to atherosclerosis. In this connection the generation of free radicals from natural and synthetic antioxidants able to prevent LDL peroxidation was studied by EPR [71].

Probucol - 4,4'-[(1-methylethylidene)bis(thio)]bis[2,6-bis(1,1-dimethyl)-phenol], is an antiatherosclerotic drug which produces a hypolipidemic effect, reduces blood cholesterol level mainly due to LDL, and does not affect much the level of triglycerides and lipoproteides of very low density. Probucol is an antioxidant of lipid-soluble azo-initiator 2,2'-azo-bis(2,4-dimethyl-valeronitrile), in dioleoylphosphatidylcholine liposomes. Using EPR spectroscopy it was shown that in human LDL ascorbate reduced probucol phenoxyl radicals. Dihydrolipoic acid alone was not able to do this. However, in the presence of both ascorbate and dihydrolipoic acid a step-wise reduction was observed. This resulted from ascorbate induced reduction of probucol, ascorbyl radicals being reduced by dihydrolipoic acid [72].

Interaction of reduced Q_{10} (ubiquinone-10) with phenoxyl radicals generated during the oxidation of probucol and α-tocopherol in models sys-tems of manganese dioxide (MnO_2) and ferric ions (Fe^{3+}) (solutions in etha-nol, isopropanol, chloroform) was studied. Q_{10} and 6-O-palmitoylascorbate reduce these phenoxyl radicals. In this case EPR signals of Q_{10} and 6-O-palmitoylascorbate free radicals were registered. In the presence of 6-O-palmitoylascorbate the Q_{10} EPR signal was observed for a longer time. Probucol in concentrations 1-2 mcmol/mg of LDL protein makes the induction period of lipid peroxidation (formation of diene conjugates) a few hours longer. Generation of probucol phenoxyl radical is induced by addition of gemin to the antioxidant-containing LDL. Since probucol radical generation requires the presence of H_2O_2 or the accumulation of some amount of lipid hydroperoxides (LOOH) in LDL, the introduction of gemin may lead to intense formation of alkoxy ($LO^•$) and hydroxy radicals These radicals cause in turn probucol oxidation. Probucol radical generation in late LDL oxidation stages is stimulated by addition of Q_{10}. The antioxidant effect of probucol is assumed to be based on its interaction with lipid radicals and Q_{10}, which is the mediator of free radical state transfer to probucol localized in the LDL central hydrophobic region. At the same time, one should admit Q_{10} involvement in the interaction of antioxidants and their free radical intermediates with metal ions inducing LDL peroxidation to be quite possible [71].

The effect of lipid-reducing compounds bezafibrate and clofibric acid on nitric oxide binding to human hemoglobin Fe^{2+} (Hb-NO) was revealed by the EPR method. In the presence of bezafibrate and clofibric acid, the EPR spectrum and absorption spectrum (in the Soret region) of Hb-NO display the same basic characteristics as those in EPR spectra in the presence of (inositol hexakisphosphate) and 2,3-diphosphoglycerate. Having an allosteric effect, these compounds reduce the oxygen affinity for Fe^{2+} human hemoglobin [73, 74].

Human serum albumin is an essential protein transporter in the organism. In EPR studies of protein transportation process the use was made of non-covalent complexes of HSA with some water-soluble polymers (hematoporphyrin and hematoporphyrin derivative) as photosensitive agents in photodynamic therapy of cancer. The results show that long-chain fatty acids and tetrapyrroles can compete for anionic ligands of HSA binding sites. EPR was used to examine spin labeled fatty acids binding to human and bovine serum albumin. The content of 5-doxyl stearate and 16-doxyl stearate binding sites on human serum albumin and bovine serum albumin was equal, whereas association constants were different. Analysis of 5-DS and 16-DS EPR spectra allows a quantitative distinguishing between slow macromolecular rotation and fast anizotropic motion [75].

On the average, the blood plasma Fe^{3+} level in IHD patients is reduced by 42% and 34% in men and women, respectively [76]. Depending on the severity of disease the Fe^{3+} content lowers and the Cu^{2+} ceruloplasmin level rises. Thus, with stable stenocardia the plasma Fe^{2+} concentration is 14.8 ± 0.7 mcmol/l and 8.2 ± 0.7 mcmol/l with progressive stenocardia ($P < 0.001$). A similar picture is observed with myocardial infarction [6, 7].

The blood plasma Fe^{3+} content determined biochemically is higher than that from EPR data. In men and women this difference is 40% and 33%, respectively. This can be explained as follows: in contrast to that of healthy persons, blood plasma transferrin of IHD patients is faster saturated with Fe^{3+} upon intravenous administration. Moreover in IHD patients the ceruloplasmin ferroxidase activity transforming Fe^{2+} to Fe^{3+} seems to be lowered [9].

The erythrocyte MetHb level in IHD patients is increased (Figure 3-3). Its level dependent on the severity of disease is 0.8 ± 0.03 and 1.3 ± 0.05 g/l in stabilized and progressive stenocardia, respectively ($P < 0.001$). In progressive stenocardia higher doses of nitrates are prescribed [56, 77].

To elucidate the reason for the higher MetHb level in IHD patients the erythrocyte MetHb content was measured on the first day of admittance to the hospital after 14 days of sustack-forte treatment (6.4 mg x 4 daily) (Figure 3-4). According to the EPR data the treatment led to a mean 43% increase in erythrocyte Met Hb concentration (table 3-11).

Thus nitrate treatment increase the erythrocyte MetHb content in IHD patients and strengthens tissue hypoxia [78, 79].

As supposed previously, α-Tq is the end product of α-tocopherol (α-T) which plays an important role in many physiological functions [80]. Usually the blood vitamin E concentration in IHD patients is lowered.

Table 3-11. Erythrocyte PMC content in IHD patients before (-) and after (+) a two-week administration of nitrates of prolonged action (M ± m)

Erythrocyte PMC	(-)	(+)
MetHb, (g/l)	0.9 ± 0.07	1.3 ± 0.10
Cu^{2+} SOD, (10^{-5} M)	0.6 ± 0.04	0.8 ± 0.05
FR, (10^{-6} M)	1.4 ± 0.06	1.5 ± 0.01
α-Tq, (10^{-6} M)	1.6 ± 0.07	1.7 ± 0.07

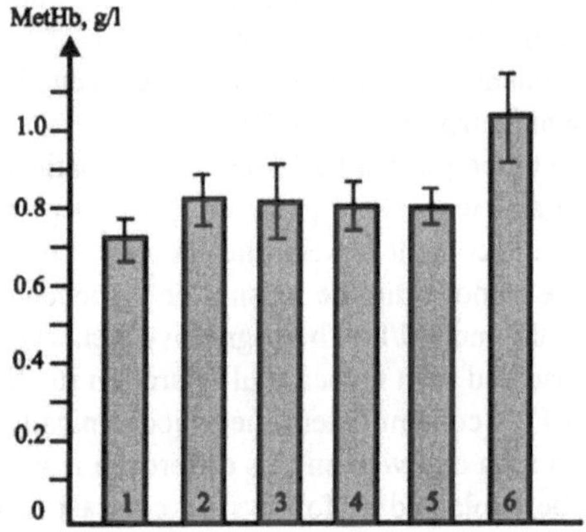

Figure 3-3. Erythrocyte methemoglobin concentration (g/l) in donors (1) and in patients with iron-deficient anemia - 2, chronic gastritis - 3, duodenal ulcer - 4, gastric ulcer - 5, ischemic heart disease – 6.

An EPR study of the effect of α-T on the blood α-Tq concentration in IHD patients was carried our in order to confirm α-T transformation in IHD patients (table 3-12) [81].

Table 3-12. Plasma and erythrocyte α-Tq concentration (10^{-6} M) in IHD patients before (-) and after (+) α-T administration for 3 days (M ± m)

Group of patients	α-Tq in plasma lipids		α-Tq in в erythrocyte lipids	
	(-)	(+)	(-)	(+)
Without vitamin E administration (placebo)	1.5 ± 0.02	1.5 ± 0.07	1.5 ± 0.06	1.5 ± 0.06
With vitamin E administration	1.4 ± 0.05	2.1 ± 0.08	1.6 ± 0.07	1.8 ± 0.08

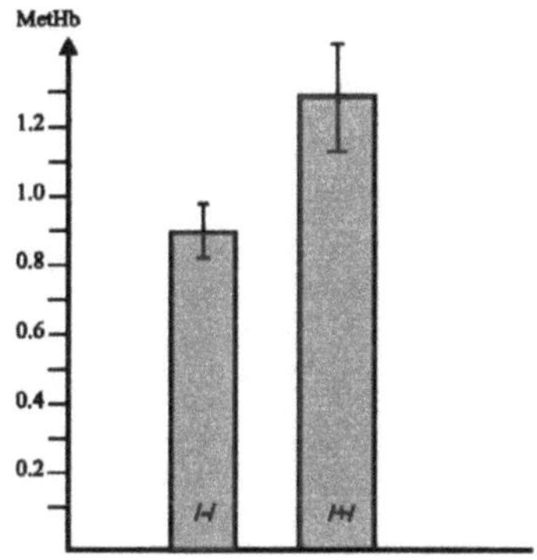

Figure 3-4. Erythrocyte methemoglobin concentration (g/l) in IHD patients prior to (-) and after (+) a two-week administration of nitrates of prolonged action.

The patients were subdivided into two homogeneous groups. Group 1 patients were treated with 2.0 ml of 30% buttery vitamin E solution administrated intravenously twice a day for 3 days. Group 2 acted as control with daily administration of 2.0 ml of physiologic salt solution. Blood samples were withdrawn and analyzed on the onset and after completion of three-day vitamin E administration. The results showed an increased α-Tq concentration in plasma and erythrocyte lipids in Group 1 patients. After aortic cross-clamping surgery accompanied by ischemia, the administrated α-T had gone to the α-Tq fraction. α-Tq is not only an indicator of lipid peroxides-vitamin E interaction. It is also involved in biochemical processes of the organism and seems to be an "active" vitamin E ingredient [81].

According to EPR data, in patients with Phase II myocardial infarction the α-Tq content in plasma and erythrocyte lipids increases compared with that in Phase I and III (Figure 3-5) [9].

This may be connected with the regeneration of myocardial tissue and the formation of cicatrix in the cardiac muscle. Consequently, administration of an antioxidant such as vitamin E is necessary to accelerate recovery.

In IHD patients treated with nitrates for releasing pain in the region of heart, the fecal Hem-NO content is usually 42% higher (after two-weeks of therapy) (Figure 3-6) [30]. This indicates that the nitrates are not fully absorbed in blood. Some portion is inactivated in the intestine by nitrate- and nitrite-reducing microorganisms.

The administrated nitrates are known to be converted to nitrogen monoxide (NO). It has been found that in human subjects nitroglycerol [82] and isosorbide dinitrate [83] liberate NO which combines with hemoglobin to form the Hb-NO complex. This suggests that the above clinical effect cannot be triggered by nitrates due to their being bound in the Hb-NO complexes and used by intestinal microflora. Possibly this provides a mechanism responsible for tolerance to nitrates observed in many ischemic patients.

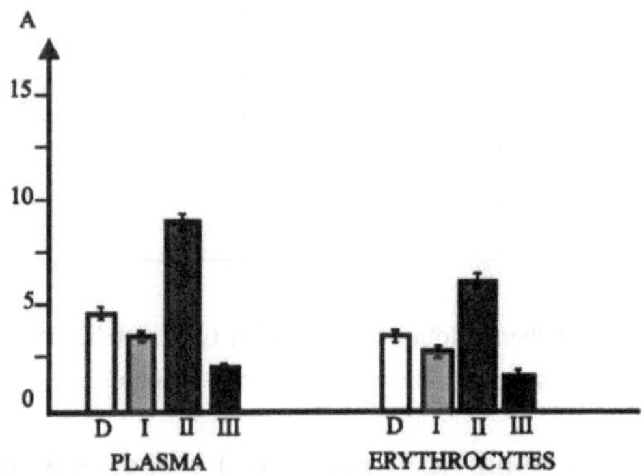

Figure 3-5. α-Tocopherolquinone semiquinone concentration (10^{-6} M) in plasma and erythrocyte lipids in donors (D) and in patients with Phase I, II and III myacardial infarction.

The EPR method was used for assessment of functional viability of a heart conserved with a goal of using as a transplant. The signals of cytochrome-C-oxidase iron-serum center at $g_1 = 1.99$, $g_2 = 2.03$ were studied [84, 85]. The clinical picture of many diseases is accompanied by cardio-vascular deficiency involving tissue hypoxia [78]. This leads to a compensatory increase in glutathione reductase activity and to GSH generation to protect Hb from oxidation in MetHb. In cases of clearly defined heart failure (HF_{III}), free radical level lowers seemingly to general attenuation of enzymatic processes. The erythrocyte FR concentration in patients with phase HF_I increases and then decreases as this complication grows progressively worse (Figure 3-7) [86].

Heart diseases are not infrequently affected by shock. Thus, stenocardia and its extreme version myocardial infarction may involve cardiogenic shock with lethal termination. Therefore investigation of the mechanism of shock origination and treatment is given a great attention. Nitrogen monoxide is believed to be one of main compounds involving in the formation of this pathology.

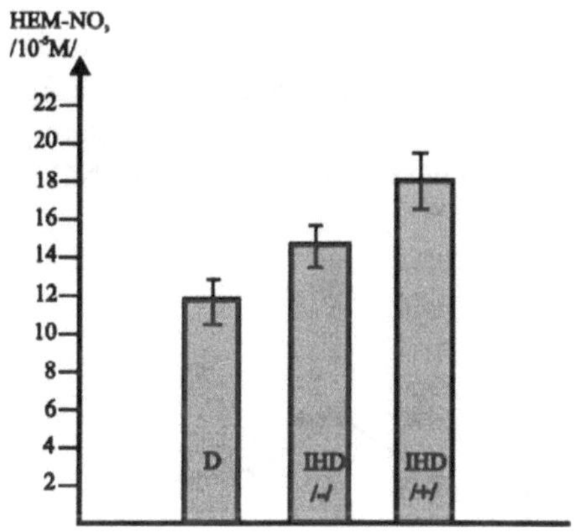

Figure 3-6. Fecal Hem-NO concentration (10^{-5} M) in donors and in patients prior to (-) and after (+) a two-week administration of nitrates of prolonged action.

Nitric oxide levels in the urine of both normal and lipopolysaccharide (LPS)-induced mice were measured by EPR spectroscopy using spin trapping. The two groups of animals were injected subcutaneously with a metal-chelator complex N-methyl-D-glucamine dithiocarbamate-ferrous iron, $[(MGD)_2/Fe\text{-}NO]$ able to bind to NO and form a water-soluble $[(MGD)_2/Fe\text{-}NO]$ complex. Two hours after the injection a triplet signal corresponding to the $[(MGD)_2/Fe\text{-}NO]$ complex was detected in the EPR spectra of the urine of both normal and LPS-induced mice. The concentrations of the $[(MGD)_2/Fe\text{-}NO]$ complex in the urine of normal and LPS-induced mice were 1.3 and 35 mcM, respectively.

Administration of N-monomethyl-L-arginine (NMMA; 50 mg/kg) inhibits *ex vivo* signal intensities of the $[(MGD)_2/Fe\text{-}NO]$ complex in the urine of the two groups of mice. Injection of ^{15}N-arginine (10 mg per mouse) results in a composite urine EPR spectrum consisting of a triplet corresponding to the $(MGD)_2/Fe\text{-}^{14}NO]$ complex, and a doublet related to the urine $[(MGD)_2/Fe\text{-}^{15}NO]$ complex. These data confirm that in the urine of mice NO is produced via arginine-nitric oxide. So, the spin trapping method provides a non-invasive means for measuring both constitutive and induced NO synthase activity in animals under physiological pathophysiological conditions where nitric oxide is overproduced [87]. In another study [88] a Fe^{2+} - sodium-N,N-(diethyldithiocarbamate trihydrate complex, $Fe(DETC)_2$ was used as a nitric oxide trap. In this way the intracellular NO production in liver tissue of mice under septic shock was investigated *in vivo* using spin trapping technique. Among all the tissues examined the most intense NO signal in the EPR spectra was that arisen from liver homogenates. The NO content in kidneys as well as in brain and lungs was 60% and 90% lower, respectively. After 1 spin trap post-injection hour the EPR signal was stabilized, which indicated the attainment of a steady state [88]. An interesting aspect of this study is an *in vivo* EPR monitoring of NO production.

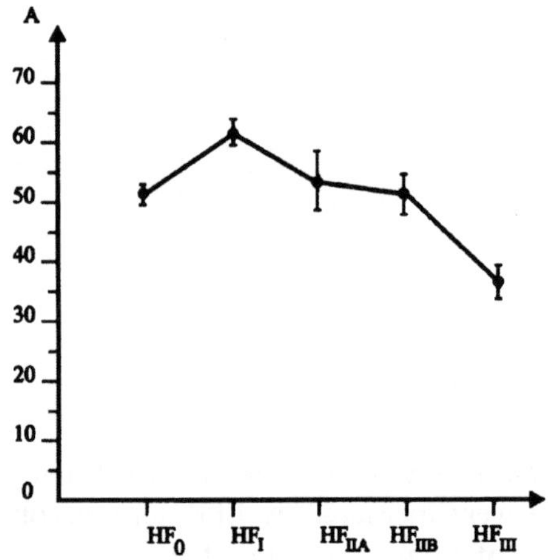

Figure 3-7. Erythrocyte free radical concentration (10^{-6} M) in patients with heart failure.

Nitrates entering the organism together with water, food and drugs are distributed by the blood flow. Twenty five per cent of the nitrates are brought through the salivary glands into the mouth cavity where 20% are reduced to nitrites by microbes. As a result, 80% of the whole daily dose of nitrites is formed in saliva [89].

Saliva of 41 IHD patients who were administrated organic prolonged nitrates (6 tablets a day) in complex treatment (sustack-forte oral formulation, 6.4 mg) was examined by EPR spectroscopy. Of 41 patients, 10 (29%) were with acute myocardial infarction and 29 (71%) with functional class II-III stenocardia. 22 Volunteers acted as control. Saliva (5-10 ml) was collected in a dry sterilized ampoule in the morning before taking food and after careful gargling of the mouth cavity. Free radical signals were observed in Saliva EPR signals of both patients and donors. g-Factor was at 2.0012-2.0040 (2.0026 ± 0.0001) in volunteers and at 2.0017-2.0029 (2.0022 ±

0.0001) in IHD patients ($P < 0.05$), line half-width being 1.15 ± 0.04 and 1.31 ± 0.03 mT ($P < 0.01$), respectively.

Free radical reactions clearly reflect even slightest changes in the homeostasis of IHD patients. Circulatory deficiency in coronary vessels sufficiently changes the metabolic processes in cardiomyocytes. In cells there are two ways of O_2 consumption: via oxidase and oxygenase. The former consists in enzymatic oxidation of substrates in tricarbonic acids, electron and proton transportation through respiratory chain to cytochromoxidase. Oxygen acts as final electron acceptor and forms H_2O under full reduction. This pathway involves ADP phosphorylation and ATP resynthesis. The other pathway is also based on the oxidation of substrates to organic compounds, however, via atomic or molecular oxygen. In oxygenase reactions oxygen is reduced to a variety of oxygen radicals, among which active forms such as superoxide anion radical, hydrogen peroxide and hydroxy radical are of primary importance [89]. Thus, the observed diffe-rences in saliva free radical characteristics can demonstrate on the molecular level myocardial changes in patients suffering from chronic coronary deficiency.

Using EPR spectroscopy saliva nitrites were determined by reduction to nitric oxide and further NO reaction with hemoglobin leading to the Hem-NO complex which gives rise to a characteristic EPR signal [90].

Hemoglobin solution was prepared from donor blood erythrocytes by hemolysis with an equal volume of distilled water and a 0.4-fold volume of chloroform. After vigorous shaking and centrifuging at 5000 rev/min for 1 h a reference (10 g/l) hemoglobin solution was prepared from the supernatant layer. Saliva was divided into the precipitate and the supernatant layer after centrifuging at 3000 rev/min for 5 min and nitrite content was assessed separately in each layer [91].

For this purpose 1 ml of hemoglobin solution (10 g/l) and 2 mg/ml of sodium metabisulfite ($Na_2S_2O_5$) were added to 1 ml of saliva. After stirring the mixture a sample for EPR examination was prepared. In intact saliva the Hem-NO EPR signal was fixed in 48% of patient and donors, whereas in the signals corresponding to the supernatant layer and the precipitate occurred in 32% and 33% of individuals. Inconstancy of EPR signal registration is suggested to be related to hemo-globin iron oxidation on exposure to air oxygen. Methemoglobin formed in this case does not add nitric oxide though gives rise to an EPR signal with $g = 6.0$ [91].

To verify this hypothesis the saliva MetHb content depending on air oxygen exposure time was determined. With this goal in view 28 saliva samples were recorded 4 times every 30 min. During this period the ampoules containing the reaction mixture were kept at room temperature and free access of air oxygen (table 3-13) [92]. As seen from the table 3-13, hemoglobin is oxidized to methemoglobin in the course of time. This accounts for inconstancy of the saliva Hem-NO signal registration.

Table 3-13. The effect of air oxygen on hemoglobin-methemoglobin oxidation (M ± m)

Initial MetHb content, g/l	MetHb Content after 30 min, g/l	MetHb Content after 60 min, g/l	MetHb Content after 90 min , g/l
1.0 ± 0.14	1.6 ± 0.19	1.7 ± 0.24	1.9 ± 0.39

In this connection, the $FeSO_4$-$Na_2S_2O_5$ system was used for nitric oxide trapping. In this study a complex containing ferric, sulfur and NO ions formed. In contrast to hemoglobin, this complex is not oxidized by air oxygen at room temperature. The complex was prepared as follows: into a dry sterilized ampoule sodium metabisulfite (5 mg), iron sulfate (3 mg) sodium ascorbate (10 mg) and the liquid to be examined were placed. Preparation of the sample was started 30-40 sec after solution of the reagents. Nitrites, if present, were reduced by sodium ascorbate to nitric oxide which reacts with the $FeSO_4$-$Na_2S_2O_5$ system to form a complex corresponding to an EPR signal at $g = 2.03$. As excess reagents were used, the signal amplitude depended on the nitrite concentration in the reaction mixture, which was calculated using calibration curves plotted on the known nitrite concentrations [92].

The EPR method was also used in studying the saliva nitrate-reducing ability estimated as follows. The saliva to be examined was divided into two portions. In one portion the nitrite concentration was measured as described above, whereas to the other portion crystalline sodium nitrate ($NaNO_3$) was added. Then the saliva was thermostatically controlled at 37°C for 30 min. Owing to nitrate-reducing activity of saliva, NO_3^- is reduced to give a nitrate ion (NO_2^-). In donors the pre- and post-incubation saliva NO_2 level was 0.07 ± 0.02 and 0.31 ± 0.08 mmol/l ($P < 0.001$), respectively, whilst in IHD patients these values were 0.10 ± 0.05 and 0.34 ± 0.12 mmol/l ($P < 0.05$) [91, 92].

The saliva nitrate-reducing properties are most likely to be conditioned by the presence of oral microflora able to reduce nitrates to nitrites. Human nitrate and nitrite load with food and water is 30-290 and 0.1-370 mg/day, respectively. On this background daily oral administration of a moderate doses of organic nitrates as drugs does not affect much the human organism. Organic nitrates seem to be inactivated by powerful nitrate-and nitrite-reducing systems promoting nitrate-tolerance in IHD patient's [92].

Dynamics of saliva nitrate-reducing ability of both ischemic patients treated with prolonged nitrates and volunteers was assessed by EPR spectroscopy. For this purpose excess crystalline sodium nitrate was added to saliva (15 ml) before and after nitrite content measurements. The ampoule was placed into a thermostat (37°C) and a sample was taken every 20 minutes of thermostatic control (140 min). Into each sample the above mentioned reducing system arousing an EPR signal at g = 2.03. The results showed that in both patients and volunteers the saliva NO_2 content had increased reaching a maximum value after 140 and 100 min, respectively. A statistically significant increase compared with the initial NO_2 content was observed after 40 min in the two groups. Thus, the data obtained indicate an increase in the saliva nitrate-reducing ability in ischemic patients [92, 93].

The EPR technique was used to study the blood and saliva of ischemic patients treated by fetal tissue graft subcutaneous injections into the abdominal wall fat tissue. Fe^{3+} and Cu^{2+} ions and free radicals were detected in blood plasma and MetHb was fixed in erythrocytes. The spectra of saliva indicate the presence of Mn^{2+} ions and endogenous nitrites (at g = 2.03). Nitrate-reducing capability is also suggested. It was found that plasma iron and copper content showed an 18% and 12% decrease, respectively, by the 42 experimental days, erythrocyte MetHb content decrease was 57%. Saliva endogenous nitrite concentration increases by the 14th day, whereas nitrate-reducing capability decreases [92].

Physiological activity of intestinal microorganisms is determined microbiologically in most cases. Some enzymatic systems and ferric-serum proteins of microorganism showed paramagnetic properties, [94, 95] which were examined, in the intestinal content by the EPR method [30].

Microflora reduces nitrates to nitric oxide and is involved in the synthesis of Hem-NO nitrosyl complex formed from NO and hem-containing groups in the gastric-intestinal tract. Hem-NO is excreted with feces in which it was registered by EPR spectroscopy [9]. Removal of a portion of nitrates as Hem NO from the intestine appears to be a factor promoting nitrate tolerance in ischemic patients.

Intestinal biocenosis disorders are corrected using dairy produce rich in viable microorganisms (eubiotics). They are also used in treating dislipoproteinemia in patients with coronary atherosclerosis. Data on fecal paramagnetic centers in patients treated with eubiotics have been reported recently. Fecal paramagnetic centers were studied by EPR spectroscopy in 70 ischemic patients (47 men and 23 women, 57 ± 1 years of age on the average). The patients were divided into subgroups depending on IHD type: acute myocardial infarction 20 patients, unstable stenocardia (both progres-sive and incipient) and stable stenocardia of II-II functional classes - 25 patients in each subgroup. 12 Volunteers acted as control. The eubiotic-treated subgroups consisted of 23 ischemic patients of age ranging from 38 to 70 years. The patients took dairy products "Bifivit" containing *B bifidum* and "Acydolact" containing *Lactobacillus* on empty stomach at a daily dose of 100 mg twice a day (in the morning and in the evening) for 14-21 days. The intestinal microflora content and fecal paramagnetic centers of all the patients were examined before the start and after the completion of the course of treatment [96, 97].

Fecal microflora disbiotic alterations were evaluated by quantitative and qualitative shifts in the microflora composition basing on the disbacteriosis form classification used in the clinic [98]. The degree of dysbacteriosis was expressed in points (from 0 to 4).

The following PMC were analyzed: Fe^{3+} at g = 4.3 with a spectral shape characteristic of its localization in albumin, Mn^{2+} with a six-component superfine structure, and Hem-NO. The Fe^{3+} signal at g = 4.3 is related to aerobic intestinal microorganisms *E.coli* and *Citrobacter* [9]. The enzymatic systems of aerobic and anaerobic representatives of intestinal microflora contain metals of variable valence as a co-factors. Non-hem Fe^{3+} ions provide structural components of microbial dehydrogenases [99], nitrate reductases and hydrogenases localized in *Clostridium* and *E.coli* [100], hydroxylases [101]. Furthermore, the EPR signal at g = 4.3 may arouse from lactoferrin present in feces [101].

In Table 3-14 the results of examination of fecal PMC of ischemic patients and volunteers are presented [102].

Table 3-14. Fecal PMC contents in ischemic patients (M ± m)

Group	N	Fe^{3+}, (mcmol/l)	Hem-NO, (10^{-5} M)	Mn^{2+}, (10^{-5} M)
Ischemic patients	70	95.0 ± 19.0	1.2 ± 0.1	8.3 ± 1.0
Volunteers	12	112.3 ± 9.4	1.4 ± 0.1	8.8 ± 0.5

As seen from Table 3-14, the Hem-NO content in ischemic patients is lower than in the control group. Statistically reliable differences between Fe^{3+} and Mn^{2+} concentrations have not been found either in the control or in the test group. However, ischemic patients show a tendency towards decreasing these values.

In Table 3-15 fecal PMC mean values in ischemic men and women are presented [102].

Table 3-15. Fecal PMC contents in ischemic patients depending on sex (M ± m)

Sex	N	Fe^{3+}, (mcmol/l)	Hem-NO, (10^{-5} M)	Mn^{2+}, (10^{-5} M)
M	47	117.9 ± 11.2	1.6 ± 0.1	9.5 ± 0.7
W	23	100.9 ± 17.2	1.2 ± 0.1	6.9 ± 0.7

Thus, higher fecal Hem-NO and Mn^{2+} contents are observed in men ($P < 0.05$ and $P < 0.02$, respectively).

Table 3-16 shows fecal PMC mean values in ischemic patients depending on ISD form [102].

Table 3-16. Fecal PMC contents in ischemic patients depending on IHD form (M ± m)

IHD form	Sex	N	Fe^{3+}, (mcmol/l)	Hem-NO, (10^{-5} M)	Mn^{2+}, (10^{-5} M)
Acute MI	M	15	138.7 ± 22.9	1.8 ± 0.2	9.9 ± 1.1
	W	5	86.6 ± 20.9	1.2 ± 0.3	4.8 ± 0.5
Unstable	M	13	88.3 ± 19.0	1.6 ± 0.3	9.4 ± 1.6
stenocardia	W	12	121.7 ± 29.0	1.6 ± 0.3	7.9 ± 1.1
Stable	M	19	121.8 ± 15.6	1.6 ± 0.1	10.1 ± 1.4
stenocardia	W	6	71.2 ± 25.5	1.2 ± 0.1	8.5 ± 1.2

From this Table, fecal Mn^{2+} concentration in women with acute myocardial infarction is reliably lower than that in the group with both unstable and stable stenocardia ($P < 0.05$ and $P < 0.02$, respectively).

In table 3-17 fecal mean PMC concentrations in subgroups of ischemic patients of various age are presented [102].

Table 3-17. Fecal PMC contents in ischemic patients depending on age (M ± m)

Age (years)	Below 39	40-49	50-59	60-69	Above 70
N	5	13	23	25	3
Fe^{3+}, mcmol/l	134.5 ± 37.8	111.0 ± 21.4	123.2 ± 18.5	94.5 ± 12.0	76.7 ± 30.6
Hem-NO, 10^{-5} M	1.9 ± 0.3	1.6 ± 0.2	1.7 ± 0.2	1.5 ± 0.1	1.2 ± 0.3
Mn^{2+}, 10^{-5} M	6.8 ± 2.0	10.6 ± 1.4	8.1 ± 1.0	9.4 ± 1.0	5.0 ± 1.3

Table 3-17 shows a decrease in the fecal PMC level with increasing age of ischemic patients ($P < 0.02$).

Assessment of the intestinal microflora composition in IHD patients showed disbiotic changes in 65 individuals (93%). Stage I and II disbacteriosis has been revealed in 28 (40%) and 16 (23%) patients; in 9 patients (13%) considerable shifts in the quantitative and qualitative intestinal microflora composition (stage III-IV disbacteriosis) have been observed. This gave grounds to examine fecal PMC contents depending on the degree of disbacteriosis by EPR spectroscopy (table 3-18) [102].

Table 3-18. Fecal PMC contents in ischemic patients depending on the intestinal disbacteriosis (DB) stage (M ± m)

DB stage (points)	n	Fe^{3+}, (mcmol/l)	Hem-NO, (10^{-5} M)	Mn^{2+}, (10^{-5} M)
0	5	106.9 + 28.8	1.9 + 0.2	12.3 + 2.5
1	28	113.0 + 15.9	1.8 + 0.2	9.0 + 0.9
2	16	105.9 + 20.7	1.3 + 0.2	7.8 + 1.1
3-4	9	77.8 + 16.7	1.2 + 0.2	6.9 + 1.0

As seen from table 3-18, with increasing the stage of disbacteriosis the fecal PMC concentration lowers.

In disbacteriosis with prevailing coccal forms in intestinal microflora (over 25% of total amount of microbes) the Fe^{3+} and Hem-NO content decreases (table 3-19) [102].

Table 3-19. Fecal PMC content in disbacteriosis with prevailing coccal forms (M ± m)

Group	n	Fe^{3+}, (mcmol/l)	Hem-NO, (10^{-5} M)	Mn^{2+}, (10^{-5} M)
Excess cocci	13	60.6 ± 6.2	1.1 ± 0.2	8.1 ± 1.3
Normal amount of cocci	55	115.2 ± 12.2	1.7 ± 0.1	8.7 ± 0.7

Coccal form concentration (% of total amount) in feces of men correlates to Hem-NO level (r = -0.4) and DB stage (r = -0.4). Fe^{3+} linearly correlates to the activity of cholesterol-reducing strains of colon bacillus (r = +0.60, P < 0.05 in all cases). The inverse dependence between the number of bifidobacilli and Mn^{2+} level is typical for woman feces (r = -0.60, P < 0.01).

Table 3-20 shows fecal PMC mean concentrations in ischemic patients before and after the course of eubiotic treatment [102].

Table 3-20. Contents of fecal PMC in ischemic patients before and after eubiotic treatment (M ± m)

Group	n	Fe^{3+}, (mcmol/l)	Hem-NO, (10^{-5} M)	Mn^{2+}, (10^{-5} M)
Before treatment	23	123.1 ± 19.2	1.5 ± 0.1	9.9 ± 1.1
After treatment	23	121.7 ± 17.2	1.9 ± 0.2	9.3 ± 0.8

From this table it follows that after the course of treatment with enriched dairy produce, fecal Hem-NO content was 27% higher on the average. In all patients of this subgroup the intestinal microflora composition was examined before and after the course of eubiotic treatment (table 3-21) [102].

Thus, as a result of eubiotic treatment the disbacteriosis degree reliably decreases and the B. Bifidum content increases. Intestinal microflora composition was improved in 14 patients (61%), remained practically unchangeable in 5 patients (22%). Microbiological characteristics were impaired in 4 patients (17%).

Table 3-21. Dynamics of the intestinal microflora composition in ischemic patients treated with eubiotics

Group*	*B bifidum*, lg CFE/g	*Lactobacillus*, lg CFE/g	*E. coli*, mln/g	Intermediate Stage DB
Before treatment	7.4 ± 0.2	6.9 ± 0.3	158.0 ± 28.3	2.2 ± 0.2
After treatment	8.1 ± 0.2	6.64 ± 0.2	134.2 ± 20.4	1.6 ± 0.2

* n = 23

The role of certain intestinal bacilli in the synthesis of Hem-NO *in vitro* was studied by EPR spectroscopy. Pure cultures of *B. bifidum, Lactobacillus* and *E.coli* were isolated from feces. The strains were anaerobicaly incubated on semiliquid nutrient media for 24 h at 37°C with addition of aqueous solutions of hemoglobin and Hb combinations with 10% sodium nitrite or nitrate. A suspension of bacterium and nutrient was examined by EPR prior to and after thermostatic control. Hem-NO presence was detected by a characteristic EPR signal [9]. Sterilized nutrient media and hemoglobin solution were used as the control [102].

No Hem-NO signal was fixed in the spectra of initial suspensions. In the spectra of control samples Hem-NO signal was absent before and after incubation. Hem-NO complex was not fixed in post-incubation *E.coli* culture either.

In *B. bifidum culture* the Hb-NO signal T conformer appeared in the presence of hemoglobin and Hb-$NaNO_3$ mixture. With *Lactobacillus* suspension the signal was fixed with incubation in hemoglobin or sodium nitrate solution or in a Hb-$NaNO_3$ mixture. Superfine structure splitting was not complete in this case.

No reliable difference in Fe^{3+} level was revealed either in ischemic patients or in the control or with various disbacteriosis forms. There is a tendency toward decreasing the fecal Fe^{3+} content as the age increases or in patients with clearly defined disbacteriosis. This seems to result from the inhibition of activity of microbial Fe^{3+}-enzymes or from the displacement of intestinal microorganisms containing the above enzymes (*E.coli, Enterobacilli*). Evidently this also accounts for a sufficient decrease in fecal Fe^{3+} concentration in patients with a predominance of conventionally pathogenic enterococci (*Streptococcus feacalis*). The linear correlation between Fe^{3+} concentration and *E.coli* cholesterol-reducing activity observed in men suggests *E.coli* cholesterol-transforming enzymes to contain non-hem iron. It is non-hem iron that incorporated into *E.coli* hydrogenases [100]. This is responsible for cholesterol metabolism.

Mn^{2+}concentration are reduced in patients with clearly defined disbacteriosis and in elderly individuals. Manganese also enters into the composition of bacterial enzymes, dehydrogenases, in particular [101], proteases, and carboanhydrases [100, 103]. The low fecal Mn^{2+} content may be related to negligible concentration of manganese-containing microbes. Mn^{2+} presence in feces has been proved to be caused by anaerobes [9]. Evidently, this may account for the reduced Mn^{2+} concentration in women feces. *B. bifidum* anaerobe level in intestinal microflora is reliably lower in women than in men.

Fecal Mn^{2+} content in women with acute myocardial infarction is reduced too. There are no reliable changes in the intestinal microflora composition in patients with various IHD forms and the decreases fecal Mn^{2+} content is caused by other factors (possibly by the level of enterokinase and alkaline phosphotase intestinal enzymes).

Hem-NO signal in EPR spectrum is formed in the intestine content in the presence of hem-groups (hemoglobin, myoglobin, catalase, cytochromes), nitrates or nitrites, nitrate- and nitrite-reducing enzymes or enzyme-containing microorganisms. Nitrate- or nitrite-reducibility is exhibited by many intestinal microbes, intestinal bacilli, enterococci, bacteroids, blue pas bacillus, *etc.* [104]. Nitrite-reducing reaction is catalyzed by protein, ferredoxy-nitrite reductase containing Cu^{2+}. Nitrate reductase ranks among metalloflavanoids and contains molybdenum ions. The enzymes display paramagnetic properties and can be fixed by EPR spectroscopy [105].

The decrease in Hem-NO concentration in the control group is likely to be due to the fact that ischemic patients constantly administrated nitrates and nitrites as antiangina medicines [30]. Fecal Hem-NO concentrations in IHD women are reliably reduced. A higher fecal Hem-NO level in men is possibly related to smoking [31] (tobacco smoke contain nitrosamines).

Reliable correlation between fecal Hem-NO content and the number of intestinal bacilli (*Enterococcus*) indicates the existence of interrelation between the nitrate-reducing activity of intestinal microorganisms and the generating Hem-NO level. The reduced fecal Hem-NO content in patients with clearly-defined disbacteriosis and elderly individuals is related to suppression of the above activity. This suggestion be supported by the inverse correlation between fecal Hem-NO level and disbacteriosis degree in men. Improvement of the microbiological pattern expressed by reliable lowering of disbacteriosis degree in eubiotic-treated patients is accompanied by a 27% increase in fecal Hem-NO content. This seems to be explained by the fact that injured intestinal microflora is not able to reduce nitrates to nitrites and NO.

B. bifidum and *Lactobacillus* are obligatory members of large intestine microflora, constituting up to 50% of the whole microbial mass [106]. These species contribute much to a fecal Hem-NO signal registered in EPR spectrum. Therefore, the increase in the fecal Hem-NO content in eubiotic-treated patients is related to the increasing *B. bifidum* amount in large intestine lumen.

These species are also met in small intestine, but in minor amounts (to 10^5 per 1 ml). Organic nitrates of prolonged action are absorbed in small intestine and brought to the blood flow through the portal vein system. Some amount of organic nitrates brought into the intestine are suggested to be use for the synthesis of Hem-NO complex during metabolism and withdrawn from the system circulation. This appears to be one of the reasons for nitrate-tolerance.

Fecal Hem-NO EPR signal is not detected in the presence of nitrate ion. Having a clearly defined toxic effect, NO_2^- is assumed to inhibit the enzymatic activity of bacteria. Unexpectedly, incubation with *E.coli* does not lead to the appearance of Hem-NO, although this bacillus is able to reduce nitrates [31]. Evidently, *E.coli* does not use nitrite in Hem-NO synthesis.

3.3 HYPERTENSIVE DISEASE

In patients with hypertensive disease the plasma and erythrocyte free radical concentration is considerably higher than in healthy individuals (table 3-22) [9].

Table 3-22. The plasma and erythrocyte free radical concentration in humans with and without hypertensive disease (M ± m)

PMC	Healthy individuals n = 94, (10^{-6} M)	Patients n = 45, (10^{-6} M)
Erythrocyte FR	1.2 ± 0.02	1.3 ± 0.06
Plasma FR	0.5 ± 0.01	0.6 ± 0.03
Erythrocyte α-Tq	1.7 ± 0.03	1.3 ± 0.06
Plasma α-Tq	2.3 ± 0.04	1.9 ± 0.06

In patients with hypertensive disease the plasma and erythrocyte free radical concentration increases and the α-Tq content reduces with the disease severity (table 3-23) [9].

In events of stage IHD the plasma and erythrocyte FR concentration remains practically normal, whereas with stage II it increases [9]. The neurogenous theory of hypertensive disease origin suggests active participation of the sympathetic part of the nervous system in the origination and development of pathogenic processes [56, 107]. This means that blood emission of catecholamines enhances the plasma FR concentration and reduces the Cu^{2+} ceruloplasmin level. This is the universal response to the sympatho-adrenal system activation.

Table 3-23. The plasma and erythrocyte FR concentration depending on disease severity

(M ± m)

PMC	I stage HD, n = 9, (10^{-6} M)	IIA stage HD, n = 20, (10^{-6} M)	IIB stage HD, n = 16, (10^{-6} M)
Erythrocyte FR	1.0 ± 0.09	1.4 ± 0.10	1.4 ± 0.09
Plasma FR	0.5 ± 0.07	0.6 ± 0.07	0.6 ± 0.02
Erythrocyte α-Tq	1.6 ± 0.21	1.4 ± 0.06	1.1 ± 0.06
Plasma α-Tq	2.3 ± 0.13	1.7 ± 0.07	1.7 ± 0.07

As hypertensive disease develops the blood FR content approaches that observed in IHD (cf. [108]). In the HD development an essential part is played by disorders of redox processes and hypoxia of tissues accompanied by glutathione reductase activation. As HD develops the α-Tq concentration in plasma lipids and, especially, in erythrocytes reduces. This is likely to be related to a decrease in the α-T concentrations in patients [56, 107].

Paramagnetic centers in the plasma, erythrocytes and blood lipids and components in the Baikal-Amur Railway Road builders suffering from hypertonic disease have been studied. The results show a reduced FR level in erythrocytes and blood lipids and a decrease in FR and plasma peroxidation lipids in sick builders compared with the control group. The initial HD symptoms occurred on the background of enhanced FR levels [109].

The EPR method provided data supporting the hypothesis of NO-induced hypertensive activity of nitro and nitroso compounds [110]. The presence of NO in animal and human tissues is used in a search for NO-producing species.

Nitric oxide (endothelium-relaxing factor) controls the vascular tonus in patients with hypertension [111]. NO is endogenously formed mainly from L-arginine under the effect of NO-synthase [112]. This has provoked extensive research of NO metabolism in human [73, 113, 114]. At the same time methods for the determination of blood and tissue NO concentrations have been developed [115, 116].

However, there are certain difficulties in NO registration due to its short half-life [117]. In this respect the EPR method is much more advantageous compared with other procedures. An EPR-based procedure for quantitative determination of nitrite, the oxidation product of NO, in human plasma was elaborated. NO is strongly bound to hemoglobin to form an adduct giving a specific EPR spectrum. The EPR signal amplitude allows determination of even nanomolar NO concentra-tions.

The determination procedure is as follows: human plasma to be examined was passed (for the conversion of nitrite to NO) in a flow of nitrogen through a column filled with bovine Hb covalently bound to agarose and treated with excess dithionite. This transforms all Hb into the reduced state. Then the column contents is placed into the EPR spectrometer cavity and the EPR signal intensity of the Hb-NO complex is recorded. In normal individuals the venous plasma nitrite level ranged from 0 to 0.6 mcmol/l. In patients with forearm or leg ischemia the plasma level nitrite increased considerably [118].

Vascular wall relaxation is a physiological effect induced by sunlight. The effect is like that of endogenous relaxation (nitric oxide). Using the EPR method the Hb-NO contribution to the mechanism of this phenomenon under exposure to He-Cd and He-Ne laser has been studied. The results show that

Hb-No is able to reversible dissociate and release NO under exposure to only He-Cd [119].

Nitrovasodilators such as sodium nitroprusside ($Na_2[Fe(CN)_5NO]$), nitrates and nitrites react with Hb to form Hb-NO or MetHb containing Fe^{3+}. The formation of the above hemoglobin derivatives depends on the presence of oxygen, pH, and the nature of substrates involved. No Hb-NO EPR signal appears when nitrovasodilators are incubated with Hb in air or in a gas medium containing 5-10% of oxygen. The signal can only be detected when the oxygen content is below 1%. But even in this case the Hb-NO level was 3- to 10-fold lower than that observed in the absence of oxygen (pure nitrogen) [120].

The metabolism of nitroprusside *in vivo* is considered to occur mainly in the liver [121] rather than in the blood as suggested previously. The metabolism of nitroprusside with hepatocytes or subcellular fractions is based on one-electron reduction with cytochrome P_{450} reductase to the corresponding metal-nitroxyl radical.

Nitric oxide is not a very strong oxidant for most organic compounds. However, secondary reactions of NO with polyvalent metals or reactive oxygen compounds considerably enhances its oxidative potential. Contribution of NO to cell injury is not well-understood yet. During interaction with oxygen compounds the NO toxicity changes and sufficiently increases in the presence of superoxide anion radical. The latter reacts with NO to form finally a strong oxidant, peroxynitrite anion ($ONOO^-$).

Since this anion is relatively stable, its protonated form, peroxynitrous acid, HOONO decomposes spontaneously to form NO_3 ions. Exposure of human blood plasma to peroxynitrite in the presence of 3,5-dibromo-4-nitrosobenzene-sulfonate (DBNBS), leads to binding to NO_3^- to form a strongly immobilized adduct with the nitroxide radical. The formation of the above adduct is related with the presence in plasma of blood proteins such as serum albumin, fibrinogen, IgG, α-1-antitrypsin and transferrin. Peroxynitrite induces in the blood plasma the formation of protein radicals centered on the tryptophan residue [122].

Peroxynitrite formed in the interaction of NO with superoxide anion radical reacts with protein tyrosine residues to form nitrotyrosines. Nitrotyrosines are met in tissues and fluids of patients with lung diseases, atherosclerosis, bacterial and viral infections, in cases of neuro-degenerative states, aging, chronic inflammation, CO poisoning, tobacco smoke intoxication.

Until recently, registration of nitrotyrosine was used for peroxynitrite detection. The mechanism of nitrotyrosine formation can be independent on peroxynitrites as well. Nitrogen oxide reacts with a stable tyrosyl radical involved in the catalytic mechanism of ribonucleotide reductase suppressing the tyrosine radical EPR signal. Nitrogen oxide is supposed to form a radical adduct with organic radicals such as phenoxy radical of the tyrosine residue. According to EPR data inhibition of ribonucleotide reductase and suppression of its tyrosyl radical is a reversible process. This is due to the fact that nitrogen oxide radical forms with tyrosine radical a complex which decomposes to give a radical pair.

Nevertheless, the oxidation of this comp-lex induces the formation of a modified tyrosine residue involving tyro-tyrosin. Like ribonucleotide reductase, prostaglandin-H-synthase contains the tyrosine residue oxidized to the corresponding phenoxy radical during the catalytic process. Nitrosine is generated from the tyrosyl radical NO-adduct formed in the course of catalytic production of prostaglandins by prosttaglandin-H-synthase-2. The latter was treated with a system generating nitrogen oxide in the presence of diethylamine nonoate [123].

A new route to peroxybitrite-dependent hemoglobine oxidation leading to the production of free radicals at $g = 2.004$, has been suggested [124].

Endogenous carbon monoxide (CO) is produced by enzyme heme oxygenase. CO regulates c-HMP through activation of guanylate cyclase sharing this biological role with nitrogen oxide. The hemoglobin tetramer in circulating erythrocytes scrambles endogenous NO. From physiological viewpoint endogenous CO enhances the oxygen affinity of other hem sites thus displacing the oxygen dissociation curve to the left. This means that CO reduces oxygen transport to tissues. Owing to the extremely high hemoglobin affinity of nitrogen oxide ($NO > CO > O_2$), the oxygen present in the hemoglobin tetramer partially bound to NO is considered to cleave with more difficulty than that partially bound to CO.

The extremely high NO concentration also displaces the oxygen dissociation curve to the left and simultaneously increases the MetHb level, which reduces the oxygen retaining capability of hemoglobin. If this is true, endogenous nitrogen oxide should also enhance tissue hypoxia and induce death of cells. Endogenous NO generated *in vivo* by cytokines or nitrovasodilatators reduces (by less than 1%) the total hemoglobin in circulating blood and is bound to erythrocyte hemoglobin tetramer α-subunits (Hb α-NO).

It is established that NO displaces to the right the erythrocyte Hb oxygen dissociation curve without formation of MetHb in this case. This is due to breakage or weakening of the Hb iron - proximal histidine binding in Hb α-subunit. As a result a characteristic Hem-NO EPR spectrum with a triplet hyperfine structure of pentacoordinate hem appears [125].

The effect of trimesoyl-tris-(3,5-dibromosalycilate) (TTDS) and trimesoyl-tris-(phosphate methyl) (TTPM) on the oxygen dissociation rate has been studied [126]. TTDS slows down the oxygen dissociation rate and increases the carbon oxide association rate. The TTDS-HbNO spectrum does not show hyperfine structure, an indicator of Hb conformational change. On the contrary, TTPM increases the oxygen dissociation rate and decreases the CO association rate compared to that observed for HbNO. In addition, in the TTPM-HbNO spectrum there is a triplet providing evidence for breakage of the α-chain proximal His-Fe bond [126].

During the last time much attention is given to treatment of arterial hypertension with angiotensin-converting enzyme (ACE) inhibitors such as captopril. This medicine is potent in treating hypertension and heart deficiency and exerts an anti-inflammatory effect. It was also used for the treatment of rheumatoid arthritis [127].

Captopril has protective properties against injuries caused by arrhythmic ischemia/reperfusion. Other ACE drugs do not show properties of this kind [128]. The captopril ability of trapping free radicals is being extensively studied as they are directly involved in the mechanism of cell and tissue injury in events of ischemia/reperfusion. In particular, this drug is able to remove hydroxy radicals [129], hypohalite radicals ($OCl^{•}$) and hypochlorous acid (HCl) [130, 131].

Concerning participation of captopril in oxygen radial scavenging has been discussed. It is reported on the one hand that captopril is able to scramble the above oxygen radicals [132, 133], and on the other hand, that captopril does this rather inefficiently [134].

Nifedipine (1,4-dihydro-2,6-dimethyl-4-(2-nitrophenyl)-3,5-pyridine-dicarbo-xylic acid dimethyl ester) is a calcium channel blocker widely used in treating hypertension [135].

As shown by EPR spectroscopy, nifedipine inhibits Mn ions intracellular penetration through calcium channels [136]. According to EPR data, nifedipine is photolabile, because when illuminated with daylight or mercury lamp, it induced the generation of nitroxide radicals. The unusually easy photochemical activation of nifedipine is likely to be stimulated by photosensitive nitro groups interacting with suitably positioned hydrogen or pyridinyl ring methyl groups [135].

It is reported that the patients treated for a long time with nifedipine die earlier and more often than those not using this drug [137, 138]. It is quite possible to be caused by a slight photochemical activation of nifedipine inducing nitroxide radical generation.

Many aspects of ethyology and pathogenesis of hypertensive disease have not been clearly understood till the present time. One of the reasons for hypertension development is considered to be hereditary (possibly, congenital) defect of cell membrane structure.

In this connection an EPR spin labe-ling study was performed to investigate erythrocyte membrane fluidity in spontaneously hypertensive rats.

The results show the membrane fluidity to be considerably decreased. These studies allow the EPR spectroscopy to be used in diagnosis of spontaneous hypertension [139].

3.4 BLOOD DISEASES

In patients with iron-deficient anemia the activity of erythrocyte paramagnetic centers changes. With increasing the disease severity the MetHb content increases and that of Cu^{2+} lowers. In this case the erythrocyte free radical concentration increases up to (1.4 ± 0.03) x 10^{-6} M whereas in normal individuals it is (1.2 ± 0.02) x 10^{-6} M (table 3-24) [9].

The increase in the glutathione reductase system activity favoring generation of erythrocyte free radicals reduces MetHb and strengthens the erythrocyte membrane.

Table 3-24. The erythrocyte PMC content in IDA patients depending on the disease severity state (M ± m)

PMC	Severity of disease		
	Easy	Moderate	Heavy
MetHb, (g/l)	0.71 ± 0.04	0.82 ± 0.05	1.03 ± 0.04
Cu^{2+} SOD, (10^{-5} M)	0.91 ± 0.07	0.80 ± 0.07	0.86 ± 0.08
FR, (10^{-6} M)	1.40 ± 0.05	1.32 ± 0.03	1.55 ± 0.09
α-Tq, (10^{-6} M)	1.12 ± 0.02	1.74 ± 0.04	1.02 ± 0.06

The blood plasma Fe^{3+} content in IDA patients shows a 69% decrease in man and a 79% decrease in woman (Figure 3-8).

This value is lower the more clearly expressed the severity of disease 6.8 ± 0.60 and 3.2 ± 0.33 mcmol/l in easy and heavy states, respectively. As the Fe^{3+} concentration increases the Cu^{2+} ceruloplasmin content decreases (1.96 ± 0.08) x 10^{-5} and (2.65 ± 0.23) x 10^{-5} M, respectively [30, 76]. Along with the reduced blood plasma iron concentration, an increased fecal iron content is observed (Figure 3-9) [9, 140]. Fe^{3+} transferring frequently appears on the erythrocyte surface in 100% of men and 70% of women with iron-deficient anemia. At the same time, normal individuals and patients with chronic gastritis, gastric and duodenal ulcer, rheumatoid arthritis, osteoarthrosis and IHD the erythrocyte Fe^{2+} content is practically the same. In iron-deficient anemic patients some early erythrocyte forms carrying 25.000-50.000 transferrin molecules appear [10].

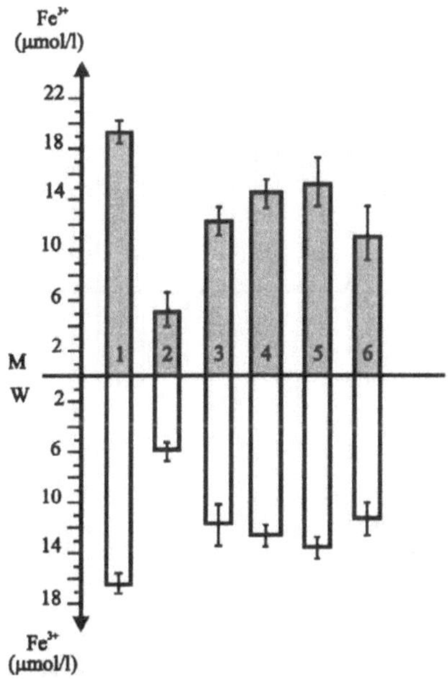

Figure 3-8. Plasma ion Fe^{3+} concentration (mcmol/l) in donors (1) and in patients with: iron-deficient anemia (2), chromic gastritis (3), duodenal ulcer (4), gastric ulcer (5), ischemic heart disease (6). M – man, W – women.

One of the iron deposition sites is protein ferritin. The pathways of generation of radicals in ferritin are not well understood. The formation of radicals in the H-chain of recombinant human ferritin and horse spleen ferritin was studied by EPR. The results revealed the potential role of tyrosyl radicals in the process of iron oxidation in ferritin [141]. The same method was used to study the human ferrochelatase function depending on the character of the enzyme amino acid terminus [142].

In moderate iron-deficient anemia the saliva Mn^{2+} content increases to nearly $(9.3 \pm 1.34) \times 10^{-6}$ M, thus showing that the more clearly expressed anemia, the more intense dystrophic processes, the lower manganese-containing enzyme activity, and which reducing to $(7.00 \pm 1.12) \times 10^{-6}$ M as the disease develops. Fe^{3+} is present in the saliva of all men with iron-deficient anemia and in 73% of iron-deficient women. At the same time in normal individuals saliva Fe^{3+} is met only in 50% of men and 23% of women. An unusual signal with g = 4.3 and half-width of 10 mT was found in the saliva of iron-deficient anemic patients (50% of men and 14% of women) (Figure 3-10) [143].

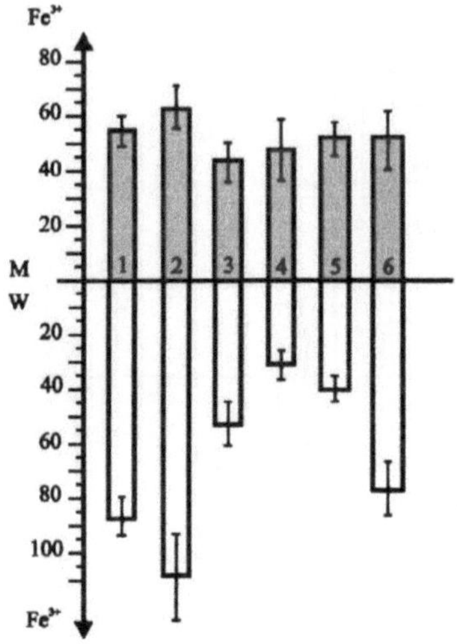

Figure 3-9. Fecal ion Fe^{3+} concentration (mcmol/l) in donors (1) and in patients with: iron-deficient anemia (2), chromic gastritis (3), duodenal ulcer (4), gastric ulcer (5), ischemic heart disease (6). M – man, W – women.

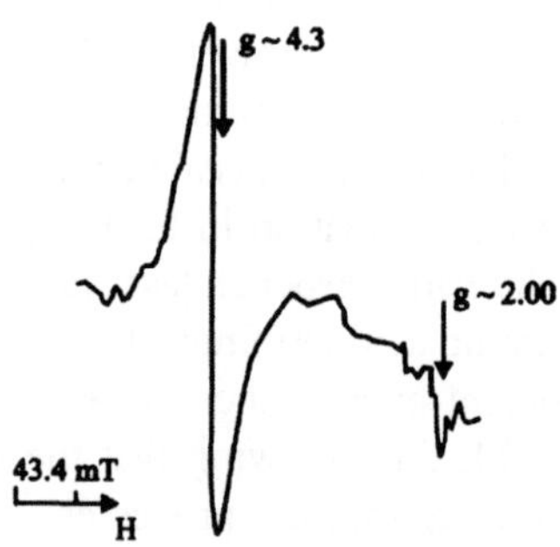

Figure 3-10. Saliva EPR spectrum of a patient with iron-deficient anemia (a sick man K., case history N 15657). Registration conditions: T = 77°K, SHF power – 5 mW, magnetic field scanning velocity - 50 mT/min, HF modulation – 0.5 mT, time constant – 0.3 sec.

This paramagnetic center is of unidentified nature. It is suggested to be caused by Fe^{3+} ion bound to chelate-forming ligands. In the saliva of iron-deficient anemic patients there is a Hem-NO signal (Figure 3-11) related to the presence of nitrate- and nitrite-restoring microorganisms in the mouth cavity.

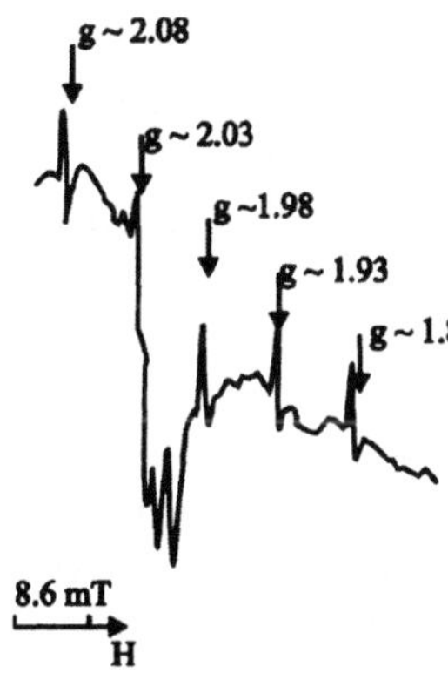

Figure 3-11. Saliva EPR spectrum of a patient with iron-deficient anemia (a sick woman S., case history N 15028). Registration conditions: T = 77°K, SHF power – 5 mW, magnetic field scanning velocity - 50 mT/min, HF modulation – 0.5 mT, time constant – 0.3 sec.

In the blood serum of patients with thalassemia, a Mediterranean anemia, there is an EPR signal with g = 6.0 never observed in patients with thalassemia minor. This EPR signal corresponds to a hem-serum albumin complex [144].

The problem of response to injury of infectious or non-infectious origin has been existed during the whole history of investigation of living organisms. The response manifests itself by a change of the organism reactivity, structure and metabolism [145]. As early as in the XIXth century I. I. Mechnikov pointed at the starfish defend reaction to spine irritation expressed as formation of a limiting leukocyte bank. Later on it has been shown that leukocytosis is a universal non-specific response of the organism to injury of different nature [146].

Investigations carried out in the second half of the XXth century revealed in leukocytes a number of species responsible for non-specific response of the organism, such as lysozyme, lactoferrin, myeloperoxidase, non-enzymatic cationic proteins. Leukocyte activation leads to the formation of free radicals in accompanied by the so called "oxidative burst" ("respiratory burst") [147].

The organic hydroxidases appearing in this case are cytotoxic. Leukocyte polymorphonuclear myeloperoxidase is primarily responsible for organic hydroperoxidase decomposition and cell defense [148]. (120). EPR and optical spectroscopy show neutrophil myeloperoxidase iron to convert from the high-spin state Fe^{2+} to low-spin in the presence of nitrite ion, the g-factor values changing from 6.84, 5.02 and 1.95 to 2.55, 2.31 and 1.82, respectively. EPR spectra fix even small differences in myeloperoxidase isoenzyme 1 compared with isoenzymes 2 and 3. This method allows detection *in vivo* the reaction of nitrites with intact neutrophil myeloperoxidase by monitoring NO production [149, 150].

Hydrogen peroxide and anion radical $O_2^{-\bullet}$ react *in vitro* in the presence of transition valence metals to form hydroxyl radical, an extraordinary oxidant essential in human phagocyte microbiological activity and tissue damage. Neutrophils or mononuclear human phagocytes either are able to form endogenous $HO^\bullet$ or need exogenous metals such as iron. This is not clearly understood until now mainly due to insufficient sensitivity of methods for the determination of phagocyte $HO^\bullet$ radicals. $HO^\bullet$ radicals can be detected by EPR spectroscopy using various spin probes such as nitrone or nitroso compounds. These compounds react with short-living $HO^\bullet$ radicals to generate long-living nitroxide radicals (spin adducts) determined by EPR spectroscopy using in most cases DMPO as a spin trap. DMPO well traps both superoxide and hydroxyl radical to form a fairly stable spin adducts. However, the spin adduct DMPO $O_2^{-\bullet}$ (2,2-dimethyl-5-hydroperoxy-1-pyrrolidinyloxyl–DMPO/OOH) decomposes to give a spin adduct DMPO-$HO^\bullet$, (2,2-dimethyl-5-hydroxy-1-pyrrolidinyloxyl–DMPO/HO). Therefore in the presence of both $O_2^{-\bullet}$ and $HO^\bullet$ it is not possible to determine them separately. This problem can be solved by addition of dimethylsulfoxide which reacts exclusively with HO radical to form CH_3 radical. The latter is scrambled by a spin trap in the form of DMPO/$CH_3^\bullet$. The EPR DMPO/$CH_3^\bullet$ method provides a specific indicator of $HO^\bullet$ radical registration in the presence of systems generating oxygen radicals in stimulated human phagocytes (neutrophils, monocytes and monocyte derivatives – macrophages) [151].

Using EPR spectroscopy the generation of oxygen radicals and the role in leukocytes were studied [152, 153]. Eosinophils also produce hydrogen peroxide, superoxide and $HO^\bullet$ which are of importance in host defense [154].

Taurine formed as a result of cysteic acid decarboxylation enters into the composition of paired bile acids and is essential in fat metabolism. The antioxidant properties of hypotaurine, the taurine precursor, were proved by EPR method using spin probe DMPO. Hypotaurine competes with DMPO for hydroxyl radicals. The biological role of hypotaurine and taurine in neutrophils was studied in terms of their antioxidant properties and subcellular location within the cell [155].

Sodium sulfite is metabolized by human polymorphonuclear leukocytes in two alternative pathways. One of these is an enzymatic way dependent on sulfite oxidase. The other is a non-enzymatic way involving the intermediate formation of sulfur trioxide anion radicals. According to EPR data the "oxidative burst" is initiated by phorbol myristate acetate that significantly stimulates the sulfate formation. In healthy individuals a considerable variation in the activity of sulfite oxidase in polymorphonuclear leukocytes is observed [156].

As shown by EPR spin trapping in monocyte-derivative macrophages nitrogen monoxides involved in the generation of $HO^{\bullet}$ radicals [157].

Using the same method it has been found that binding of mono-, di- and trisaccharides to membrane lectine reduces the rotational motion of membrane proteins and lipids. As a result of interaction of polisaccharides with lymphocytes the membrane fluidity decreases considerably which makes the membrane almost rigid. The membrane fluidity decreases depen-ding on the dose of saccharides, on the multivalency of ligands and is sensitive to the presence of EDTA and sodium azide. Binding of two different carbohydrate ligands on the lectin-containing membrane surface exerts a synergistic effect on the membrane fluidity [158-160].

According to EPR data, cell lines of human fibroblasts generate oxygen radicals when stimulated with serum-treated zymosan, N-formyl-methionyl-leucyl-phenylalanine, leukotriene B4 or 12-O-tetradecanoylphorbol 13-acetate. As known, all the above compounds stimulate respiratory burst of phagocytes (in phagocytic cells) [161].

Oxygen-derived active species and, in particular, superoxide radicals are generated and excreted by activated granulocytes. These radicals play an essential role in host defense against bacterial and fungal infections. The formation of radicals peaks soon after neutrophil stimulation and decays within minutes. Nevertheless, neutrophil function is expected to last considerably longer. Kinetics of anion radical formation by stimulated human neutrophils was investigated by three independent methods: EPR, chemiluminescence and ferricytochrome C reduction. Under the appropriate experimental conditions stimulated neutrophils are able to produce anion radicals $O_2^{-\bullet}$ within several hours [162].

An agonist-activated phospholipase D/phosphatidic acid (PA) and phosphohydrolase (RAH) pathway of "oxidative burst" generation was demonstrated in human neutrophils. Phosphatidic acid and diradylglycerol (DG) formed in this way, participate in activation of the "oxidative burst" followed by anion radical $O_2^{-\bullet}$ generation. To investigate the role of the phospholipase D (RAH) pathway in neurophil activation the use was made of a series of cationic amphiphilic compounds (sphingosine, propanolol, chlorpromazine and desipramine) and antibiotics (clindamycin, trimethoprim and roxithromycin), all of which inhibiting the "oxidative burst". By EPR spectroscopy using ([3H]-O-alkyl- lysoPC) labeling the content of phosphatidyl in neutrophil intact cells was examined. The [3H]-phosphatidic acid and [3H]-diradylglycerol released in this case could quantitatively be determined only after addition of either chemo-attractant or phorbol miristate acetate. All these agonists showed dose-dependent inhibition of [3H]-diradylglycerol generation which correlated well with inhibition of O_2 radi-cal generation. However, the compounds cannot directly inhibit the NADP.H oxidase in cells. In many cases inhibition of [3H]-DG corresponds to an increase in the [3H]-PA concentration. This implies PAH to be the locus of inhibition. With a decrease in the content of phosphatidic acid the anion radical O_2 generation is either inhibited or ceased. $O_2^{-\bullet}$ release and DG generation are independent on the structure of propranolol stereoisomers. PAH is suggested to be not involved in inhibition of this type. No effect of the inhibitors on PA mobility was revealed by means of EPR monitoring using spin labeled PA in a model membrane system. These results suggest diradylglycerol generated in the phospholipase D/RAN pathway to activate and maintain the "oxidative burst" [163].

Polymorphonuclear leukocytes appear in infected areas of the organism and adhere via diapedese and chemotaxis. These leukocytes kill pathogenic bacteria thus producing active oxygen radicals. However, the latter are toxic not only to microorganisms, but to the host cells and even to leukocytes [164, 165].

"Oxidative burst" during phagocytosis of human neutrophils is related to the generation of superoxide anion radical. This radical is converted to hydrogen peroxide and hydroxyl radical which can be harmful to the organism. The radicals were studied by EPR spectroscopy using DMPO spin trap. Overproduction of superoxide anion radicals (by chronic inflammation, for example) is endogenously suppressed by superoxide dismutase or other drugs of similar activity [166].

Polymorphonuclear leukocytes are damaged by radicals formed on their own. Leukocyte damage by these radicals can be inhibited using phenyl-N-*tert*-butyl nitrone (PBN) spin label [167-169].

3.5 HEMOSTASIS DISORDER

Hemostasis is an evolutionary developed defense reaction manifesting itself as hemorrhage arrest in cases of blood vessel wall injury. The hemostasis system represents a combination of blood components (plasma and cells) and vessels.

The process of hemostasis involves extravascular tissue, vessel wall, blood coagulation plasma factors, and other blood formed elements. In this process an important role is played by endogenous physiologically active substances (adrenalin, noradrenalin, serotonin, *etc.*), local changes in ionic relations and blood pH, ξ(zeta)-potential (difference of potentials between dispersion medium, dispersion phase and the vessel wall). Thrombocytes also take part in the hemostasis vascular-thrombocytic mechanism. On the wound surface they adhere to injured endothelial cells and tissue fibers, thus forming a hemostatic thrombocytic thrombus. The structure and functions of thrombocyte membranes are essential in blood coagulation process.

Three subfractions of human thrombocytes different in density and function were studied using EPR spin label measurements and fluorescence polarization. The anisotropy of plasma membranes of low-density thrombocytes is higher than that of intermediate and low density thrombocytes. This high plasma membrane anisotropy correlates with a higher cholesterol/phospholipid molar ratio in the plasma membranes of low density thrombocytes. As compared with other subfractions the low density thrombocytes show a very small increase in cAMP after activation with aggregation inhibitor prostaglandin E1. After thrombin stimulation the low density thrombocytes show the highest percentage of inositol phosphate accumulation. A high correlation between functional parameters and the architecture of thrombocyte plasma membranes is observed [170].

The EPR method using paramagnetic analogs of endogenous phospholipids was employed to study phospholipid distribution in thrombocyte plasma membranes in the process of aging. Asymmetrical distribution and translocation kinetics of spin labeled phosphatidylserine and phosphatidylcholine in fresh thrombocyte plasma membranes were different. After keeping for seven days in a freshly prepared thrombocyte culture the spin labeled phosphatidylserine fast penetrated to the inner side of the thrombocyte plasma membranes. However, the spin labeled phosphotidylcholine mainly remains on the external side. After stimulation of the thrombocyte culture by the calcium ionophore A23187 the inhibition of translocase activity and modification of membrane stability persists up to the nineth day and disappears between the day 10 and 12. After 7-9 days a strong cyctoskeleton proteolysis and a noticeable decrease in intracellular ATP concentration take place. After 9 days thrombocytes cannot be activated any longer. This indicates complete suppression of β-N-acetyl glucosaminidase secretion and vesicle formation after stimulation of aged thrombocytes by ionophore calcium (A23187). During thrombocyte aging *in vitro* there are metabolic and membrane changes analogous to those observed in activated thrombocytes. All these phenomena simultaneously occur between the seventh and nineth day. The EPR data emphasize the importance of maintaining the plasma membrane asymmetry to increase the hemostatic effectiveness of transfused thrombocyte concentrates [171].

In order to investigate specific changes in the fatty acid composition of cell membrane phospholipids an original method based on the transfer of pure phospholipid macromolecular fragment to membranes has been developed. For this purpose phosphotidylcholine (PC) and phosphotidylethanolamine (PE) subclasses containing unsaturated fatty acids with C_{18}:2n-6 and C_{22}:6n-3 in the sn-2 position were incorporated into human thrombocyte membranes. This was performed using endogenous phosphotidylinositol/PC transfer protein (PI/PC-TP) and phospholipid transfer protein from maize (L-TP), respectively. The PI/PC protein transfer carriers catalyze phospholipid exchange between thrombocyte membranes and unilamellar vesicles containing 1,2-diacylglycerophosphocholine (diacyl-GPC, 16:0/18:2-GPC or 16:0/22:6-GPC). The proportion of unsaturated fatty acids 18:2n-6 and 22:6n-3 in diacyl-GPC constantly increases from 11% to 17% and from 0.8% to 10%, respectively, At the same time the PE and PI fatty acid compositions are unchangeable. According to electron-spin resonance of 5- and 16-nitroxy stearic acids, the diacyl-GPC-enrichment in 22:6-3n and 18:2-6 does not lead to changes in the membrane fluidity.

Analogously, 18:2-6 and 22:6-3 esterified in 1,2-diacylglycerophosphoethanolamine (diacyl-GCP) were incorporated in thrombocyte membranes by an exchange process when donor vesicles had a phospholipid composition equivalent. Thus, the L-TP and PI/PC-PP catalytic effect can be used for studying the modulation of membrane biological activities by changes in fatty acid compositions of specific phospholipid classes or subclasses [172, 173].

Electrostatic forces enhancing the thrombocyte adhesion to a wound surface are of importance in the mechanism of interaction of thrombocytes to injured vessel wall during hemostasis. Adhesion facilitates slowing down of bleeding.

Thrombocyte adhesion is followed by aggregation in the wall injury site. ATP is released from injured endothelial cells and from erythrocytes and thrombocytes and further transformed to ADP under the influence of cellular adenosinetriphosphotase. ADP induces a reversible thrombocyte aggregation. A number of hypotheses have been put forward for elucidating the aggregation mechanism. According to one of these the ADP thrombocyte-aggregating activity is conditioned by three free negative valences in the ADP molecule. Two valences bind a calcium ion whereas the third one, together with the same valence of neighboring ADP molecule, adds another calcium ion. Thrombocyte aggregation results from the formation of calcium "bridges" between thrombocyte-nucleotide complexes. However, many thrombocyte aggregation mechanisms are not well understood yet.

EPR spectroscopy using three different traps: two nitroso – DBNBS and 2-methyl-2-nitrosopropane (MNP), and a nitrone – DMPO was applied to investigate the role of L-arginine/nitric oxide pathway in human thrombocyte aggregation. In this study the effect of spin trap concentration on the collagen-induced human thrombocyte aggregation was compared with the antiaggregation effect exerted by L-arginine. It was found that the nitroso spin traps prevented thrombocyte aggregation more effectively than L-arginine. At concentrations below 300 mM/l DMPO did not affect much the collagen-induced aggregation. Thrombocyte activation with a low collagen concentration (17 μg/l) in the presence of DBNBS and MNP leads to EPR-detectable spin-adducts. Some of them are different from the spin adducts forming in the interaction of free radicals and nitric oxide with the spin traps. Only one DBNBS adduct observed in the spectrum of water-washed thrombocytes in the presence of superoxide dismutase was similar to a product obtained by a reaction of NO with DBNBS. DBNBS and MNP adducts were also generated during thrombocyte activation in the presence of Ca^{2+} and a cytosol(e)-depleted L-arginine preparation from washed thrombocytes to which L-arginine was added. N-ω-methyl-L-arginine (MeArg, 100 mcM) inhibits the formation of these DBNBS and MNP spin adducts. This suggests the latter to originate from a product of NO synthase. The formation of these DBNBS and MNP spin adducts in thrombocyte suspensions was enhanced in the presence of superoxide dismutase. However, their formation was suppressed by methylene blue and hemoglobin, endothelial-derived relaxing factor (EDRF) inhibitors. These data provide evidence for the existence of the endogenous L-arginine NO pathway in human thrombocytes [174].

The role of erythrocytes and leukocytes in hemostasis is related to the presence of most blood coagulation factors. When the blood vessel wall is injured these factor get involved in fibrination reaction. During hemostasis thrombocytes are held in the fibrinous network thus facilitating a thrombus formation and growth. Erythrocyte rheological properties and aggregational ability have been studied by EPR spectroscopy using spin labels [175].

The rate of thrombus formation depends on the interaction between thrombocytes and erythrocytes. However, the mechanism of this process remains unclear. Thrombocyte aggregation can be caused by nanomolar levels of hemoglobin released from injured erythrocytes. At the molecular level this process is not receptor-based. It involves oxyhemoglobin oxidation by hydrogen peroxide derived from thrombocytes with subsequent generation of unknown radical products detected by EPR spectroscopy. MetHb and Hb-CO do not induce thrombocyte activation or radical formation. Hemoglobin-induced thrombocyte aggregation is completely blocked by catalase or radical "scavengers" [176].

The interaction of human thrombocytes and endothelial cells with neutrophils and monocytes was studied by EPR spectroscopy using spin probes. Sialic acid receptors located on the thrombocyte surface were also examined by this method [158].

Thrombocytes stimulated by arachidonic acid or collagen are activated with nanomolar quantities of H_2O_2. This effect is mediated by hydroxyl radicals extracellularly formed by a Fenton-like reaction. H_2O_2-induced thrombocyte aggregation results in serotonin release and thromboxane A2 production. This process is inhibited by hydroxy radical "scavengers" and by iron chelator desferrioxamine. The intermediate HO radicals were detected by EPR spectroscopy using spin probes. The role of hydroxyl radicals was supported in experiments with exogenously added iron. A free or EDTA-bound Fe^{2+} ion activates thrombocytes pretreated with deoxyribose, mannitol or catalase. Fe^{3+} ion acts in a similar manner only in the presence of a reductant. $HO^{\bullet}$ radical activation depends on accompanying release of arachidonic acid and is blocked by phospholipase A2 inhibitors mepacryne and aristolochic acid and by (Na^+/K^+ antiporter inhibitor ethylisopropylamiloride). In contrast, neomycin and staurosporin were ineffective. This indicates that phospholipase C, protein kinase C were not involved in the initial activation phase. Neither radial formation nor arachidonic acid release were aspirin-blocked. In whole blood thrombocyte activation is induced by hydrogen peroxide generated upon stimulation of neutrophils by N-formylmethionyl-leucyl-phenylalanine. Thrombocyte activation and $HO^{\bullet}$ radical formation is blocked by the NADP.H oxidase inhibitor diphenyliodonium, catalase and mannitol. This suggests that the reactive oxygen species act as "second messengers" during the initial phase of thrombocyte activation process [177].

Introduction of albumin-bound docosahexaenoic acid (22:6n-3), but not linoleic acid (18:2n-6) into cellular phospholipids inhibits thrombocyte aggregation induced by thromboxane analog U46619. [3H]U46619 specific binding to thromboxane A2/prostaglandin H2 (TXA2/PGH2) receptors, as well as specific binding of the antagonist [3H]SQ29548 to these sites were also decreased in these modified cells. Over 80% of the 22:6n-3 incorporated in these cells was esterified in the various endogenous phospholipid classes. The remaining was found in neutral lipids and in the unesterified fatty acid pool. Membrane fluidity parameters were measured by electron spin resonance of 5- and 16-nitroxy-stearic acids. As a result it was established that the molecular species of phosphotidylcholine (PC) bound to 22:6n-3 in the position 2 can affect were not significantly different in membranes enriched with 1-palmitoyl-2-docosahexaenoyl-glycerophosphocholine (16:0/22:6-GPC) relative to those enriched with 16:0/18:2n-6-GPC, arguing against a generalized perturbation of the membrane due to 22:6n-3 incorporation. We conclude that molecular species of PC with 22:6n-3 at the sn-2 position can affect TXA2/PGH2 receptors. Selective inhibition of thrombocyte membrane fluidity by 22:6n-3 - containing phosphotidylcholine is discussed [172, 173].

Changes in membrane fluidity caused by recombinant tissue-type plasminogen activator were detected by means of h+1/0 parameter calculated from EPR spectra for [2-(14-carboxytetradecyl)-2-ethyl-4,4-dimethyl-3-oxazolidinyloxy] (16-DOXYL-Ste) and [2(3-carboxypropyl)4,4-dimethyl-2-tridecyl-3-oxazolidinyloxy] (5-DOXYL-Ste) incorporated into the lipid bilayer (h+1 and h0) are the heights of the low-field and middle-field lines of the spectra, respectively). For this purpose washed thrombocytes labeled with 16-DPXYL-Ste and 5-DOXYL-Ste free radicals were incubated with recombinant tissue-type plasminogen activator. Interaction of the above activator contained in both stimulated and non-stimulated thrombocytes increases the membrane lipid bilayer rigidity. This is suggested by a decreased $h+1/h_0$ value. This can be explained by either conformational changes of the membrane receptors due to the attachment of recombinant tissue-type plasminogen activator with subsequent rearrangement of the lipid matrix of thrombocyte membrane. Or by direct association of recombinant tissue-type plasminogen activator with membrane phospholipids. Partial embedding of the protein molecules into the lipid bilayer restrains the lipid mobility [178].

In order to examine which of the two Pro47-Asp87 domains of epidermal growth factor is involved in the reaction of recombinant tissue-type plasminogen activator with thrombocytes this domain was expressed in *E.coli*. Thus the peptide fragment was produced from a plasmid expression vector as a fusion protein with β-galactosidase Met1-Val444 at high yield in eight clones of *E.coli*. The fusion protein was purified and hydrolyzed with formic acid under mild conditions. Then the Pro47/Asp87 peptides identified by immunoblotting using specific antibodies to tissue plasminogen activator, were isolated by HPLC. According to EPR data, after incubation with thrombocytes spin labeled with 5- or 16-doxylsteraic acid the Pro47/Asp87 peptide fragment reduced the membrane lipid bilayer rigidity analogously to intact tissue plasminogen activator. The data obtained suggest that the epidermal growth factor-like domain of tissue plasminogen activator directly interacts with thrombocytes [179].

Thrombin is a blood coagulation proteolytic enzyme catalyzing the hydrolytic cleavage of fibrinogen to fibrin-monomers. Thrombin is related to peptide-hydrolases, serine-proteinase subclass. Fibrin-monomers are polymerized to fibrin constituting the thrombus basis.

The primary structures of bovine and human α-thrombins are highly homologous. Therefore, X-ray diffraction analysis does not provide sufficient information about their structural differences. To compare the dynamic conformations of bovine and human α-thrombins the use was made of EPR spectroscopy. The active sites were labeled by fluorosulfonylphenyl inhibitor and fluorophore which probed the catalytic serine residue in the 10-15 A region. According to EPR data, the nitroxide moieties in human and bovine thrombin are completely mobilized, the steric motion in bovine thrombin being restrained. Dansyl fluoride and *para*-nitrophenyl anthranilate phosphores reveal differences in λ^{m}_{max} and λ^{x}_{max} values of the complexes with bovine and human α-thrombins, respectively [180].

Hirudin, a specific thrombin inhibitor, irreversibly blockades the "recognition" site of high molecular substrates outside the active center. Interactions of hirudin analogs with bovine and human thrombin were studied by EPR (using spin labeling and fluorescence spectroscopy [181].

A certain role in hemostasis is played by plasma fibronectin, a high-molecular adhesive glycoprotein. Its macrochain contains two subunits (250 kDA) bound by disulfide bond within the carboxyl terminals. One of them (SH1 site) is located between the DNA-binding and cell-binding domains, while the other one (SH2 site) is located within the carboxyl-terminal fibrin-binding domain. Using the EPR method it is shown that these sulfhydrile groups of soluble fibronectin do not react with sulfhydryl reagents. Also they fail to interact with collagen, fibrinogen, heparin, hyaluronic acids, calcium ions, EDTA, deoxycholate or methylamine solution [182].

Fibronectin molecular dynamics was investigated with the help of EPR method (^{15}N, ^{2}H-maleimide spin label) and fluorescent spectroscopy (coumarinylphenyl maleimide label). Upon absorption of fibronectin to a gelatin-coated surface, the SH_1 site, located between the DNA-binding and the cell-binding domains, is partially exposed, while the SH_2 site, located within the carboxyl-terminal fibrin-binding domain, remains buried and unreactive. The procedures for the preparation of the selectively labeled fibronectins are described in detail. The fluorescence technique indicates that the environments of SH_1 and SH_2 sites are hydrophobic. The EPR results show that heparin or high salt induces an increase in the domainal flexibility in both SH_1 and SH_2 regions, perhaps through the disruption of domain-domain interactions in the fibronectin molecule, and that heparin is more effective in this respect than the polysalts. The observed heparin effect is reversible by addition of calcium ions in the SH_2 regions but not in the SH_1 regions. At temperatures above 44°C, both types of homologous regions containing the free sulfhydryl groups undergo denaturation and aggregation. These data suggest that the EPR method is useful for mapping the spatial arrangement of structural domains in the protein molecule [183].

To correct hemostasis dipyridamole (curantyl) is often used. Dipyridamole dose-and time-dependently inhibits the production of oxygen radical from human stimulated by opsonized zymos and formyl-methionyl-leucyl-phenylalanine. Dipyridamole directly inhibits the generation of active oxygen radicals at therapeutic concentrations. Possibly dipyridamole may be used as a potential "scavenger" of active oxygen metabolites because it inhibits very quickly biochemically formed radicals $HO^{\bullet}$ and $O_2^{-\bullet}$. At therapeutic concentrations dipyridamole inhibits the generation of oxygen active forms in human polymorphonuclear leukocytes [184].

Chemical modifications of thrombocyte membranes and erythrocyte ghosts, induced by aspirin (acetylsalicylic acid), acetyl chloride or salicylate, have been investigated using fluorescence quenching and EPR spectroscopy. The results show that only aspirin can induce disorders in the lipid-protein matrix and membrane protein conformation. The aspirin action in thrombocyte and erythrocyte membranes decreases the distance separating the membrane tryptophan and bound 1-anilino-8-naphthalenesulfonate molecules. This significantly increases the maximum efficiency of energy transfer. The decrease in the ratio of amplitudes of low-field peaks in EPR spectra from weakly to strongly immobilized fractions of maleimide spin label (4-maleimido-and the increase in the relative rotation correlation time of iodoacetamide spin label [4-(2-iodoacetamido)- 2,2,6-6-tetramethylpiperidin-1-oxyl) indicate that aspirin effectively immobilizes membrane proteins in the lipid bilayer. At the same time, acetylchloride and sodium salicylate fail to produce a noticeable effect. Thus, aspirin affects the membrane protein structure inducing reorganization of both lipid and membrane protein assembly. The interaction of aspirin with thrombocyte and erythrocyte membranes may lead to local conformational changes with impairment of thrombocyte functions. This new type of aspirin-induced protein chemical modification is suggested to involve the generation of a reactive salicylic residue during fast degradation of aspirin under physiological conditions [185].

The effect of pentoxifylline and other hemorheologically active drugs on human erythrocyte membranes was studied by EPR spectroscopy. On external addition of the drug, an increase in the fluidity in the regions of phospholipid head groups of the erythrocyte membrane bilayer is observed. The fluidity is dependent on the incubation time suggesting a drug-membrane protein interaction. On the other hand, the acyl chain motion in the hydrophobic end of the phospholipid group is reduced in the presence of the drug and does not depend on time. These pentoxifylline-induced changes in the membrane fluidity at different depths of the membrane may correlate to erythrocyte deformability [186].

EPR spectroscopy was used to examine the effect of WEB 2086, 3-[4-(2-chlorophenyl)-9-methyl-6H-thieno[3,2-f][1,2,4]triazolo- [4,3-a][1,4]-diazepin-2-yl]-1-(4-morpholinyl)-1-propanone, an antagonist of thrombocyte-activating factor, on the generation of active oxygen radicals by human polymorphonuclear leukocytes. This agent inhibits the generation of oxidative metabolites by these cells following stimulation by and formyl-methionylleucylphenylalanine in a time-dependent fashion. WEB 2086 directly inhibits the generation of active oxygen radicals by these cells washed prior to stimulation may be a "scavenger" of active oxygen radicals. WEB also "scavengers" active oxygen radicals generated by opsonized zymosan-stimulated human polymorphonuclear leukocytes. As established by EPR, this action of WEB 2086 may be exerted against hydroxyl radicals $HO^{\bullet}$ and $O_2^{-\bullet}$. Thus, it follows that WEB 2086 may inhibit the generation of active oxygen radicals by human polymorphonuclear leukocytes and directly influences the scavenging of these radicals [187].

3.6 DISEASES OF GASTRO-INTESTINAL TRACT

The development of gastric and duodenal chronic diseases is accompanied by some changes in the host PMC content. In events of peptic ulcer, especially, duodenal ulcer, the erythrocyte free radical concentration is slightly increased. Activation of the glutathionereductase system seems to be caused by hypoxia favoring chronic development of peptic ulcer [188].

In patients with chronic gastritis and ulcer the plasma Fe^{3+} concentration is decreased, whereas that of Cu^{2+} is increased (cf. [189]). Opinions on the iron concentration in the blood plasma of patients are contradictory (lower values are most often reported). At reduced gastric juice acidity, the Fe^{2+} - Fe^{3+} oxidation is decreased [190]. The same is also observed in gastritis, ulcers and cancer of stomach. Concomitant injuries of other departments of the gastro-intestinal tract (enteritis, colitis, cholecystitis, pancreatitis, *etc.*) are accompanied by disorders in iron utilization and metabolism. The results of EPR and spectrophotometry show the duodenal b-type cytochrome to be involved in the process of intestinal iron absorption [191]. The decrease in the blood plasma Fe^{3+} content in ulceric patients is induced first of all by concealed bleeding.

In the stomach contents of patients with gastritis, gastric and duodenal ulcer an EPR Hem-NO signal was observed. Its concentration in healthy individuals is higher than in patients with chronic gastritis and peptic ulcer (71% and 66%, respectively) (Figure 3-12) [192, 193].

Hem-NO exerts an activating effect on guanylate cyclase thus increasing tissue c-GMP and stimulating cell proliferation. This supports the gastric wall integrity and prevents ulceration [194, 195]. C-GMP inhibits thrombocyte aggregation and produces a vasodilatory effect. The above diseases are characterized by a reduced blood flow and the appearance of thrombocyte aggregations [190]. Certain amount of carcinogenic nitro- and nitroso-compounds are brought into the gastro-intestinal tract with food. The activity of nitrate- and nitrite-reducing microorganisms is most likely to be compensatory reaction preventing the development of oncologic diseases. Malignant tumors are often met in patients with chronic gastritis and peptic ulcers. This suggests Hem-NO to be a compulsory component of gastric contents. Hem-NO is supposed to be involved in the development of chronic gastritis and peptic ulcer by the following pathogenic mechanism. Changes in the stomach microflora, in particular, a decrease in the content of nitrate- and nitrite-containing microorganisms leads to a drop of the Hem-NO level [196]. This inhibits cell proliferation, decreases mucous wall restoration, strengthens thrombocyte aggregation and reduces vascular tonus regulation.

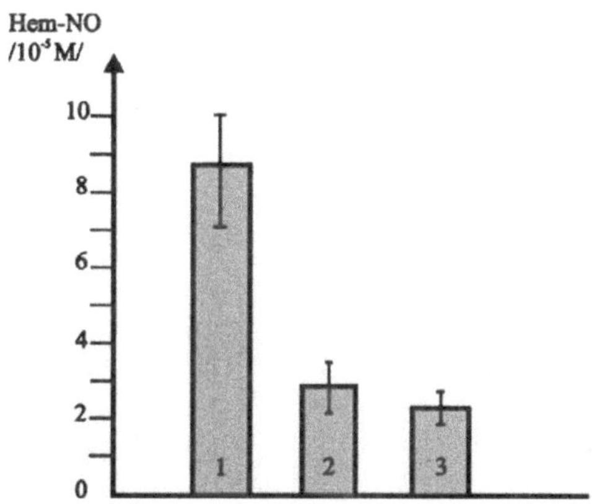

Figure 3-12. Intestinal Hem-NO concentration (10^{-5} M) in donors (1) and in patients with chronic gastritis (2) and gasric-duodenal ulcer (3).

In some patients with injury of the gastro-intestinal tract a decrease in the α-Tq content in plasma and erythrocyte lipids is well defined (Figure 3-13).

In patients suffered from ulcer of stomach the saliva Mn^{2+} concentration is extremely low (6.3 ± 0.42) x 10^{-6} M. In events of chronic gastritis and peptic ulcer the fecal Mn^{2+} content is lowered too (Figure 3-14).

The Mn^{2+} ion was detected in upper departments of the gastro-intestinal tract (saliva, pancreas, bile). High concentrations of Mn^{2+} ion were also found in intestinal anaerobes of the *Bacteroides* genus.

Assessment of the stomach acid-forming function is of importance in both diagnosis and treatment of the disease. The available methods to determine stomach acidity (gastric intubation, intragastric pH-metry, *etc.*) are painful. Therefore, the development of novel diagnostic approaches is rather essential. The latest achievements in EPR spectroscopy involving examination of intact animals look rather promising [197, 198].

Spin labeled pH-sensitive proteins were studied by EPR spectroscopy. For this purpose human serum albumin was modified by two different spin pH-sensitive labels of the imidazole series containing an alkylating or acylating group. Then the pK values of radicals covalently bound to protein were measured by the EPR method ($pK_1 = 3.2 \pm 0.1$, $pK_2 = 4.75 \pm 0.1$). The resulting spin pH-probes may be used for measuring local pH values by the EPR method within the 1.8-6.2 pH range [199]. A method based on peroral administration of pH-sensitive nitroxide seems to be of great promise in clinical practice [200]. The method enables intragastric pH monitoring. These tentative data suggest that the EPR method will provide a valuable noninvasive tool for a prolonged pH characterization *in vivo*.

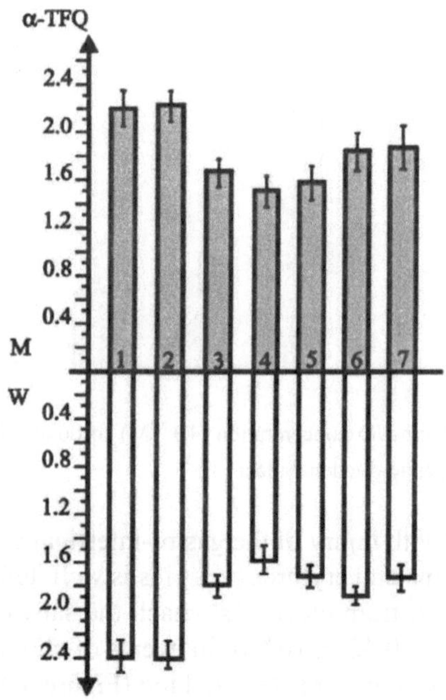

Figure 3-13. Plasma lipid tocopherilquinone concentration in donors (1) and in patients with iron-deficient anemia (2), chronic gastritis (3), duodenal ulcer (4), gastric ulcer (5), ischemic heart disease (6), hypertensive disease (7). M – men, W – women.

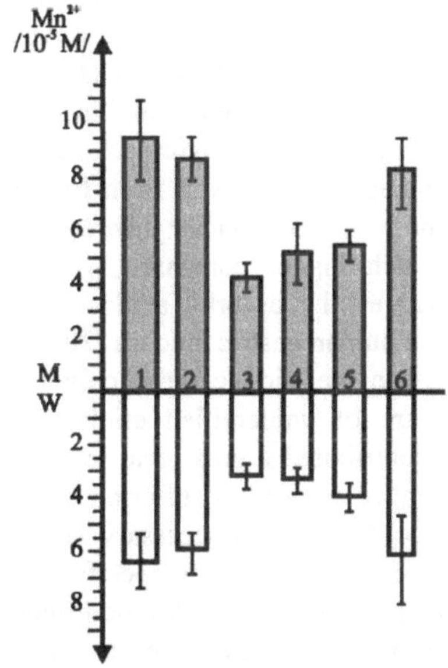

Figure 3-14. Fecal Mn^{2+} ion concentration (10^{-5} M) in donors (1) and in patients with iron-deficient anemia (2), chronic gastritis (3), duodenal ulcer (4), gastric ulcer (5), ischemic heart disease (6). M – men, W – women.

The acid-forming function of stomach can be reduced by a wide range of drugs (antacids, cholinolitics, H_2-histamine receptor blockers, "proton pump" blockers). However, mechanism of action is not well-understood for most of them. The effect of proton pump blocker omeprazole on the $K/[H^+]$ value of ascorbic acid (K = the ascorbic acid-ascorbic free radical equilibrium constant) has been studied. Changes in the concentration of ascorbate free radical in guinea pig serum were examined after intraperitoneal administration of omeprazole. It was found that the serum $K/[H^+]$ value increased significantly after administration of omeprazole. The drug turned out to amplify the EPR signal intensity of ascorbate free radical *in vitro*. The results suggested omeprazole to act like oxygen radicals [201].

Recently a new antiulcer agent, rebamipide 2-(4-chlorobenzoylamino)-3-[2(1H)-quinolinon-4-yl] propionic acid, has been synthesized. EPR spectroscopy using spin trap DMPO showed that rebamipide had antioxidant properties. This agent is active as $HO^{\bullet}$ radical "scavenger" and inhibitor of superoxide production in neutrophils [202]. Rebamipide inhibits active oxygen generation by neutrophils stimulated with *Helicobacter pylori* extracts.

This suggests rebamipide to protect gastric mucosa against inflammation [203].

The high ascorbic acid content of diet is suggested to protect against gastric cancer. This may be due to its action as a "scavenger" of free radicals forming in the gastric mucosa, which finally leads to DNA damage. Using EPR spectroscopy ascorbyl radicals were found in gastric mucosa of 82 patients taken vitamin C earlier, of whom 37 had *Helicobacter pylori*-associated gastritis. This appears to result from free radical scavenging by ascorbic acid. In patients with *Helicobacter pylori* gastritis the ascorbyl radical concentration is much higher compared with subjects with normal histology. These data suggest that ascorbic acid is acting as a "scavenger" of free radicals generated in human gastric mucosa [204].

Crohn's disease and non-specific ulcerative colitis are called the diseases of the century. They are of unidentified etiology. Industrialization and "western" way of life contributed a great deal to spreading these diseases. The ever increasing number of patients suffering from Crohn's disease and non-specific ulcerative colitis makes it necessary to search for the origin and mechanisms of these diseases as well as to specify and enlarge the scope of drugs to treat them. The pathogenesis of chronic inflammations of the intestine such as Crohn's disease and nonspecific ulcerative colitis, is suggested to involve the formation of oxygen radicals ($O_2^{-\bullet}$ and $HO^{\bullet}$) generated by phagocytes. It was found that sulfasalazine (SASP) and its metabolites, such as 5-amino-salicylic acid (5-ASA), and acetyl-5-amino-salicylic acid (AC-5-ASA), but not sulfapyridine (SP) and acetylsulfa-pyridine (Ac-SP) act as direct $O_2^{-\bullet}$ and $HO^{\bullet}$ "scavengers" *in vitro*. The *in vivo* anti-inflammatory effects of sulfasalazine and its metabolites are, at least partly, due to the direct oxygen-centered scavenging activity of these drugs. Upon oral administration of sulfasalazine and 5-aminosalisic acid their concentrations in the intestine my reach a level high enough to have a scavenging activity for oxygen radicals [205].

Gluten enteropathy represents a classical example of thin intestine disease with malabsorbtion syndrome. In patients genetically indisposed to this disease the proximal intestine mucous membrane may be injured by proteins of cereals such as wheat, rye, barley and oats. At first gluten is bound to specific enterocyte receptors and later it reacts with interepithelial lymphocytes and with those of its own platelet of the intestine mucous membrane. Thus, an afferent unit of pathogenesis is formed. Further the afferent unit comes into effect in the form of specific immunic products such as lymphokines and antibodies which injure enterocytes. The viscoelastic properties of wheat dough is determined by prolamin proteins. Using EPR spin labeled spectroscopy glutens from various wheat varieties were investigated in a temperature range from 5 to 90°C. The spectra were of a composite type, resulting from at least two populations of spin labels largely differing in molecular mobility. The correlation time of the less mobile nitroxide radicals was determined by saturation transfer EPR. Upon heating there was a transfer from the slow to the fast moving population of radicals, and an increase of mobility of this last category that followed the Arrhenius law. The effect of temperature on molecular flexibility was reversible. This was not the case for purified, polymerized glutenin subunits extracted from gluten. Urea created similar modifications on gluten as heat [206, 207].

EPR Spectroscopy was employed for the investigation of liver pathology. A stable free radical was revealed in the urine of acute hepatic patients with Dubin-Jonson's syndrome. The radical is not related to melanin and absent in other liver diseases [208].

A liver glycogen storage disease (GSD) is associated with severe dyslipidaemia which can induce cell membrane alterations. Using EPR spectroscopy with two fatty acid nitroxide probes (5NS and 16NS) the membrane erythrocyte fluidity in 25 GSD patients aged 1-27 years and 15 controls aged 1-28 years has been studied. The erythrocyte membrane of GSD patients appeared less fluid compared with the controls. The patients showed a relative increase in saturated fatty acids (SFA) and a decrease in polyunsaturated fatty acid (PUFA). These data indicate a reduced PUFA/SFA ratio in GSD cases. This should be taken into account for the management of the dietary treatment of GSD patients [209].

Liver is of primary importance in alcohol metabolism. Ethanol induces disbalance between the peroxidant and oxidant liver systems and an oxidative stress, thus injuring the liver. The formation of reactive oxygen compounds [C_2H_5OH] by liver microsoms and an increase in their content as a result of cytochrome induction contribute a great deal to ethanol induced oxidative stress. In chronic drunkards the lipid peroxidation intensity is dose-dependent. The liver microsomal system oxidized ethanol to 1-hydroxyethyl radicals found *in vivo* as well. 1-Hydroxyethyl radicals are covalently bound to proteins to form adducts having immunologic properties leading to generation of antibodies. Antibodies are found in the blood of patients with chronic alcoholic cirrhosis [210, 211]. 1-Hydroxyethyl radicals induce DNA disjointing, DNA basis alkalizing and hemoprotein reduction. Therefore the formation of 1-hydroxyethyl radicals from ethanol may be responsible for ethanol-induced hepatotoxicity. Two pathways of 1-hydroxyethyl radical formation are suggested. The one pathway requires iron catalysis and is catalase-inhibited and, possibly, involves $HO^\bullet$ radical formation. The other pathway does not involve $HO^\bullet$ radical formation. This pathway is a results of one-electron ethanol oxidation to 1-hydroxyethyl radicals in the cytochrome P_{450}. Human liver microsoms catalyze the ethanol oxidation to 1-hydroxyethyl radicals either by NADP.H or by NAD.H which participate as reductants. The effectiveness of NAD.H may be significant due to the increased NADP.H/NAD^+ redox ratio in the liver as a consequence of ethanol oxidation by alcohol-dehydrogenase. The formation of 1-hydroxy-ethyl radicals by human liver microsoms is catalyzed by oxidants derived from the interaction of iron with superoxide or peroxide anion radicals. In this process there is a close association between the formation of 1-hydroxyethyl radicals and superoxide production. Cytochrome b_5 seems to play a role in the formation of 1-hydroxyethyl radicals most likely owing to their effect on the production of superoxide anion radicals [212].

As shown by EPR spectroscopy and potentiometry, the fragments of hepatitis B virus proteins can specifically bind Cu^{2+} ions using arginine lateral NH_2 group donor sites [213, 214].

A highly sensitive method to determine HbsAg has been offered. The method is based on assessment of the peroxidase activity using EPR measurements of the stability of nitroxide radicals. The method was tested on donors, patients with hepatitis B and those with neither B nor C [215].

Liver plays a broad spectrum of roles. It takes part in blood system coagulation as well as in fat, carbohydrate, protein and pigment metabolism. One of the most important liver functions is detoxication of harmful exo- and endogenous substances with the help of many enzymes, the structure and function of which in the liver is under investigation.

EPR spectroscopy was used in studying the His25Ala mutant of human liver heme oxygenase and its complex with heme. In this complex the Fe^{3+}-containing heme is in the penta-coordinate state. This is distinct from the same complex in the normal type enzyme in which the heme is hexa-coordinate. The ferrous His25Ala heme oxygenase complex exists as an equilibrium mixture between tetra-coordinate and penta-coordinate forms. Although the His25Ala mutant protein shows no heme oxygenase activity, the heme participates in binding carbon monoxide [216].

Calphostin C is a specific protein kinase C inhibitor. Using EPR spectroscopy an enzymatic light-independent mechanism of calphostin C free radical formation was established. Free radicals are generated in liver microsoms and their formation is determined by the presence of NADP.H, functional microsomes, and protein and calphostin C concentration. Cytochrome P_{450} reductase is shown to be responsible for the calphostin C free radicals. They are likely to belong to perylenesemiquinone [217].

EPR spectroscopy showed that cytochrome P_{450} 2E1-containing human microsoms from clone MV2E1-9 produced superoxide ion-radicals. Microsome treatment by viral DNA increases the rates of formation anion radicals O_2-and hydrogen peroxide by one and two orders of magnitude, respectively [218].

The *in vitro* metabolic activation of flutamide, a nitroaromatic anti-androgen, in liver microsoms was studied by EPR. Flutamide s oxidatively transformed by cytochrome P_{450} into reactive metabolites covalently bound to microsomal proteins. Covalent binding requires oxygen and NADP.H, and is decreased by the nucleophile glutathione and by the cytochrome P_{450} inhibitors SKF 525-A, piperonyl butoxide and troleandomycin (an inhibitor of the cytochrome P_{450} 3A subfamily), and increased by dexamethasone (an inducer of the cytochrome P_{450} 3A subfamily). [219].

During the latest years viral hepatitis was successfully treated using interferons of various types. The term "interferon" has appeared as a result of observations of patients after viral infection. It has been found that in the convalescence stage they are to some extent protected against other viruses. However, the reason for the so-called interference remained not understood. And only as late as 1957 A. Azeks and J. Lindeman found the factor responsible for this phenomenon which they called "interferon". At present interferon is subdivided into two types. The one involves interferon-α and interferon-β, whereas the other involves interferon-γ. This subdivision is based by different biological roles of these types. Interferons-α and -β, are mainly characterized by antiviral activity, whereas interferon-γ represents a universal endogenous immuno-stimulating agents. Interferons strengthen antigen cell-binding, HLA antigen expression, lymphocyte response to mitogenes, target-cell lysing, increases immunoglobulin formation, macro-phage phagocytic activity and cooperative interaction with T- and B-cells. Interferons inhibit the growth of tumor cells and suppress intracellular reproduction of bacteria and protozoa, facilitate T-lymphocyte differen-tiation. In high doses all interferons inhibit rather than stimulate immunic reactions. Interferon of one type potentiate the effect of interferon of another type. The formation of interferon-α *in vivo* is effected via B-lymphocytes and macrophages whereas that of interferon-β involves fibroblasts, epithelial cells, macrophages, interferon-γ-T and NK lymphocytes [220]. Mechanisti-cal studies of interferon are in progress.

Using electron spin resonance techniques the effect of α-interferon on fluidity of the plasma membrane lipid bilayer in human lymphopblastic cell lines has been studied. These cells are sensitive to the antiproliferative action of α-interferon. α-Interferon decreases the fluidity of the plasma membrane lipid bilayer of lymphoblastoid cells [221].

In Burkitt's lymphoma patients, interferon-α produces dose-dependent membrane hyperpolarization, decreases plasma membrane viscosity, modulation of the microfilament system, and alters the synthesis of cAMP, cGMP and prostaglandins in interferon-α-sensitive Daudi cells. No such changes are observed in interferon-α-resistant cells. Using EPR spectrometry and flow cytometry it was found that indomethacin, pentoxifylline, or the cGMP induced sodium azide (NaN_3) had no effect on the interferon-α induced membrane potential changes, or on membrane viscosity changes. Indomethacin may influence interferon-α synthesis, but not it's antiproliferative actions. Relatively high doses of pentoxifylline slightly inhibit proliferation of Daudi cells and synergize with interferon-α. Thus, EPR spectroscopy presents a new screening method to rapidly explore certain types of interferon [222].

The EPR method was also used for a study of liver porphirin and bilirubin metabolism. Coproporphyrinogen oxidase catalyzes the conversion of coproporphyrinogen III to protoporphyrinogen IX. This enzyme contains and copper needed for its activity. It was studied by EPR, UV, visible and atomic emission spectroscopy. Human coproporphyrinogen oxidase contains no metal centers and is not stimulated *in vitro* by iron or copper. Addition of these metal ions to cultures expressing the protein has no effect [223].

The nature of EPR signals derived from bilirubin-IX α has been studied on the samples treated with free radical generating and inhibiting systems, such as Fe/EDTA, superoxide dismutase, mannitol/ascorbate, CO, KCN, *etc.* In all the cases, stable signals originated from semiquinone radicals (g = 2.0012) and superoxide radicals ($g_{\|}$ = 2.041, $g_{\perp}$ = 2.0040) were registered. Bilirubin is suggested to act as "active oxygen sin" in mammalians [224]. Bilirubin enhances membrane lipid fluidity. The effect of bilirubin on motion parameters of human erythrocyte membranes was investigated using spin label EPR spectroscopy [225].

In patients with mechanical jaundice the liver tissue EPR signal intensity is many times higher than that in normal individuals [226].

Among gastroenterologic problems a prominent place is occupied by gallbladder diseases. The incidence of chronic noncalculous cholecystitis and cholelithiasis is ever increasing. Calculus generation presents an evolutionary phase of earlier forms of biliary tract pathology such as dyskinesia and chronic noncalculous cholecystitis. Therefore early diagnosis and therapy of functional and inflammatory biliary tract diseases are of primary importance in prophylaxis of cholelithiasis. Among numerous existing diagnostic methods there are no optimal ones assuring correct diagnosis of chronic noncalculous cholecystitis. The reliability of most criteria in diagnosing an inflammatory process in the biliary tract leaves much to be desired. In the initial stage of gallbladder inflammation no wall thickening, deformation, periprocess or other ultrasonically or X-ray-detectable alterations are observed. Therefore laboratory examination are of primary importance for early revealing of inflammatory elements in withdrawn samples (blood, bile, duodenal contents). EPR-detectable free radicals and paramagnetic centers of metalloenzymes play an essential role in living cell metabolism.

Using EPR spectroscopy 81 patients with chronic noncalculous cholecystitis (53 in acute phase and 28 in remission stage) and 29 patients with acute calculous cholecystitis (men to women ratio 1 : 1.2, average age 46 ± 1.3 years) were examined. The control group consisted of 20 practically healthy individuals (men - 31%, women - 69%). The presence of endogenic nitrites in saliva, duodenal contents and gallbladder bile collected during cholecystectomy (acute calculous cholecystitis) was fixed by EPR spectroscopy. For this purpose sodium metabisulfite, iron sulfate and sodium ascorbate were added to the above samples. A characteristic EPR signal with g = 2.03 was recorded. Nitrate-reducing capability was assessed on samples to which sodium nitrate was introduced prior to incubation (37°C, 30 min). The presence of nitrate reductase system was indicated by increasing the signal intensity at g = 2.03 compared with the initial signals. $FeSO_4$, ascorbic acid or *para*-phenylenediamine solutions were added to bile to determine the ferroxidase, ascorbatoxidase and catecholaminoxidase activity of copper-containing biliary protein [227, 228].

The most informative diagnostic criterion of cholecystitis exacerbation is the presence of Fe^{3+} and FR in plasma, FR in erythrocytes, and FR, Cu^{2+} and Mn^{2+} in portion B of the duodenal contents (see table 3-25) [228].

Table 3-25. Criterion of exacerbation activity in cholecystites according to EPR data (M ± m)

Group	n			In duodenal contents (portion B)		
		Fe^{3+} in plasma, (mcmol/l)	FR in plasma, (10^{-6}M)	FR in erythrocytes (10^{-6}M)	Cu^{2+}, (10^{-5}M)	FR, (10^{-6}M)	Mn^{2+}, (10^{-5}M)
ACC	29	17.7 ± 0.9	0.75 ± 0.1	1.17 ± 0.1	5.1 ± 0.6	4.8 ± 0.1	6.2 ± 0.8
CNC (exacerbation)	48	15.9 ± 0.9	0.68 ± 0.1	0.84 ± 0.0	1.4 ± 0.1	0.96 ± 0.1	4.77 ± 0.6
Normal individuals	20	12.4 ± 0.7	0.3 ± 0.0	0.55 ± 0.0	0.87 ± 0.1	0.68 ± 0.1	8.37 ± 0.5

Notes: FR – free radicals, ACC- acute calculous cholecystitis, CNC –chronic non-calculous cholecystitis.

As seen from table 3-25, the highest Fe^{3+} concentration is observed in cases of acute calculous cholecystitis and exacerbation of chronic non-calculous cholecystitis. In patients with chronic non-calculous cholecystitis in the exacerbation and remission phase the Fe^{3+} content is reliably enhanced.

The data on blood serum non-heme iron content in patients with bile secretion system pathology are contradictory. An increase in plasma Fe^{3+} of concentration (table 3-26) [228] in these patients and in those with exacerbation of both calculous and noncalculous cholecystites possibly can be explained by Fe^{3+} release as a result of inflammatory destruction of cells including blood formed elements. Lipid peroxidation intensification releases Fe^{3+} from hemoglobin thus increasing the transferrin Fe^{3+}-saturation. The consistency of Fe^{3+}-bound antioxidant activity of transferrin and ceruloplasmin decreases with increasing the transferrin Fe^{3+}-saturation [229].

As seen from table 3-26, the plasma and erythrocyte free radicals level in cholecystitic patients is higher than in practically healthy individuals. Lipid peroxidation appears to be one of the most probable injuring factors in the genesis of gallbladder inflammation. Depending on the disease severity, the highest plasma FR content in observed in acute calculous cholecystitis [228].

Table 3-26. Plasma PMC content in patients with chronic non-calculous chole-cystitis (CNC) and acute calculous cholecystitis (ACC) (M ± m)

Group	n	Fe^{3+}, (mcmol/l)	Cu^{2+}, (10^{-5} M)	FR, (10^{-6} M)
CNC (exacerbation)	54	15.93 ± 0.89	1.31 ± 0.09	0.68 ± 0.02
CNC (remission)	27	13.34 ± 0.78	1.35 ± 0.06	0.65 ± 0.03
ACC	35	17.73 ± 0.91	3.10 ± 0.09	0.75 ± 0.05
Normal individuals	20	12.36 ± 0.74	1.30 ± 0.09	0.30 ± 0.03

Blood plasma free radicals are conditioned by free radical adrenalin form. The EPR spectrum contains a signal with $g = 2.0024–2.0029$ and line half-width 0.6-0.8 mT [9, 230]. Adrenalin presents "an emergency hormone" responsible for mobilization of all the functions under extreme conditions. In emotional and pain stress, ischemia and hypoxia the blood catecholamine content is enhanced. Adrenalin increases the amount of plasma free radicals and activates the erythrocyte glutathionereductase system.

The erythrocyte FR level is the higher, the more severe cholecystitis occurrence. The highest erythrocyte FR concentration is observed in patients with acute calculous cholecystitis ($P < 0.001$). In chronic noncalculous cholecystitis in the exacerbation phase this factor was somewhat lower (see table 3-27) [228].

Table 3-27. Erythrocyte PMR centers in patients with chronic non-calculous chole-cystitis (CNC) and acute calculous cholecystitis (ACC) (M ± m)

Group	n	MetHb, (g/l)	Cu^{2+}-SOD, (10^{-5} M)	FR, (10^{-6} M)
CNC (exacerbation)	54	0.76 ± 0.11	1.27 ± 0.08	0.84 ± 0.03
CNC (remission)	27	0.64 ± 0.10	1.16 ± 0.10	0.65 ± 0.06
ACC	25	0.78 ± 0.12	3.00 ± 0.08	1.17 ± 0.10
Normal individuals	20	0.61 ± 0.11	1.11 ± 0.13	0.55 ± 0.03

Using EPR spectroscopy plasma and erythrocyte PMC centers were determined in ACC patients before and 6-7 days after successful surgery. In spite of normalization of biochemical indexes and total blood count, the PMC concentration remained high even before discharge from hospital. Immediately after surgery the ACC patients showed a decreased plasma Fe^{3+} content. This can be explained by blood loss during surgery. FR, Mn^{2+} and Cu^{2+} were fixed in the gallbladder bile and duodenal content of ACC and CNC patients. At the same time, the spectrum of bile withdrawn as a result of gallbladder puncture during cholecystectomy also shows methemoglobin Fe^{3+} signals at g = 6.0 and Fe^{3+} at g = 4.3 of a shape characteristic of Fe^{3+} localization in albumin. However, the MetHb Fe^{3+} signal was observed not in all ACC patients. At the same time this signal usually appeared in the bile spectrum of patients with both simple calculous and destructive-phlegmonous cholecystitis [228].

The bile albumin Fe^{3+} signal at g = 4.3 was always fixed in ACC patients. The signal intensity was higher gallbladder in destructive ACC compared with non-destructive ($P < 0.05$) (table 3-28) [228].

Table 3-28. PMC concentration in the duodenal content (portion B) of CNC patients and in the bile of ACC patients (M ± m)

Group	n	MetHb, (g/l)	Fe^{3+}, (mcmol/l)	Cu^{2+}, (10^{-5} M)	FR, (10^{-6} M)	Mn^{2+}, (10^{-5} M)
CNC (exacerbation)	48	-	-	1.39 ± 0.07	0.96 ± 0.08	4.77 ± 0.61
CNC (remission)	27	-	-	1.11 ± 0.08	0.71 ± 0.06	6.89 ± 0.70
ACC	16	5.92±0.15	22.17±0.73	5.07 ± 0.63	4.78 ± 0.08	6.21 ± 0.83
Normal individuals	10	-	-	0.87 ± 0.06	0.68 ± 0.07	8.37 ± 0.56

Apart from iron the duodenal content contains copper. The bile copper-containing protein of apparently serum origin is suggested to be localized in ceruloplasmin [231]. However, no correlation has been found between the Cu^{2+} content of blood plasma ceruloplasmin and duodenal juice of ulceric patients [30]. Therefore, it was found reasonable to examine the duodenal protein properties by EPR spectroscopy. It turned out that the duodenal copper-containing protein does not show ferroxidase activity but exerted an ascorbatoxidase and, in some patients, catecholaminoxidase effect. Thus, this protein is not functionally identical to blood plasma cerulloplasmin.

The highest bile Cu^{2+} concentration was observed in ACC patients: 3.5 times as high as in CNC patients in the exacerbation phase ($P < 0.001$). Alteration in the bile Cu^{2+} level is dependent on the severity of gall-bladder inflammatory process. The bile Cu^{2+}-containing protein is not functionally corresponding to the Cu^{2+}-containing ceruloplasmin. Nevertheless, the former is suggested to be responsible for binding oxygen free radicals. This protein also accounts for an increase in the bile Cu^{2+} concentration especially in acute and chronic calculous cholecystitis. The Cu^{2+} level of duodenal portion B in all CNC patients in both exacerbation and remission phase is statistically reliably different [232].

Duodenal Mn^{2+} ions evidently are localized in phosphatase, DNA-ase and RNA-ase enzymes [233]. The gallbladder Mn^{2+} content in ACC patients is lower than in normal individuals and CNC patients in the exacerbation phase [234, 235]. Sometimes this is related to depression of the synthetic and excretory liver function. The duodenal Mn^{2+} concentration is of lowest value in patients with chronic noncalculous cholecystitis in the exacerbation phase (see table 3-28) ($P < 0.05$).

Bacterial infection plays a leading part in cholecystitis etiology. Colon bacilli, streptococci, staphylococci, enterococci are often met in bile. Most of them show nitrate-reducing ability. They can transform nitrates to nitrites and, finally, to nitric oxide. Owing to this fact the presence of microflora in bile can be detected by EPR spectroscopy. In this way the presence of nitrites as well as the nitrate-reducing ability of bacilli in the duodenal content, bile and saliva of patients with cholecystitis were revealed and determined.

After reduction with ascorbic acid to nitric oxide, nitrites were detected by a characteristic EPR signal at $g = 2.03$ recorded in the saliva and duodenal content in 80% of normal individuals and 50% of patients with chronic cholecystitis. In the gall-bladder bile drawn under sterile conditions during cholecystectomy colon bacilli, staphylococci, streptococci and klebsiella are present. However, no nitric oxide EPR signal was fixed in the spectrum of the saliva and duodenal content of ACC patients. As these patients were treated with antibiotics, the latter were suggested to suppress the nitrate-reducing ability of microflora. Like duodenal content and bile, microorganisms are present in saliva as well, and the saliva EPR spectrum also shows the signal at $g = 2.03$. In this connection, we have examined the saliva EPR spectrum after addition of antibiotics in doses analogous to their blood concentrations in treating acute cholecystitis. As a result the nitric oxide EPR signal in the saliva spectrum disappeared. Thus, it is quite possible that the absence of the signal at $g = 2.03$ in the bile spectrum and rare appearance in the spectra of saliva and duodenal contents of patients with cholecystitis are connected with antibiotically induced depression of microflora nitrate-reducing properties.

The biosynthesis and metabolism of bile acids are influenced by Mn^{2+}, Fe^{3+}, Cu^{2+} ions involved in the formation of active centers of redox enzymes (oxydases, hydroxylases, superoxide dismutases, *etc.*). Iron is incorporated in hemoproteides responsible for molecular oxygen activation and participation in cholesterene oxidative degradation. Using EPR spectroscopy and HPLC the content of serum bile acids and their correlation with blood PMC have been studied. A correlation between the concentrations of cholic acid, Cu^{2+}, plasma FR and superoxide dismutase have also been found (r = +0.4, +0.5 and +0.6, respectively). The blood plasma Fe^{+} content depends iniversely on chenodesoxicholic acid (r = -0.4), whereas desoxicholic and lithocholic acids are in direct proportion to plasma free radicals (r = +0.4 and +0.5, respectively) [228, 236].

EPR spectroscopy was also used in studying the efficiency of treatment of Mencke "silver hair" syndrome which manifests itself, in particular, as hypoceruloplasminemia and abrupt drop in blood serum cooper content approaching 25% of normal level. Peroral copper administration leads to an increase in albumin copper, but not in ceruloplasmin copper. At the same time, intravenous administration induces an increase in the ceruloplasmin copper content and enhances the blood ceruloplasmin concentration [237].

Erythrocyte membrane fluidity was compared in 15 normal women and 15 women with anorexia nervosa using EPR spectroscopy. The patients were studied at hospital admission and again after 1 month of refeeding. At admission, the fluidity from the hydrophobic part of the erythrocyte membrane, was decreased in the sick women. After 1 month of refeeding, fluidity increased. One of the possible mechanisms of membrane fluidity variations could be the effect of cholesterol on the membrane structure. Increased cholesterol levels in anorexic patients could reduce fluidity. These alterations in membrane physical properties could explain some of the neurobiological abnormalities observed in patients with anorexia nervosa [238].

3.7 LUNG DISEASES

Using EPR spectroscopy blood paramagnetic centers of patients with non-specific chronic lung diseases were studied. The increased ceruloplasmin/transferrin (CP/TF) ratios and erythrocyte superoxide dismutase level indicate an activated antioxidant system in these patients. PMC concentration is dependent on both the form of chronic lung disease and severity of inflammatory process [239].

Blood plasma PMC level in patients with bronchial asthma is dependent on the form and severity of illness [240].

An important role in the pathogenesis of many lung diseases is played by oxygen radicals with a unpaired electron [241], in particular, in obstructive lung diseases. The EPR method using DMPO was employed to investigate the products of xanthine oxidase activity in bronchoalveolar fluid of both patients with obstructive lung disease and healthy non-smoking individuals. The results show that xanthine oxidase is involved in mechanisms responsible for the formation of superoxide anion radicals in bronchoalveolar fluid, thus contributing to oxidative stress in this pathology [242].

Occupational exposure to coal mine dust causes coal workers pneumoconiosis and other pulmonary diseases. Their mechanisms remain unclear. The hydroxyl radicals $HO^{\bullet}$ may play an important role in the pathogenesis of coal pneumoconiosis. The catalytic effect of bituminous coal mine dust in the generation of radicals $HO^{\bullet}$ from hydrogen peroxide was investigated using spin trap EPR spectroscopy. In the presence of DMPO as a spin trap $HO^{\bullet}$ radicals were effectively inhibited by deferrioxamine and catalase, but only partially inhibited by superoxide dismutase.

At the same time the radical generation is enhanced by metal chelators DETAPAC and EDTA. The $HO^{\bullet}$-generating potential of all the coal dusts showed a positive correlation with the surface iron content of coal mine dusts. The coal dust-induced lipid peroxidation is in agreement with the surface iron content. It is suggested that higher concentrations of surface iron in coal mine dust may intensify the generation of $HO^{\bullet}$ radicals which may play an important role in the development of coal pneumoconiosis [243].

Stable coal radicals of polynuclear aromatic hydrocarbons were detected by EPR spectroscopy in the lung tissue of autopsied coal miners. These radicals were also found in the lung tissue of 98 coal miners with and without coal pneumoconiosis, cancer of the lung and a history of cigarette smoking. Concentrations of the radicals were also measured in the lungs of non-miners served as the control. The content of lung stable coal radicals was found to be related to the length of coal mining work, coal pneumoconiosis or lung cancer severity, and cigarette smoking.

The mean concentration of stable coal radicals in the lung tissue of miners with 30 years of coal mining exposure was $(5.3 \pm 1.3) \times 10^{17}$ spins/g versus controls who had a nondetectable level (less than 10^{15} spins/g).

An increase in disease severity was accompanied by a progressive increase in stable coal radical concentration. A stable coal radical concentration of $(4.8 \pm 0.7) \times 10^{17}$ spins/g was found for simple coal pneumoconiosis (with moderate coal macules) versus $(7.8 \pm 4.6) \times 10^{17}$ spins/g lung tissue for complicated coal pneumoconiosis (with progressive massive fibrosis) [244].

By EPR spectroscopy using spin trapping silicosis-inducing crystalline silica was shown to be involved in the generation of oxygen radicals (hydroxyl, superoxidase and singlet oxygen). Radicals $HO^{\bullet}$ are generated from H_2O_2 (Fenton's reaction) [245] resulting in cell injury and DNA damage [246]:

$$O_2^{-\bullet} + M^{(n+1)} \rightarrow M^{(n-1)+} + O_2$$

$$M^{(n-1)+} + H_2O_2 \rightarrow M^{n+} + HO^{\bullet} + HO^{-}$$

$$O_2^{-\bullet} + H_2O_2 \xrightarrow{M^{n+}/\,M^{(n-1)+}} HO^{\bullet} + O_2 + HO^{-}$$

Asbestosis ranks among pneumoconiosis varieties. Asbestos is a well-known carcinogen in the development of lung cancer and mesothelioma. However, its mechanism remains unclear. Asbestos is reported to release oxygen radicals from neutrophils and macrophages. In neutrophils and macrophages an anion radical $O_2^{-\bullet}$ were revealed by EPR spectroscopy with the aid of spin trapping techniques in the presence of DMPO. Mechanism responsible for $O_2^{-\bullet}$ release in the presence of asbestos is not understood yet [247]. Nevertheless, it has been shown that the harmful asbestos effect consists in catalyzing the formation of free radicals [248].

A combination of asbestos, lung and pleura cancer is met considerably more often than in other pneumoconiosis. Oxygen radicals induced by crocidolite from human polymorphonuclear leukocytes were studied using both an EPR spin trapping method with DMPO, and a luminol-dependent chemiluminescence method. The results confirm the generation of $HO^{\bullet}$. Superoxide dismutase, catalase, and DMSO inhibit the production of $O_2^{-\bullet}$ and HO radicals in a dose-dependent manner. It is proposed that HO radicals play key role in the mechanism of carcinogenesis in patients [249].

By EPR spectroscopy using DMPO as a spin trap, $O_2^{-\bullet}$ production was proved upon stimulation of dimethylsulfoxide-differentiated HL60 cells by crocidolite opsonized with fresh or refrigerated serum, as well as by PMA. Crocidolite, unopsonized or opsonized with frozen-thawed or heat-inactivated serum, did not induce $O_2^{-\bullet}$ release. The addition of iron chelators or superoxide dismutase inhibited $O_2^{-\bullet}$ release completely. Neither undifferentiated nor PMA-differentiated HL60 released $O_2^{-\bullet}$ upon stimu-lation with opsonized crocidolite [250]. These radicals were shown to induce permeability alterations and cell injury in alveolar epithelial monolayers. The formation of hydroxyl radicals in hydrogen peroxide-treated cells was proved by electron spin resonance measurements using the spin trap DMPO. Chelation of the cellular iron by deferrioxamine inhibits radical formation and helped protect the cells from oxidative changes [251].

The formation of respiratory tract pathology is greatly influenced by smoking. The cigarette smoke induces cancer, emphysema and other diseases even many years after giving up smoking. The high concentration of polynuclear aromatic hydrocarbon free radicals detected by EPR spectroscopy in the lungs of heavy smokers points to much probability of developing cancer of lungs in the future. Cigarette tar contains stable semiquinone radicals which can be directly observed by EPR. Aqueous extracts of cigarette tar, which contain these radicals, reduce oxygen to anion radical $O_2^{-\bullet}$ and thus produce both H_2O_2 and $HO^\bullet$ radicals [252, 253]. Reactive free radicals contained in cigarette smoke reduce antimicrobial activities of polymorphonuclear leukocytes thus injuring the lung tissue. Superoxide dismutase and an iron chelator, diethylenetriaminepentaacetic acid (DETAPAC), totally protect the lung cells against injury. Pretreatment of polymorphonuclear leukocytes with antioxidants such as α-tocopherol and dihydrolipoic acid produces a similar effect. Anion radicals $O_2^{-\bullet}$ are probably generated when cigarette smoke comes in contact with the upper and lower respiratory tract lining fluid when they come in contact with cigarette smoke. Antioxidant therapy is suggested to improve the lung patients state. Antioxidants such as α-tocopherol or dihydrolipoic acid might be successful in preventing cigarette smoke deleterious effects [254].

Apart from the above free radicals, the cigarette smoke contains nitrogen oxide and its oxidation products (NO_2, NO_3). These all exert an adverse effect on the respiratory tracts. A Hem-NO signal was observed by EPR spectroscopy in bronchoalveolar lavage cell pellets obtained from human volunteers exposed to NO_2, for 20 min every other days [255].

As known ozone facilitates the formation of free radicals in biological systems [256]. Vitamin E protects cells *in vitro* and *in vivo* against the effects of ozone. Ozone reacts by two quite different mechanisms to produce radicals; one involves an ozone-olefin reaction resulting in splitting an R radical out of an olefin with the structure R-CH=CH_2. The R radical then reacts with O_2 to transform to a peroxyl radical ($ROO^\bullet$). Both carbon- and oxygen-centered radicals can be detected by the electron spin resonance spin trap method. The other mechanism involves a reaction of ozone with electron donors. If the electron-donor is GSH or its ion GS, this reaction produces the thiyl radical ($GS^\bullet$) and $O_3^{-\bullet}$ The ozone radical anion then reacts with a proton to form the hydroxyl radical and $O_2^{-\bullet}$ [257]:

$$O_3^{-\bullet} + H^+ \rightarrow HO^\bullet + O_2.$$

The hydroxyl radical is determined by the EPR method using a spin adduct DMPO [258]. All this indicates ozone to be one of the most potent toxic agents. The noxious effect of ozone on living organisms appears mainly in the lungs and blood. A considerable role in the mechanism of its action is ascribed to free radicals. The EPR method was applied to indirectly demonstrate the free radical effects and the significance of ascorbate in ozonizing the blood. The effect of ozonizing the blood *in vitro* is expressed in triggering a "local" protective mechanism of the blood cells, while *in vivo* it is to point out the consumption of ascorbate in "scavenging" of free radicals [257].

The nature of melanin-containing centers in bronchoalveolar smears and blood of persons responsible for cleaning up after the accident at the Chernobyl Atomic Energy Station was determined using EPR spectroscopy [259].

The concentrations of free radicals in the whole blood of the patients with cor pulmonale were determined using the EPR method. The results show that the levels of free radicals, serum malonic dialdehyde and superoxide dismutase activity were higher compared with those of the control group. These indicate that there is impairment in free radical metabolism of the patients suffering from cor pulmonale [260].

It is known that the breathing muscle fatigue syndrome contributes some deal to pulmonary insufficiency in patients with lung diseases. This may lay the foundation of the development of new approaches to treating pathologies of this type. The results of recent EPR investigations in this field have shown that oxygen free radicals generated during active muscle work, contribute to the development of muscle fatigue [261].

The EPR method allows the diagnosis of destructive changes in patients with pulmonary and osteoarticular tuberculosis [262].

Using EPR spectroscopy at 77°K signals at g = 4.2 and g = 2.049 arisen from Fe^{3+} and Cu^{2+}, respectively, were revealed in the pleural exudate of both oncologic and non-oncologic patients. A significant difference in the pleural liquid Cu^{2+}/Fe^{3+} ratio of these patients enables malignancy assessment [263].

3.8 DISEASES OF KIDNEYS

Kidneys excrete metabolic and products, excess salts, water, foreign and toxic substances. They are responsible for monitoring the blood composition and provide for the inner medium constancy. An important function of kidneys is oxygen supply to the renal cortical layer. The tissue oxygen tension in the kidney cortex and outer medulla *in vivo* in mice was measured by EPR oximetry. In the cortex region the oxygen tension was higher compared to that in the outer medulla. An intravenous injection of endotoxin decreases the oxygen tension in the cortex and increases in the medulla region. In intact kidneys Functional Magnetic Resonance (MR) images showed the cortex region to be better supplied with oxygenated hemoglobin compared with the outer medulla. After administration of endotoxin an immediate increase in the oxygenated blood level is observed in the outer medulla region. Pretreatment of mice with N-monomethyl-L-arginine prevents both the changes in tissue oxygen tension and distribution of oxyge-nated hemoglobin. The formation of nitric oxide has a critical role to play in renal medullary hemodynamics. For monitoring renal oxygen tension and changes in the oxygen distribution a combination of EPR with MR imaging of kidneys was proved to be very useful [264].

Excess free radicals in combination with antioxidant system insufficiency produce an advert effect on tissue functions. Oxygen free radicals and red blood cell damage in acute renal failure were examined by the EPR method. The results revealed a correlation between free radical level and erythrocyte hemolysis. [265]. Renal ischemia is accompanied by an increased oxygen radical production in cortical tissue as established by the EPR method in rabbits with renal ischemia [266].

Glomerulonephritis is caused by damage of renal nephron glomerulus. Using the EPR method the structure of lipid bilayer in patients with chronic glomerulonephritis was determined. Studies were carried out in 30 patients with nephrotic syndrome (NS) and 30 patients without this syndrome and 15 controls. Patients with chronic glomerulonephritis revealed disorders of lipid bilayer structure. These disorders could be responsible for decreased thrombocyte aggregation observed in patients with chromic glomerulo-nephritis in comparison to healthy controls [267].

IgA-glomerulopathy (Berger disease) regarded as a chronic glomerulonephritis variety, is characterized by excess production of immunoglobulins A (IgA). An efficient renal blood flow in patients with Berger disease after administration of standard doses of isosorbide-5-mononitrate (NO donor) was examined by EPR spectroscopy. Blood nitric oxide levels are significantly increased after isosorbide administration and decreased after drug discontinuation. Blood NO concentration is closely connected with the effective renal blood flow, but not with the glomerular filtration rate. Proteinuria levels significantly decrease after drug administration and return to baseline levels thereafter. Values of filtration are decreased after administration of this drug compared to basal levels. These results suggest the effects of nitric oxide on glomerulus endothelium [268].

Superoxide dismutase in renal tissue biopsy specimens obtained from patients with Berger's disease (13 cases) and mesangial proliferative glomerulonephritis (9 cases) was examined by the EPR method using spin labeling. Normal renal tissues obtained from the kidneys of malignant patients (6 cases) served as controls. In the two groups of patients the copper and zinc form of SOD activity was lowered proportionally to the severity of kidney tissue damage. A lower level of SOD activity, whether as a cause or a consequence of the disease process, might induce a decrease in the anion radical "scavenger" reaction of the superoxide, thus causing the tissue to become more vulnerable to oxidative stress [269].

Chronic renal failure is a symptomatic complex resulting from gradual damage of nephrons in cases of any progressive disease of kidneys. Superoxide dismutase activity in serum samples of 23 patients with chronic glomerulonephritis and 6 patients with chronic renal failure was measured by EPR spectroscopy using spin trapping. 10 healthy individuals were examined as controls. Among 23 patients with chronic glomerulonephritis, there were 12 patients with Berger disease and one patient with membranous nephropathy diagnosed by immunofluorescence of renal biopsy specimens.

Other chronic glomerulonephritis patients were diagnosed by its clinical criteria. The serum activity of SOD in patients with chronic glomerulonephritis and chronic renal failure was significantly higher than those in healthy adults. The serum SOD activity in patients with chronic renal failure was also higher than that in patients with chronic glomerulonephritis. Marked by high levels of serum SOD activity were observed histologically in the advanced stage of Berger disease. Thus, the increase in serum superoxide dismutase activity may reflect renal injuries in patients with chronic glomerulonephritis and chronic renal failure [270]. EPR was used to show that stable oxidizing components of blood plasma accumulate in uremia. The oxidant was detected by its capacity to oxidize the spin trap DBNBS. The oxidant dialyzable from plasma, had an upper molecular weight limit of about 3,000 Daltons and was stable over many months. Physiological plasma concentrations of vitamin C, vitamin E and reduced glutathione are unable to inhibit the oxidizing capacity of uremic plasma. Thus, uremia is associated with accumulation of an endogenous oxidizing activity at much higher concentrations than in subjects with normal renal function [271].

However, in experiment no relation between the increase in renal oxygen radical levels and chronic renal failure development has been found. There was no detectable renal production of oxygen free radicals in rats with chronic renal failure subjected to 3/4 nephrectomy. Kidney parenchymal content of other oxidants and the oxidant/reductant ratio are similar in control and uremic animals. The plasma redox capacity and ascorbate levels are elevated in uremic rats. Early in the course of chronic renal failure, there is no excessive production of oxygen free radicals. There is accumulation of reductants, primarily plasma ascorbate in uremic animals [272].

In the terminal stage of chronic renal failure the plasma Cu^{2+} ceruloplasmin EPR signal intensity is decreased. A decrease in the superoxide dismutase activity and an increase in the MetHb EPR signal intensity observed with increasing blood osmotic pressure and reducing total renal function indicate a considerable disorder in antioxidant processes in the kidney cells. In patients with renal function failure the erythrocyte FR content is lowered too [273, 274].

EPR spectroscopy using spin labeling has not practically been applied to diagnosis studying of various diseases up to now. Sorption processes of widely accepted fatty-acid spin probe (16-doxyl stearate) by human blood components in different renal pathologies were examined with a goal to elaborate diagnosis technique. The probe introduced into the whole blood is absorbed mainly (90%) by serum albumin. The 16-doxyl stearate binding constants are quite different in donors and uremic patients. This suggests that in uremic patients the sites of albumin affinity to higher fatty acids are occupied by the products of pathologic metabolism [275].

16-Doxyl stearate distribution between the free and the bound states (probe: protein ratio = 3 : 1) was used as a probe express-parameter. In this way 250 patients were examined. In renal diseases without of nitrogen-releasing function (chronic pyelonephritis and glomerulonephritis) the binding parameter ranges from 1.2 to 1.7 in donors and is significantly higher (up to 3.2) in chronic renal failure patients under programmed hemodialysis.

In patients underwent successful transplantation of kidney taken at autopsy the distribution parameter is close to normal. This parameter is abruptly increased upon rejection-induced deterioration of transplanted kidney function. After arresting of acute rejection crisis by large doses of corticosteroids the distribution parameter is brought back to normal.

EPR studies of renal diseases were also performed with other types of spin labels such as lauric acid (7-doxyl laurate) and benzo-g-carboline (BgC). These spin labels react with other blood serum binding sites than 16-doxyl stearate does. 7-Doxyl laurate is sorbed by binding sites of medium fatty acid serum albumin. In this case the parameter to be measured is an apparent dissociation constant (K).

The benzo-g-carboline spin label shows high affinity for not only albumin, but for blood lipoproteins as well. It was used in measurements of non-bound spin label concentrations in the serum diluted to equal albumin content (250 mcM), total spin label concentration being 60 mcM.

Summarized data of comparative tests for different spin labels are presented in table 3-29 [275].

As seen from table 3-29, the parameters of spin label binding by blood serum in CRF patients differ much from those in healthy donors. The donor – CRF patient difference for spin label 7-doxyl laurate is significantly higher than that for 16-doxyl stearate. A donor serum-induced decrease in spin label 7-doxyl laurate is observed in the presence of typical uremic toxins - hippuric acid and sulfate indoxyl. No localization of spin label benzo-g-carboline binding sites on lipoproteins was found. Evidently, in CRF the spin label trapping ability is markedly enhanced thus indicating a quantitative enlargement of the lipoprotein fraction in the blood uremic patients. Thus, the use of spin labels in both renal disease diagnosis and detoxication therapy looks very promising [275].

Table 3-29. Blood serum spin label binding parameters in patients with chronic renal failure (CRF)

CRF Stage (number of observations)	n	Measured spin label parameter		
		7-DL (K, mcM)	16-DS (K, mcM)	BgC (Cf, mcM)
Compensatory	5	134 ± 12	-	1.25 ± 0.20
Intermediate	4	163 ± 21	0.13 - 0.15	0.95 ± 0.25
Terminal	5	320 ± 40	0.15 - 0.16	1.55 ± 0.15
Control	15	127 ± 7	0.12	2.35 ± 0.20

Many nitroheterocyclic compounds (xenobiotics) are potential radiosensitive and antimicrobial agents. However, their use in therapy is strongly limited by their toxic action including mutagenous and carcinogenic activity. The toxicity of these compounds is caused by intracellular reduction to active metabolites. The first step of activation involves the formation of free radical $NO_2^{-\bullet}$, that is rapidly reoxigenated by molar oxygen to related compounds to form superoxide anion radical. In the course of this process called a redox cycle the generation of a constant superoxide flow leading to other active oxygen intermediates occurs. The intermediates are strong oxidants (hydrogen peroxide and hydroxyl radical) able to weaken the cell function. Oxygen energy consumption during a redox cycle causes cell and tissue hypoxia that reduces $NO_2^{-\bullet}$ radical to nitroso derivatives and other reactive intermediates and stable end products. Some intermediates are strong electrophiles and can be readily covalently bound to proteins and nucleic acids. This is the way of cell injury responsible for the antimicrobial activity of heterocyclic nitro derivatives. However, the redox cycle may be the reason for the high toxicity of this class of drugs. Numerous potential pathways of xenobiotics activation mainly involve primary reduction by flavoproteins such as xanthinoxidase, aldehyde oxidase and cytochrome P_{450} reductase, by mitochondrial redox chain components, or diaphorases. In spite of the fact that most investigations in this field were undertaken with pure enzymes and subcellular fractions, principle pathway responsible for activation *in vivo* has not been elucidated yet. 5-Nitrofuran preparations (furazolidon) are mainly used as prophylactic additives to farm animal feed to suppress gastrointestinal and renal microorganisms. Nevertheless, the ability of these products to cause disease in poultry and, possibly, in human is well-known [276, 277]. Therefore the residues of nitroheterocyclic compounds and their toxic metabolites in the tissue of farm animals and poultry fed on a diet containing the above additives, may be a risk factor to human.

The gastrointestinal tract provides the first drug barrier in the host organism. CaCo-2 cell line (human colorectal carcinoma cells) provides a good model for investigating drugs of the nitroheterocyclic series. Mechanism of action of a nitrofuran, N-[5-nitro-2-furfurylidene]-3-amino-2-oxazolidinone (furazolidone) was investigated by noninvasive EPR spectroscopy. One-electron reduction of furazolidone led to the formation of a free radical intermediate that could be monitored in dense cell suspensions. The effects of enzyme inhibitors on the kinetics of radical production and decay were used to estimate the enzyme-induced reductive activation of nitrofurans. The relative contribution of enzymes such as xanthinoxidase, aldehyde oxidase, DT-diaphorase, mitochondrial chain components to the reduction of nitrofurans is insignificant. The microsomal cytochrome P_{450} reductase proved to be most active in this respect. Thus, the EPR method makes it possible to distinguish the bioactivation sites of redox active drugs in intact cells and can be used in therapy [278].

A new antinephritic drug, 6-(1H-indol-3-ylmethyl)-5-methoxy-3-(2-methylpropyl)-2-(1H)-pyrazinone-4-oxide (OPC-15161) does not react with 1, 1-diphenyl-2-picrylhydrazyl (DPPH) and is only slightly reactive with $O_2^{-\bullet}$, but it easily reacts with hydroxyl radical. The scavenging ability of this drug against the hydroxyl radical is dependent on its concentration. The life-span of free radicals generated by the reaction of the drug (OPC-15161) with $HO^\bullet$ is 30 min. The structure of this new radical species corresponds to the chemical formula below [279]:

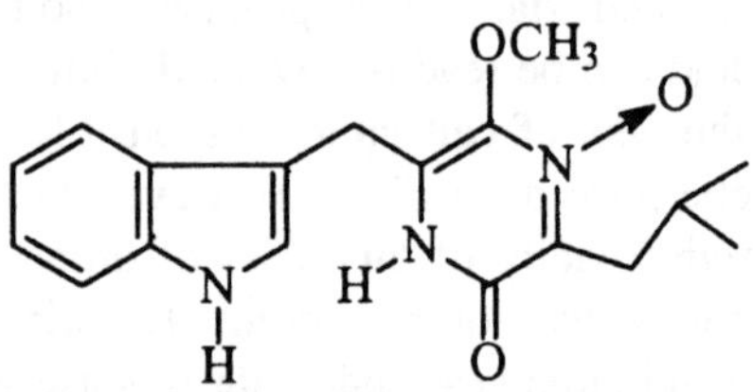

OPC - 15161

Alkaptonuria is a hereditary disease characterized by tyrosine exchange disorder and a large amount (up to 4-8 g/day and more) of homogentisic acid (2,5-dihydroxyphenylacetic acid) excreted with urea. Darkening of urea (till a nearly black color) on exposure to air, especially upon alkalining or heating, is a feature of this disease. When urine samples from alkaptonuria patients are allowed to stand, they turn black, presumably owing to the oxidation of homogentisic acid to a melanin-like substance. The pigments formed by polymerization of the components in the urine from a patient with alkaptonuria and homogentisic acid can be characterized as follows. The absorption spectra and electron spin resonance signals of these pigments are similar to those of eumelanins. Irradiation of the pigments with nitroblue tetrazolium reduced the tetrazolium; which was partially inhibited by superoxide dismutase. Irradiation of Ehrlich ascites carcinoma cells with the pigments from homogentisic acid or urine caused cell lysis. The fact that this lysis was inhibited by catalase, allows a conclusion that it was mediated by H_2O_2. A similar pigment was also extracted from the tissue from an alkaptonuria patient. It is suggested that the degeneration of tissue *in vivo* may be due to the deposition of melanin-like pigments in the tissues, probably in combination with metal ions [280].

In complete renal failure the only radical cure (though not infrequently successful) is kidney transplantation. In order to elucidate the reason for transplant rejection a study was performed for evidence of the production of free radicals in the blood of 14 patients underwent kidney transplantation. The assessment was done by the measurement of vitamin E (an index of lipid peroxidation) and of myeloperoxidase (a marker of neutrophil activation) in the systemic blood. Early (2 min) and late revascularization (30 min) of the kidney was respectively associated with a significant decrease of 35% and 40% of the initial level of plasma vitamin E. This consumption was parallel to the decrease in the vitamin E/total lipids ratio, a better indicator of tocoferol status. Heparin administration preceding renal artery clamping resulted in a two-fold significant increase of baseline plasma myeloperoxidase (MPO) level (523 ± 214 ng/ml). At kidney reperfusion, MPO concentration rose again and reached a maximum value of 1653 ± 882 ng/ml, indicating the presence of considerable neutrophil activation. A return to the baseline value was observed after 30 min of reperfusion. These data strongly suggest that free radical production, leading to lipid peroxidation phenomena, can occur within the early phase of kidney revascularization. Preliminary data using EPR with the spin trapping technique strengthen this hypothesis [281].

Many diseases such as syphilis, gonorrhea, trichomonoasis, chlamydiasis) are transmitted by sexual contact, thus causing injury of the urogenital system. In particular, an extracted and partially purified hydrogenase *Trichomonas vaginalis* was studied using the EPR method. The catalytic and spectroscopic properties of the enzyme indicate that it belongs to the class of [Fe]-hydrogenases, rather than the [NiFe]-hydrogenases. The hydrogenase activity is highly inhibited by carbon monoxide. Thus as little as 1 mcM CO induces 50% inhibition. The EPR spectrum of the most active fractions from chromatography, after reduction by hydrogen and partial reoxidation under argon, showed an EPR spectrum at g = 2.10, 2.04, 2.00. This unusual spectrum is characteristic of the "H-cluster", as seen in [Fe]-hydrogenases of anaerobic bacteria such as *Clostridium spp* [282]. The EPR method was also used to examine spectral characteristics of *Trichomonas vaginalis* ferredoxin [283].

Many microorganisms and chemical compounds are nephrotoxic. Orellanine, a tetrahydroxylated and di-N-oxidized bipyridine toxin is responsible for lethal nephrotoxicity of some fungi of the *Cortinarius genus*. Heavy and even lethal fungal intoxications of human have been observed during the latest years in Europe and North America. The patients showed renal insufficiency from moderate to heavy degree of severity. In some cases, recourse to hemodialysis or even kidney transplantation was necessary. No specific cure has been found until the present time. Orellanine *in vitro* inhibits protein synthesis, but this effect needs preliminary activation of the toxin [284].

The mechanism responsible for orellanine lethal nephrotoxicity was unknown until now. The properties of the toxin were studied and compared with those of other bipyridine derivatives using EPR spectroscopy. After a one-electron oxidation (e.g., photochemical oxidation upon visible light), a radical form of orellanine was observed at physiological pH under aerobic or anaerobic conditions. This radical, identified as *ortho*-semiquinone anion radical, can also be generated by oxidation with biological oxidizing agents or enzymatic systems. Production of superoxide and hydroxyl radicals was shown by the spin trapping method using DMPO as a spin trap. Bioreducing agents like GSH and cysteine involve *in vitro* the semiquinone radical and orellanine in a redox cycling process resulting in the production of glutathionyl and oxygen free radicals. This process leads *in vitro* to a large oxygen consumption and to a dramatic depletion of glutathione level. The formation of an apparently stable *ortho*-semiquinone anion radical and of reactive oxygen radical species is observed for the first time with a mushroom toxin [285]. These species are shown to be readily detected by EPR spectroscopy in fresh mushrooms Cortinariace species [286]. This enables a wide application of EPR spectroscopy in toxicological centers.

3.9 DIABETES

Diabetes mellitus is a disease caused by an inadequate action of insulin on tissues due to either a reduced level of circulating insulin or target tissue-resistance to the insulin action. Diabetes is associated with metabolism disorder, remote complications such as ophthalmic, renal, nervous and vascular damage as well as with basal membrane injury. The mechanism of diabetes pathogenesis is not well-understood in spite of numerous investigations carried out using different techniques. Thus, for example, microviscosity and polarity of erythrocyte membranes of the blood of patients suffering from diabetes were studied by EPR using spin labeled fatty acids. In this study structural changes were discovered 0.6-0.8 nm from the membrane surface in the lipid bilayer of erythrocytes obtained from diabetic patients' blood. However, no essential immobilization of the acyl chains of phospholipids was found in deeper layers [287].

The dynamic properties of erythrocyte membranes in diabetic children and of control erythrocyte membranes subjected to *in vitro* glycation have been investigated by means and EPR spectroscopy. The apparent distance separating the membrane protein tryptophan and the bound 1-anilino-8-naphthalenesulphonate molecules was decreased in the erythrocyte membranes from children with diabetes. This resulted in a significant increase of the maximum energy transfer efficiency in diabetic membranes. These changes are accompanied by an increase in the relative rotational correlation time in diabetic membranes and in the membranes subjected to *in vitro* glycation. The results suggest that that nonenzymatic glycosylation of membrane proteins *in vivo* and *in vitro* may be the major factor attributable to the alterations in the dynamic properties of erythrocyte membrane in diabetic state [288].

The membrane fluidity of intact erythrocytes from diabetic patients and sex-matched controls has been examined between 20°C and 40°C by EPR spectroscopy using the 5-doxyl palmitic acid spin label. In contrast to the normal erythrocytes, in types I and II of diabetes a significant non-linearity was found around 30°C in the fluidity-temperature plots of the lipid bilayer. The magnitude of the bilayer fluidity showed a little decrease in insulin-dependent diabetes and an increasing trend (vanishing around 37°C) in non-insulin-dependent diabetes. In addition, two protein-immobilized lipid subpopulations, showing slightly higher apparent concentrations and modified fluidity were observed in diabetes. The fluidity changes may be explained by membrane composition alterations, mainly in the fatty acid concentrations. These data also demonstrate the high potential of the spin label application in diabetic investigations [289].

The EPR method was also used for studying the behavior of insulin receptors in human tissues. Thus, the divalent cation-binding properties of the human insulin receptor tyrosine kinase domain were examined. The protein-tyrosine kinase activity of the purified cytoplasmic domain was activated nearly 10-fold by 3 mM Mn^{2+} in the presence or absence of 5 mM Mg^{2+}. Electron paramagnetic resonance spectra of the purified, acid-denatured kinase domain and assays of EDTA-treated kinase show that the purified protein does not possess residual, tightly bound Mn^{2+}. Electron paramagnetic resonance spectroscopy was used to directly measure the binding constant of the kinase domain for Mn^{2+}. The results indicate that the recombinant cytoplasmic domain of the human insulin receptor does not bind Mn^{2+} tightly in the absence or presence of Mg-ATP (K_d greater than 0.8 mM). The enzyme does not show a strong preference for Mn-ATP binding when both Mg-ATP and Mn-ATP are present. Mn^{2+} ions, although they bind weakly, induce an activating conformational change in the secondary structure of the human insulin receptor cytoplasmic domain [290].

The EPR method using the spin label 5-nitroxystearate inserted into the lipid bilayer of cell membranes has revealed insulin receptor internalization in response to insulin incubation (down-regulation) occurred in human erythrocytes as well as in human erythrocyte ghosts. Changes in spectral parameters at 37°C were followed over a 3 h time period. As a result it was found that a transient decrease in erythrocyte membrane order began within 30 min of the start of insulin incubation and within 90 min reached a minimum level of 5.25 mT, that was nearly a 0.2 mT lower than that at the start of incubation. Membrane order returned to the initial value by 2.5 h. These time-related changes in membrane order correspond well with the insulin receptor internalization process as followed by surface binding assays. Similar correlations between membrane order and receptor internalization are observed at 23°C. Erythrocytes incubated with denatured insulin, and ATP-depleted erythrocytes incubated with native insulin, do not down-regulate their insulin receptors. This is consistent with the hypothesis that an increase in membrane disorder is part of the mechanism of insulin receptor down-regulation [291].

The EPR technique using a spin label (5-nitroxysrearate) was employed to investigate cholesterol-enriched human erythrocytes and erythrocytes in patients with hepatocyrrhosis. The two groups of cells show a high cholesterol/phospholipid ratio compared to that in volunteers. Erythrocytes in hepatocyrrhotic patients and those rich in cholesterol do not reduce the regulation and the membrane order observed in intact erythrocytes. A higher membrane order inhibits the internalization of erythrocyte insulin-receptor. An increase in membrane disordering reduces insulin receptor regulation [292].

Oxidative imbalance plays an important role in many cell pathologies a well as in cell aging. Insulin-induced down-regulation of erythrocyte insulin receptors is a simplified model that can provide useful information on the cell surface regulative phenomena and on the role of the plasma membrane and cytoskeleton in such physiological processes. The free radical inducer menadione was examined in order to evaluate if this compound is able to modify (and in which manner) the down-regulation process. The menadione-induced oxidative damage is able to decrease the insulin-induced down-regulation process. This effect is accompanied by slight alterations in plasma membrane ultrastructure and by insignificant variations in plasma membrane lipid composition. EPR measurements of changes in membrane order revealed no decrease to occur during the process of down-regulation. In contrast, cytoskeletal protein assembly appears to be remarkably altered. Such changes in specific cytoskeletal elements could lead to the decrease of down-regulation phenomenon induced by menadione. This suggestion is supported by changes in electrophoretic pattern of some cytoskeletal proteins (e.g., spectrin). Using the EPR method the importance of free radicals in cell injury and cell surface receptor modification was established [293].

Non-enzymatic glycation of reactive amino groups in model proteins increases the rate of free radical production at physiologic pH by nearly fifty-fold over non-glycated protein. Anion radical $O_2^{-\bullet}$ generation was confirmed by EPR measurements with the spin trap phenyl-*t*-butyl-nitrone. Free radicals generated by glycated protein increase the peroxidation of membranes of linoleic/arachidonic acid vesicles nearly 2-fold over control. It is suggested that the increased glycation of proteins in diabetes may accele-rate vascular wall lipid oxidative modification [294]. To determine oxygen radicals which generate in Type II diabetes melitis an interesting marker in the form of dimethylsulfoxide / ascorbyl free radial (DMSO/AFR) complex has been proposed [295].

Spleen tissue paramagnetic centers were examined by EPR prior to spleen autotransplantation both experimental and in a clinic. The results show the presence of free radicals, Cu^{2+}, Fe^{3+}. The Fe^{3+} concentration did not change after antioxidant therapy [296].

Cytokines, immunological effectors of molecules, that mediate β-cell destruction, are involved with the development of insulin-dependent diabetes mellitus. The combination of cytokine with human recombinant interleukin 1-β, tumor necrosis factor α (TNF-α), and interferon-γ induces the formation of nitric oxide by human islets. This combination stimulates both the formation of the nitric oxide derivative, nitrite, and the accumulation of cGMP by human islets. The nitric oxide synthase inhibitor N-monomethyl-L-arginine prevents the formation of both cGMP and nitrite. Interleukin 1-β and interferon-γ are sufficient to induce nitric oxide formation by human islets, whereas TNF-α potentiates nitrite production. This combination of cytokines with interleukin 1-β, TNF-α, and interferon-γ also influences the insulin secretion by human islets. Pretreatment of human islets with low concentrations of this cytokine combination (interleukin 1-β 15 units/ml, TNF-α 0.7 nM, and interferon-γ 150 units/ml) appears to slightly stimulate insulin secretion. Higher concentrations (interleukin 1-β 75 units/ml, tumor necrosis factor α 3.5 nM, and interferon-γ 750 units/ml) inhibit insulin secretion from human islets. The inhibitory effect is prevented by NG-monomethyl-L-arginine. This higher concentration of cytokines also induces the formation of an electron paramagnetic resonance-detectable signal at $g = 2.04$ by human islets that is characteristic of the formation of an iron-dithio-dinitrosyl complex. The formation of this complex is prevented by N-mono-methyl-L-arginine, thus confirming that this cytokine combination induces the formation of nitric oxide by human islets. These results indicate that nitric oxide mediates the inhibitory effects of cytokines on glucose-stimulated insulin secretion by human islets. Nitric oxide is suggested to participate in β-cell dysfunction associated with insulin-dependent diabetes mellitus [297].

Investigations of the structure and properties of insulin are in progress. The R-state conformation of the Cu^{2+}-substituted insulin hexamer has been identified and a number of its derivatives have been studied via EPR, ^{1}H NMR, and UV-visible spectroscopy. The results indicate that each Cu^{2+} center of the R-state Cu^{2+}-insulin hexamer possesses a coordination site that is accessible to anions from solution. Both phenol and anionic ligands that coordinate to the Cu^{2+} ions are required to generate the necessary heterotropic interactions stabilizing the R-state structure. With phenylmethylthiolate (PMT), a Cu^{2+}-R6 adduct that displays the spectral features of blue (type 1) copper proteins is obtained. This complex is proposed to embody a pseudotetrahedral Cu^{2+}N3S(PMT) chromophore, in which N is HisB10 (imidazolyl). The remaining ligands examined give rise to Cu^{2+}-R6 adducts that possesses the spectral characteristics of normal (type 2) Cu^{2+} proteins. Under reducing conditions, Cu^{+}-T6 and Cu^{+}-R6 hexamers have been identified [298].

3.10 NERVOUS SYSTEM DISEASES

The EPR method has found many uses in clinical and experimental neurology. [299, 300].

In order to decrease oxygen toxicity of the central nervous system and to treat cancer and HIV-infected patients the use has long been made of diethyldithiocarbamate (DEDTC), a potent copper chelating agent, and an immunomodulator. The antioxidant properties of DEDTC, including its scavenging of reactive oxygen species were evaluated by EPR spectroscopy. The same method was used to examine its reducing and iron-chelating properties, and its protective effects on oxidant-induced damage of brain tissue, protein, human low density lipoproteins, and DNA. It is found that DEDTC is a powerful reductant and antioxidant since it scavenges hypochlorous acid, hydroxyl radical and peroxynitrite (HOONO). All this may provide an explanation for the apparent beneficial effects of DEDTC against oxidative stress-related diseases that have been observed in experimental and clinical studies [301].

Anion radicals $O_2^{-\bullet}$ cannot be directly fixed by EPR spectroscopy owing to their high reactivity. With advent of spin labels [302] the detection of hydroxyl and hydroperoxyl radicals ($HOO^{\bullet}$) using their stable spin adducts became possible [303, 304].

EPR spectroscopy using spin traps was applied to determine Cu,Zn superoxide dismutase activity in cerebrospinal fluid (CSF) of 4 patients with rectal carcinoma, 10 patients with endometrium cancer, 9 with cervical carcinoma, 21 with leiomyoma, 3 with ovarian cyst, 4 with inguinal hernia and 8 expectant mothers. The results showed the SOD activity to increase with age [305].

Ascorbic acid (vitamin C) attracts the attention of numerous researchers since it is not only essential to the health of human, but prevents from many diseases [306]. The acid possesses an antioxidant activity due to its ability of scavenging oxygen radicals. Therefore much information on this process can be provided by measurements of CSF and plasma ascorbic acid levels made using the EPR method. Ascorbic acid transport from plasma to CSF is controlled by choroid plexus. In this way it was possible to examine CSF and plasma of patients with acute lymphoblastic leukosis by an ingenious technique which avoids using plane quartz cuvette, in a polymethylmetacry-late capillary at room temperature. The EPR spectra of both CSF and plasma show a characteristic ascorbyl radical doublet at $g = 2.005$ and $\Delta H = 0.18$ mT. The CSF and serum ascorbyl radical concentration determined using known concentration of nitroxyl radical TEMPO is 3.5×10^{-8} and 9×10^{-9} mol/dm^{-3} [307].

In the pathophysiology of lumbosacral radiculopathy, inflammation of the nerve root is of critical importance. This inflammatory process has been shown to be associated with free radicals. EPR spectroscopy with DMPO spin trap was used to SOD activity in CSF of 31 patients with unilateral lumbosacral radiculopathy caused by a herniated disc. The SOD activity in the hernia patients was remarkably decreased. The concentration gradient of SOD activity was different between central herniation and centrolateral herniation. These findings indicate that free radicals are generated after nerve root compression. Under severe deficiency of SOD activity in CSF, serum SOD penetrates into CSF and may play an important role in protecting against nerve root involvement [308].

Patients with Alzheimer's disease show a suppressed highest cortical function leading to dementia caused by diffusive cerebral atrophy. Alzheimer's disease is related to presenile dementia of unclear pathogenesis. During the latest time free radicals have been widely believed to play a part in neuron damage and the development of age-dependent degenerative diseases. The membrane fluidity of thrombocytes isolated from 15 patients with Alzheimer's disease (AD), 11 patients with multi-infarct dementia

(MID), and 7 neurologically healthy controls was studied by EPR spectroscopy employing spin label techniques. Spin label I(12, 3) probed the shallow site (hydrophilic region) and spin label I(5, 10) the deeper site (hydrophobic region) of the thrombocyte membrane. A significant increase in membrane fluidity was observed in patients with Alzheimer's disease and multi-infarct dementia, as compared to age-matched controls. However, there were no significant differences in fluidity between AD and MID patients. The increased thrombocyte membrane fluidity observed in dementia does not seem to be specific to Alzheimer's disease [309].

Electron paramagnetic studies of heavy polyvalent metals were performed on autopsy material of 5 aged patients (mean age 80.2 years) and 15 control patients (mean age 29 years). The obtained results lead to the following conclusions:

1. The aging brain is characterized by a tendency to decrease in concentration of isolated Cu^{2+} ions, and a marked decrease in concentration of Fe^{3+} ions as well as of free radicals, whereas the concentration of Cu^{2+} clusters is significantly increased.

2. The cases showing both arteriosclerotic and senile degenerative changes are characterized by higher concentrations of Cu^{2+} clusters than the brains with dominance of the arteriosclerotic process, whereas the concentrations of isolated Cu^{2+} and Fe^{3+} ions as well as that of free radicals do not differ between the two subgroups.

3. The diminished concentrations of Cu^{2+} and Fe^{3+} ions and free radicals observed in brains of old persons and concomitant with increased concentration of multi-ion aggregates (clusters), more marked in cases of senile atrophy of Alzheimer type, seem to result from some slow-down of metabolic processes in the aging brain [310].

In Alzheimer's disease progressive dementia and gradual formation of extracellular senile plagues in the hippocamp and adjacent brain areas are observed. β-Amyloid (Aβ) oligopeptides represent an essential component of these plagues. A-β peptides consist of 40-43 amino acids, derivatives of a larger Aβ precursor β-APP. β-APP is a unit gene product existing in several forms coding plasma membrane type I integral proteins. Amyloid formation is successfully monitored by EPR spectroscopy [311]. Alzheimer's disease genetic forms called hereditary Alzheimer's disease are characterized by a comparatively early beginning and are determined by mutation of three genes. The first gene corresponds to β-APP located in chromosome 21, the second gene is in chromosome 14 and codes protein S182, and the third gene, responsible for coding protein STM2, is present in chromosome 1. Mutation in either of the three genes accelerate the processes of β-APP formation and accumulation. The functions of V-APP, S182 and STM2 are not clearly understood. In an aqueous solution β-amyloid undergoes spontaneous fragmentation to produce low molecular free radicals which can be fixed by EPR spin trapping technique using spin trap phenyl-*tert*-butylnitrone. In aqueous media β-amyloid produce free radicals without metals. For β-amyloid it takes the hours or days to produce detectable free radicals, whereas the extremely neurotoxic A β (25-35) fragment of β-amyloid peptide produces a detectable radical in minutes. A β (25-35) is a potent lipoperoxidation initiator. It rapidly quenches the membrane-bound 12-nitroxyl stearate spin probe deep within the lipid bilayer. These data enable formulating a "molecular shrapnel" model of neuronal membrane damage in Alzheimer's disease [312-314]. Thus, β-amyloid should be considered the central constituent of senile plaques in Alzheimer's disease. β-Amyloid peptide neurotoxicity is inhibited by antioxidants. It is established that the free radicals derived from A β peptides are able to inhibit glutamate uptake in cultured astrocytes. These results support the hypothesis that A β neurotoxicity in Alzheimer's disease may be due in part to A β-derived, oxygen-dependent free radical inhibition of glutamate uptake [315]. α-Hydroxyethyl radical formation induced by the action of ethanol on human brain glyoma C6 cell astrocytes and astrocytic to ethanol has been investigated by EPR spectyroscopy using spin traps [316].

The formation of free oxygen radicals in homogenates of frontal cortex from brains taken at autopsy and verified histologically to be from five patients with Alzheimer's disease or from six age matched normal controls, was investigated by EPR spectroscopy. During incubation at 37°C, in the presence of ferrous sulfate (200 mcM), samples of Alzheimer's frontal cortex produced nearly 50% more free radicals than did controls. Although these are post mortem *in vitro* observations are consistent with an increased free radical burden in tissue from patients with Alzheimer's disease, that comparable differences exist *in vivo* between Alzheimer's patients and non-demented people remains to be demonstrated [317].

Migraine is a severe mostly unilateral headache of various intensity, frequency and duration. Sometimes, migraine is accompanied by vegetative disorders and transient focal neuralgic symptoms. The role of vascular phenomena during an attack of migraine remains unclear. For a better understanding of these phenomena a study involving measurements of systemic levels of nitric oxide and endothelin-1 has been carried out. Nitric oxide levels were measured using EPR spectroscopy, whereas endothelin-1 concentrations were estimated by means of radioimmunoassay. In the pilot study seven patients (five women and two men suffered from migraine) were enrolled. Nitric oxide levels in women and men were slightly higher compared to basal values (1.56 ± 0.88 and 0.85 ± 0.46 pg/ml, respectively) and were raised after pharmacological intervention (2.91 ± 1.93 pm/ml). Plasma endothelin-1 concentrations decreased during migraine attacks with respect to interictal conditions from 4.23 ± 1.19) to (3.99 ± 1.21 pg/ml), and returned to basal values (4.44 ± 1.08 pg/ml) after relief of pain. The measurements of systemic levels of nitric oxide and endothelin-1 will provide useful information on the hemodynamic changes of cerebral blood flow regulation in migraine. This will give new insights into the mechanisms of the migraine attack [318]. NO formation by nerve cells and brain macro-phages was demonstrated by the EPR technique using spin trapping [319].

The EPR technique and total reflection X-ray fluorescence spectroscopy were employed to measure paramagnetic metal ions, free radicals (neuromelanin), and total metal content in nine areas of the brain. In addition, the extent of accumulation of metal ions by melanins incubated in homogenates of a region of the brain (putamen) was determined. The EPR spectrum of the *substantia nigra* had a prominent g = 4 EPR signal characteristic of Fe^{3+} ions in the rhombic state. Only the *substantia nigra*, and to a lesser extent the *locus coeruleus*, had a free radical signal consistent with that of neuromelanin. This signal was much more prominent in the unprocessed *substantia nigra*. However, when metal ions were removed the EPR signal of neuromelanin of the *locus coeruleus* increased much more than that of the *substantia nigra*. This suggests that the structure of the pigment may differ in these two regions. Incubating synthetic melanins with homogenates of putamen resulted in accumulation of iron, copper and zinc ions. This seems to be caused by metal binding to the neuromelanin during the isolation procedure [320].

The role of neuromelanin, a pigment that accumulates in catecholaminergic neurons during normal aging, is poorly understood in many respects. According to histochemical data, neuromelanin has a cytoprotective function under normal conditions, but it is cytotoxic at advanced ages and in patients with Parkinson's disease. EPR spectroscopy is an especially effective technique for investigating melanins [321]. This method was used to examine human *substantia nigra*. It has been unambiguously shown during the purification of neuromelanin that a portion of paramagnetic metal ions are bound to the pigment *in situ*. This made it necessary to elaborate a purification procedure to avoid contamination of the pigment by extraneous metal ions. The EPR method is also useful for identifying other natural pigments [322].

Parkinson's disease (paralysis agitans) is a brain disease that slowly grows progressively worse and manifests itself by shaking, extrapyramidal rigidity and akinesia. The origin of Parkinson's disease is unclear. Shaking is mainly caused by functional disorder of the spinal system. Distribution pattern of biogenic amines in the brain was examined by histochemical methods using electron microscopy. It has been found that extrapyramidal rigidity was induced by dophamin deficiency in subcortical structures.

Parkinsonism is also associated with *substantia nigra* malfunction and breakage of basal nuclear control of reticular-spinal tracts. All this increases the time of tonic muscle delay. Under this syndrome both the considerable decrease in dophamin levels in basal nuclei and synthesis of *substantia nigra* melanin are considerably reduced. A certain part is played by acetylcholine as well. Using the EPR technique the melanin ability to scavenge free radicals produced from tetrahydroisoquinolines has been studied [323].

In the etiology of Parkinson's disease and other CNS disorders a role of metal ions and/or neuromelanin is postulated. The interactions between neuromelanin and metal ions as well as the amount and type of metal ions in human neuromelanin in intact *substantia nigra* and in isolated neuromelanin have been studied using EPR spectroscopy. In *substantia nigra* substantial amounts of iron, zinc, lead, copper, manganese, and titanium are present at concentrations up to 4 times greater than those of non-pigmented brain tissue (basis pedunculi). The concentrations of metal ions in isolated neuromelanin were 5-260 times higher than in *substantia nigra*. There were substantial amounts of paramagnetic metals ions, especially iron, bound to neuromelanin in intact *substantia nigra*. Compared to other regions of the midbrain, the *substantia nigra* contains increased amounts of many different metal ions. These results are consistent with the hypotheses that postulate a role of metal ions in promoting oxidative reactions in pigmented neurons [324].

Free radicals are considered to be involved in the pathogenesis of Parkinson's disease as the level of radical $NO^{\bullet}$ is increased in Parkinson's disease brain. A system generating $NO^{\bullet}$ radicals *in vitro* using 3-(2-hydroxy-1-methylethyl-2-nitrosohydrazino)-N-methyl-1-propanamine as an radical $NO^{\bullet}$ donor and 2-(4-carboxyphenyl)-4,4,5,5-tetramethylimidazoline-1-oxyl 3-oxide (carboxy-PTIO) has been examined by EPR spectroscopy. The effects of the dopamine agonists pergolide and bromocriptine on the amount of radical $NO^{\bullet}$ generated has also been studied. Radical $NO^{\bullet}$ appeared to be scavenged by pergolide and, to a lesser extent, by bromocriptine. Pergolide and bromocriptine treatment of Parkinsonism *in vivo* completely protects against the decrease in levels of striatal dopamine and its metabolites. Pergolide and probably bromocriptine may also protect against dysfunction of dopaminergic neurons by direct scavenging of radicals $NO^{\bullet}$ and thus eliminating NO-related cytotoxicity. Thus the ESR spectrometry method using carboxy-PTIO may be useful for screening other drugs that can quench radical NO [325].

Idiopathic Parkinson's disease (PD) involves is accompanied by a decline in the activity of mitochondrial complex I in *substantia nigra* (1-3). EPR spectroscopy was used to investigate complex I in human *substantia nigra* and *globus pallidus*. EPR signals characteristic of the iron-sulfur centers of complexes I and II were observed with *globus pallidus*. There was no significant difference between control and PD patients. The complex 1 signals could not be clearly observed in *substantia nigra*. Instead, nitric oxide radicals in PD nigra were detected at g approximately 2.08, 1.98 due to [Hem-NO] formation. Although an EPR signal indicative of Hem-NO was observed with control nigra, it lacked the distinctive g approximately 1.98 trough observed with PD nigra. As PD is associated with a reactive gliosis, the difference in the Hem-NO EPR signal, between control and PD nigra, may result from cytotoxic radical $NO^{\bullet}$ generated by microglia in PD *substantia nigra* [326].

Down's syndrome (chromosome disease) is an oligophrenia form characterized by mental retardation in combination with a peculiar appearance of the patient. Sulfite oxidation and sulfur trioxide radical formation were studied in polymorphonuclear leukocytes isolated from healthy young, old and centenarian donors and from patients with Down's syndrome. The sulfur radical formation measured by EPR spectroscopy using spin trapping is correlated with the activity of sulfite oxidase and with the rate of sulfite oxidation to sulfate by polymorphonuclear leukocytes. Sulfite metabolism was studied both in resting, and phorbol myristate acetate stimulated freshly isolated cells. The rate of sulfur trioxide radical formation was demonstrated by use of the spin trapping agent DMPO with subsequent formation of an adduct. The intensity of adduct formation was most intense in cells with low sulfite oxidase activity, while a mixture of the adduct and of DMPO hydroxyl radical was mainly observed in cells with high sulfite oxidase activity. Furthermore, the spectra of purified sulfite oxidase show that in the presence of sulfite the enzyme can also give rise to a DMPO-OH adduct. Sulfite oxidase activity in cells isolated from healthy young and old donors was positive correlated with both rates of sulfur trioxide radical formation and sulfite oxidation to sulfate, respectively. However, sulfite oxidase activity in cells isolated from centenarians and patients with Down's syndrome seems to loose partly its rate of oxidizing sulfite to sulfate. The intensity of the sulfur centered radical adduct increased in the two latter groups of population and the radical observed was predominantly sulfur trioxide radical [327].

Amyotrophic lateral sclerosis (Charcot-Kozhevnikov's disease) is an organic nervous system leisure mainly caused by damage of anterior horn, pyramid tract, nervous roots and peripheral nerves. The etiology and pathogenesis of this disease remain unclear. On the one hand this disease is considered to be of hereditary origin, on the other hand it is regarded as an infectious disease. Histologic pattern and clinical symptoms in familial and sporadic varieties are practically indistinguishable. Therefore, investigation of amyotrophic lateral sclerosis at the molecular level will contribute much to a better understanding of processes occurring in this pathology.

Several point mutations in the gene coding for human Cu,Zn superoxide dismutase reported as being responsible for familial amyotrophic lateral sclerosis (FALS) have been distinguished. However, no direct correlation between total superoxide dismutase activity and severity of the FALS pathology has been found. In order to get a better insight into the mechanism underlying FALS phenotype, the activity and the copper-binding properties of the single mutant H46R, which is associated with a Japanese form of FALS, has been investigated by the EPR method. The results show that this mutant is structurally stable but lacks significant enzyme activity and has impaired capability of binding catalytic copper. The mutant protein can be fully reconstituted with copper *in vitro* but its ESR spectrum quite differents from that of the wild-type [328].

Familial amyotrophic lateral sclerosis responds to treatment with difficulty. Positive tendency appeared after the discovery of some FALS cases associating with superoxide dismutase-1 (SOD-1) genus mutation. SOD-1 is responsible for coding of Cu-Zn SOD-1 formation [329]. Only 10% of FALS patients display features of the familial type. SOD-1 mutation was observed in 20% of FALS patients. It is suggested that biochemical processes that lead to the damage of motor neurons in SOD1-associated FALS are like sporadic FALS cases [330].

Unlike the wild type, mutant Cu-Zn-superoxide dismutase observed in familial amyotrophic lateral sclerosis, has peroxidase Cu-dependent activity. This was proved by EPR method using spin trap DMPO. The effect of two Cu chelators (sodium diethyldithiocarbomate and D-penicillamine) on the relation between the death of motor neurons and the mutant activity has been studied. These chelators inhibit mutant peroxidase, the wild type enzyme activity in stoichiometric concentrations being unchanged. Moreover, these Cu-chelators increase the neural survival in cell culture of FALS patients. The wild type cell expression survival of Cu-Zn-SOD remains unchangeable. It is suggested that oxidative reactions catalyzed by mutant Cu-Zn-SOD initiate neuropathologic changes in familial amyotrophic lateral sclerosis [300].

Duchenne-Aran amyotrophy is characterized by a chronic degenerative process in spinal anterior horn cells, anterior roots and peripheral nerves. Histologic examinations of muscle fibers of patients suffering from this disease reveal a fascicular character of atrophy and electromyographic alterations associated with neural atrophy. Mobilization and aggregation of intramembrane particles are physiological changes observed in all cells. In erythrocyte membranes this can be induced by exocytosis, vasopressin secretion, oxidation and pH changes. The association between aggregation of intramembrane particles, protein mobility, and membrane fluidity in erythrocyte membranes of healthy controls and patients with Duchenne-Aran muscular dystrophy was investigated by the EPR method using spin labels. In control individuals a partial spectrin removal induced a decreased segmental motion of protein spin label. This indicates an increase in protein-protein interactions. Stearic acid spin labels 5- and 16-(N-oxyl-4,4'-dimethyloxazolidine) induces an increase in membrane fluidity. In patients with Duchenne-Aran's amyotrophy changes of membrane fluidity compared to those of the controls are observed. It is suggested that in Duchenne-Aran's amyotrophy the defects in the interactions between skeletal proteins, membrane and skeleton components may induce alterations in erythrocyte membranes [331].

Myotonic dystrophy, a neuromuscular disease, is associated with the presence of a myotonic phenomenon or "contracture". Patients with myotonic dystrophy experience difficulty with relaxation of muscles after strong contraction. Pathogenesis of this disease is related to injury of cell membrane permeability and changes in ionic and mediatory exchange. The levels of free radicals in the blood of six patients with myotonic dystrophy (mean age 52.8 ± 5.0 years) and seven controls (mean age 48.8 ± 6.3 years) were studied by EPR spectroscopy using DMPO spin trap. In the same groups lipid peroxides and the levels of antioxidants (vitamin E, coenzyme Q_{10}, selenium), as well as superoxide dismutase activity were examined. The SOD activity in both erythrocytes and serum was the same in the 6 patients with myotonic dystrophy and the seven controls. Free radical concentration and serum lipid peroxidation was increased in 5 of myotonic dystrophy patients and tended to increase further as the disease progressed. At the same time. The levels of serum vitamin E, Q_{10} and selenium were low. Thus, an increase in the levels of free radicals and lipid peroxides and a decrease in the levels of antioxidants play an important role in the pathogenesis of myotonic dystrophy [332].

Ataxia-telangiectasia (Louis-Bar's syndrome) is an early progressing cerebellar ataxia with symmetrical telangiectases of conjunctiva and skin. This pathology is associated with atrophy and glyosis of cerebellum, pale globe and substantia nigra, numerous telangiectases in brain substance as well as with thymus aplasia, adenohypophysis hypoplasia, bronchiectases. Ataxia-telangiectasia is supposed to be a hereditary disease, that is transmitted by autosome-recessive type. In erythrocyte membranes from an ataxia-telangiectasia patient and from his mother (heterozygote) an additional protein fraction migrating slower than spectrin has been detected. In erythrocyte membranes labeled with maleimide spin label changes in signal of the weakly immobilized spin label as related to that of strongly immobilized one (w/s) were noted. In controls the w/s values increased with age, whereas in families with ataxia-telangiectasia no significant differences were noted between children and parents. The presence of an additional protein fraction in erythrocyte membranes of ataxia-telangiectasia patient and ataxia-telangiectasia heterozygote indicates that they can be differentiated from the healthy controls [333].

Using The EPR method was used to assess the action simultaneously produced by decimeter microwaves and deresinified naphthalane on neurons. Clinical and neurophysiological trials were conducted with 133 patients suffered from vertebrogenic scapulohumeral periarthritis. It was found that the above combined efforts produce beneficial effect and correct functional activity of segmental-peripheral neuromotor system [334].

Jugular bulb venous oxygen saturation and free radical formation in the process of cerebral ischemia/reperfusion were studied in 13 patients undergoing carotid endarterectomy. Oxygen content were determined by a calorimetric assay. Free radical levels were recorded by EPR spectroscopy using spin trap adducts α-*tert*-butyl phenyl nitrone. During clamping of the internal carotid artery the oxygen saturation decreased greatly from 68% to 61% and returned to baseline (65%) when the carotid clamp was removed. Jugular venous plasma antioxidant potential also decreased from 33% to 28%. There was no concomitant change in arterial plasma antioxidant potential values. This indicates a decrease in antioxidant capacity across the cerebral circulation, which is supported by EPR data. These results provide strong evidence for oxidant production during cerebral ischemia/reperfusion [335].

3.11 DISEASES OF JOINTS AND CONNECTIVE TISSUE

Erythrocyte and plasma paramagnetic centers were studied in 110 patients with rheumatoid arthritis and 82 patients with osteoarthrosis deformans for the purpose of differentially diagnosing inflammatory and degenerative diseases of joints and connective tissue. In osteoarthrosis deformans and, especially, in rheumatoid arthritis a noticeable increase in erythrocyte free radical levels and a decrease in plasma Fe^{3+} content compared to those in normal donors are observed (Table 3-30) [336]. As the activity of inflammatory process enhances, the erythrocyte free radical level grows higher. Nothing of this kind is observed in patients with osteoarthrosis deformans. A considerable decrease in plasma Fe^{3+} concentration in patients with rheumatoid arthritis may serve as a differential criterion of this disease. The Cu^{2+}-ceruloplasmin and Fe^{3+} transferrin levels in blood plasma of RA patients is increased too [6].

Analogously, in patients with systemic lupus erythematosus the blood free radical concentration is higher than that in healthy donors and increases as the process activity grows higher [9].

In the course of seropositive RA the value of FR signal and the intensity of lipid peroxidation are higher than those in the seronegative version [337, 338]. It is known that hydroxyl radicals attack hyaluronic acid and related polymers, that leads to their strand-breakage [339].

The low erythrocyte FR level in the RA visceral form is most likely to be associated with glutathionereductase depletion (table 3-30) [336].

Table 3-30. Erythrocyte and plasma PMC content in RA and OAD patients (M ± m)

Indexes	RA	OAD	Control
	Erythrocytes		
MetHb, (g/l)	2.0 ± 0.06	1.8 ± 0.07	0.8 ± 0.03
FR, (10^{-6} M)	1.6 ± 0.06	0.8 ± 0.02	1.2 ± 0.03
Cu^{2+} SOD, (10^{-5} M)	1.0 ± 0.03	0.8 ± 0.03	0.3 ± 0.04
	Plasma		
Fe^{3+}, mcmol/l	5.6 ± 0.29	14.2 ± 0.5	16.0 ± 0.33
FR, (10^{-6} M)	0.5 ± 0.02	0.5 ± 0.02	0.50 ± 0.01
Cu^{2+} CP, (10^{-5} M)	2.22 ± 0.08	0.92 ± 0.03	1.80 ± 0.03

RA and OAD patients show and increased blood MetHb concentration. This seems to be due to the administration of anti-inflammatory drugs, that are methemoglobin producers [20]. These include not only toxic substances such as lysol, creosote, barium sulfate, β-naphthol, nitrates, nitrites, nitrogen oxide, but various medicines (phenacetin, *para*-aminosalicylic acid, hydrazine derivatives, nitroglycerol, sulfanylamides) [340]. Electron spin resonance investigations have revealed that hemoglobin in intact erythrocytes react with hydrophilic xenobiotics, such as hydroxylamine [341] and phenol [342] to form MetHb.

The use of ESR spectroscopy made it possible to verify whether Cu^{2+} complexes bound to dipeptides or tripeptides and introduced into human red blood cells, interact with hemoglobin to form MetHb [343].

Using EPR spectroscopy the effect of superoxide anion radicals on synovial fluid has been studied. Anion radicals $O_2^{-\bullet}$ are generated by microvascular endothelial xanthinoxidase in inflamed synovial tissue [344]. In rheumatoid arthritis the activity of polymorphonuclear leukocyte superoxide dismutase is decreased, therefore the forming superoxide anion radicals injure synovial fluid [345].

In RA patients the synovial fluid Fe^{3+} content is four times lower than in OAD patients. The Cu^{2+} content inversely changes in these diseases (table 3-31) [336].

Table 3-31. Synovial fluid PMC content in RA and OAD patients (M ± m)

Indexes	RA	OAD
Fe^{3+}, (momol/l)	4.5 ± 0.38	13.5 ± 0.86
Cu^{2+}, (10^{-5} M)	2.24 ± 0.11	1.20 ± 0.10
FR, (10^{-6} M)	0.5 ± 0.03	0.4 ± 0.03

The analysis of synovial fluid of 18 RA patients was carried out on admission and after 3 weeks of stationary treatment. The positive dynamics in the patient state was characterized by a statistically reliable decrease in the Cu^{2+} level and an increase in the Fe^{3+} signal amplitude in EPR spectra [346].

The molecular fundamentals of auto-immune diseases of RA type became more clearly understandable owing to investigations of the dynamics and conformational alterations in immunoglobulin molecules forming an antigen-antibody complex. Some dynamic and conformational characteristics in human IgG molecules have been studied by EPR spectroscopy using spin label 2,2,5,5-tetramethyl-3- maleimide pyrrolidin-1oxide. The value of rotational correlation time (τ) of the spin labeled IgG is 26±2 ns, what is known to be characteristic of labeling of the IgG peptide residue. The spin labeled Lys residues are subdivided into two forms with different mobility and rigidity, the ratio of mobile to rigid forms being 3:1. The im-mobility of IgG Lys residues is conditioned by their involvement in H-bonds formation. The temperature dependence of the ordering parameter S provides evidence for a significant immobility of Lys residues. When kept for a long time at 4 C, the spin labeled IgG samples are mobilized, as shown by EPR spectra. This is caused by the process of autolysis, that can be completely eliminated by addition of phenylmethylsulfonylfluoride [347].

The use of non-steroid anti-inflammatory drugs often fails to furnish the desirable result. This is caused by the fact that lactoperoxidase (LPO, EC 1.11.1.7; donor-H_2O_2 oxidereductase) catalyses the oxidation of indomethacin. As shown by the EPR method, lipid peroxidation of indomethacin follows a one-electron mechanism [348]. Cu,Zn-superoxide dismutase is thought to act as peroxidase in the presence of H_2O_2 at high pH (pH>9). However, a recent EPR study has shown this enzyme to function also at physiological pH (pH 7.4-7.6) in the presence of bicatbonate [349].

The EPR method was successfully used for the assessment of the anti-inflammatory effect and the scavenging activity with respect to free radical (hydroxyl and superoxide) of the *Arctium lappa L.* root [350].

Physiological changes in neutrophils after elimination of the effect of sialic acid, mucin and fucoidan have been investigated by the EPR method using spin probes. Binding of sialic acid markedly reduces the mobility of neutrophil membrane proteins and lipids, whereas binding of mucin and fucoidan totally restricts the mobility probably due to cross linking of the surface lectins. This also results in an increase in the cytoplasmic viscosity [160].

Elaboration of new approaches to the diagnosis of osseous abnormalities is of great importance in clinical practice. New data concerning the EPR method potential in interpreting osseous images including characterization of trabecular patterns by micromorphometry appear rather promising [351].

Generation of free radicals in mammals with bone fractures or cuts has been fixed using ER spectroscopy. The detected species were identified as alkylperoxyl, $RCH_2^{\bullet}$, and $O_2^{-\bullet}$ radicals. Evidently, these data and the above method will be of help in traumatologic and orthopedic practice [352].

3.12 MYOGLOBIN PATHOLOGY

Myoglobin (Mb) is a protein related to chromoproteides, that is present in red muscles of animals of various classes and species. The major myoglobin function is accumulation of the oxygen brought from blood into muscles in the process of breathing and the loss of the oxygen as needed [353].

Some more comprehensive information concerning the electron structure and stereochemistry the O_2 bond and ligand-binding Mb and Hb sites could be derived by substitution of protoporphyrin IX iron in the Mb and Hb complexes by cobalt. The EPR method and Raman spectroscopy have revealed that cobalt in cobaltous Mb binds O_2. As shown by using deuterium buffer, in this case a hydrogen bond between the O_2 atom and the histidine E7 distal residue is formed [354].

Proteins and the role of distal histidine residue on the interaction between the bound ligand and the myoglobin molecule have been studied by EPR spectroscopy. The results have shown that the distal histidine residue is essential in the orientation of the bound oxygen molecule. Recombinant human myoglobin mutants having the distal histidine residue replaced by Leu, Val or Gln residues have been prepared by site-directed mutagenesis and expression in E.coli. The recombinant apomyoglobin proteins were successfully reconstituted with cobaltous protoporphyrin IX to obtain cobalt myoglobin mutant proteins. Low temperature photolysis of cobalt-oxygen proteins and ferric nitric oxide complexes indicated that the nature of the photolyzed form depends on steric crowding of the distal heme pocket. The distal Leu mutant has a less restricted, less sterically crowded distal heme pocket than that of the distal Val mutant myoglobin. The EPR data demon-strate that the distal heme pocket steric crowding is not necessarily related to the side chain volume of the E7 residue [355].

Oxo-ferrylporphyrin π-cation radical [$Por^{+\bullet}$ Fe^{IY}=O] is a highly reactive intermediate responsible for the catalytic reactions of heme enzymes such as cytochrome P_{450}, peroxidase and catalase. This compound is involved in the utilization of hydroperoxides or reacts with their equivalents containing Fe^{3+}-heme proteins or model porphyrin complexes. They are formed as the oxygen atom is transferred from peroxide via hemolytic cleavage of the O-O bond [356].

$$PorFe^{3+} + ROOH \longrightarrow Por^{+\bullet}Fe^{IY}{=}O + ROH$$

Simultaneously hemolytic cleavage of the O-O bond takes place leading to the formation of oxo-ferryl species ($PorFe^{IY}{=}O$):

$$PorFe^{3+} + ROOH \longrightarrow PorFe^{IY}{=}O + RO^{\bullet} + H^{+}$$

Human Mb histidine-93 (F8), a proximal ligand of the heme iron, has been replaced with cysteine or tyrosine by site-directed mutagenesis. The resultant proximal cysteine and tyrosine mutant Mbs (H93C and H93Y Mbs, respectively) exhibit the altered axial ligand analogous to P_{450}, chloroperoxidase and catalase. Coordination of cysteine or tyrosine to the ferric heme iron is confirmed by EPR, ^{1}H NMR, UV and Raman spectroscopy data and by redox potential measurements of Fe^{3+}/Fe^{2+} couple. Mutant H93C myoglobulin is five-coordinate ferric high-spin with the proximal cysteine. H93Y myoglobin containing the tyrosine residue is of analogous structure. The reactions of the mutants with hydroperoxide show that the thiolate ligand enhances heterolytic O-O bond cleavage of the oxidant. At the same time, the phenolate ligand hardly affects the heterolysis/homolysis ratio for O-O bond scission compared to with wild-type Mb. The catalytic activity of wild-type Mb is five times lower than that of mutants [357].

3.13 CUTANEOUS DISEASES

Skin is one of accessible tissues used for diagnosis and selection of tactic of treating various diseases.

EPR spectroscopy was used to examine human normal connective tissue, skin and keloid. Most oxygen free radicals are identified in keloid, less in normal connective tissue. No free radicals are found in skin. The relation between the occurrence of oxygen free radicals and the formation of keloid is the topic of discussions. It is suggested that reduction or elimination of production of oxygen free radicals by scavenging may be of help in preventing the excessive growth of scar tissue [358].

Skin paramagnetic centers of patients with various burns have been examined. In the EPR spectrum along with the signal of melanin ($g = 2.003$,

$\Delta H \sim 0.9$ mT), there is an EPR signal (g = 2.3) corresponding to the non-heme iron nitrosyl complex indicating the hypoxic state of tissue [359].

The EPR method using nitroxide radicals has a wide application in dermatologic research for studying biophysical properties of skin lipids and epidermal membranes [360, 361]. The biostability of various chemical types of nitroxide radicals in keratinocytes, epidermis homogenate, and intact skin has been examined. Imidazoline, pyrrolidine, piperidine, and oxazolidine nitroxides may loose their paramagnetic properties in biological tissues due to their reduction to the corresponding hydroxylamine. The rate of nitroxide reduction in skin is considerably dependent on the nitroxide ring structure and substitution. The nitroxide stability in isolated human keratinocytes, mouse epidermis homogenate, and intact mouse and human skin decreases in the following order: imidazoline > pyrrolidine > di-*t*-butylnitroxide (DTBN) > piperidine > oxazolidine. Thus, imidazoline and pyrrolidine nitroxides show the highest biostability. Cationic nitroxides are reduced much faster than neutral or anionic probes, presumably due to transmembrane electron shuttle or internalization. Piperidine-type nitroxides are versatile probes for studying one-electron transfer reactions in skin [360]. As shown by EPR spectroscopy using 5-doxylstearic acid, a 1% sodium lauryl sulfate solution (surfactant) significantly affects human skin barrier function [361].

A relationship between tumor production and free radical formation has been shown by the EPR method in cultured murine keratinocytes and human skin-tumor cell line. The bioactivation of *t*-butyl-hydroperoxide, and benzoyl peroxide via one-electron oxidation or reduction was compared in freshly isolated and in cultured normal human keratinocytes. The formation of methyl free radicals during the metabolism of t-butyl-hydroperoxide was shown by the EPR spin trapping technique.

Radical formation increases under hypoxic conditions. At a low concentration (1 mM) hydroperoxides do not affect much the cell viability, whereas at a higher concentration hydroperoxides display cytotoxicity. Cultured human keratinocytes can be used as a model system for the study of the metabolic activation to free radical inter-mediates of toxic and carcinogenic compounds in the epidermis [362].

The necessity of biopsy presents the only disadvantage of using EPR spectroscopy in dermatologic research. However, with the development of new approaches examination of tissue *in vivo* became quite possible. Thus, free radicals were directly detected *in vivo* in hairless mice skin using low-frequency EPR spectroscopy after local application of anthralin. The EPR signal intensity increased approaching maximum during 1 day and decreased slowly in the following days, probably due to desquamation of the skin. Judging by the spectroscopic features (singlet with a line width of 6 mT; g = 2.0036) and by the pharmacokinetic pattern, the observed signal is related to the final products of anthralin metabolism (ether-insoluble polymeric structures-"anthralin brown"). Vitamin E and the spin trap *tert*-butylphenylnitrone decrease the amount of anthralin-derived radicals. [363].

In this case it is evident that EPR is a valuable tool for the noninvasive and direct *in vivo* monitoring of drugs-induced radical formation in the skin. These experi-mental studies are likely to find much use in the clinic practice.

3.14 GYNECOLOGIC DISEASES

Capacitation of spermatozoa is an essential procedure for fertilization. Capacitated spermatozoa have an increase in the intracellular cAMP and acrosome reaction (AR) occurs immediately. The effect of superoxide anion on the level of intracellular cAMP and on both spontaneous AR and lysophosphatidylcholine-induced AR (LPC-AR) were studied by EPR spectroscopy using semen samples collected from 10 healthy and fertile volunteers. Spermatozoa were incubated at 37°C in a medium with $O_2^{-\bullet}$ generation system: xanthine + xanthine oxidase + catalase + diethylenetriamine-pentaacetic acid + sodium formate (Ham's F-10). The intracellular cAMP was determined by (3H)-cAMP radioimmunoassay at 3 h of incubation, and the percentages of AR and LPC-AR were evaluated by the triple-stain technique at 3.5 h of incubation. The effects of SOD with different concentration were also determined. The results showed that the level of intracellular cAMP of spermatozoa increased from 14.0 to 23.2 (pmol/10^8 of spermatozoa) whereas lysophosphatidyl-choline-induced acrosome reaction increased from 4.5% and 14% to 16% and 32.5%, respectively. These processes are inhibited by superoxide dismutase. For the elucidation of the source of SOD anion radicals influencing sperm capacitation female genital tract fluids collected from 6 healthy nonpregnant donors of reproductive age, and seminal plasma, capacitated and noncapacitated spermatozoa from 10 fertile volunteers were investigated by the EPR spin trapping method. A typical EPR spectrum for SOD anion radical spin adduct was exhibited only in capacitated spermatozoa but not in vaginal or cervical secretions, uterine and fallopian tubal fluids and noncapacitated spermatozoa. These results suggested that only capacitated spermatozoa themselves are able to generate superoxide anion radicals which stimulated their capacitation in turn [364].

Measurements of the water content of human sperm were performed by EPR spectroscopy using a spin label TEMPO and a broadening agent potassium chromium oxalate. This method proved to be better than the generally accepted optical and electronic particle counter techniques [365].

Using the EPR techniques the permeability of human spermatozoa to glycerol and its activation energy were determined. For this purpose the aqueous spin label ^{15}N-tempone and the membrane impermeable broadening agent potassium trioxalatochromiate were used. The permeabilities of human spermatozoa in 1 molar and 2 molar glycerol at 20°C are (10.3 ± 0.3) and (6.0 ± 1.4) x 10^{-4} cm/min, respectively. The permeabilities of human spermatozoa in 2 molar glycerol at 30°C, 20°C, 10°C and 0°C are (8.3 ± 1.3), (6.0 ± 1.4), (2.1 ± 0.4), and (1.1 ± 0.3) x 10^{-4} cm/min, respectively. The activation energy (Ea) for glycerol permeation between 30°C and 0°C is 11.6 kcal/mol [366].

In the EPR spectrum of the placenta tissue of women in various pathologic conditions there is a narrow EPR signal of free radicals (ΔH = 0.8-1.0 mT), whereas in the women placenta and the umbilical cord of infants born in asphyxia a broad signal (ΔH 20-30 mT and g = 2.7-3.0) is observed [367]. Intact uterus tissues gives rise to EPR signals with g = 2.11-2.15. In uterus cancer the signal intensity is lowered [368, 369].

In major tumoral injuries of human uteric muscle (leiomyoma and endometriosis) a signal of ribonucleotide reductase (g = 2.05 with splitting into a double 1.7-2 mT) is observed in the EPR spectrum [370].

A decrease in iron level in pregnant women has been found by EPR spectroscopy [371]. With increasing transferrin saturation by iron the antioxidant activity of transferrin and ceruloplasmin decreases. The interrelation between serum iron and iron-binding capacity and concentrations of the lipid peroxidation metabolite malondialdehyde in normal and preclamptic pregnancies has been examined by EPR spectroscopy. It is suggested that the enhancement of lipid peroxidation induces the release of iron from free hemoglobin and the increase in iron saturation of transferrin [229] (Huber, 1996).

To elucidate the rheological difference in maternal and fetal blood, the erythrocyte deformability in 20 pairs of mothers and newborns, and in 20 nonpregnant women was measured by means of an electron spin resonance method. As a result it has been found that erythrocyte deformability is dependent on the hematocrit. The hematocrit at which the deformability was maximal showed a lower value (32% to 35%) in maternal blood; conversely, a higher value (47% to 50%) occurred in fetal blood, and an intermediate value (40% to 43%) is observed in nonpregnant control women. These results show the erythrocyte deformability of fetal blood to be much higher compared with the maternal blood. This suggests the effective oxygen supply to the fetal tissues even under the conditions of lower oxygen tension [372].

The EPR method was used to study the receptor of rat ovarian membrane luteinizing hormone of human chorionic gonadotropin (LH/hCG) on a reconstituted system of proteoliposomes. The ability of sodium cholate to extract and reconstitute the binding activity of human chorionic gonadotropin is dependent on the protein/detergent ratio. Trypsin treatment of the LH/hCG receptors containing proteoliposomes indicates that approximately 57% of hCG binding sites were oriented extravesicularly. The presence of 20% glycerol or other osmolytes in the reconstituted system increases the accessibility of LH/hCG receptors but not the activity of adenylate cyclase in proteoliposomes. Changes in the membrane properties were monitored by the EPR method using a spin label 16-, 12-, and 5-doxyl stearic acid. Reconstituted proteoliposomes contain less ordered membrane lipids than do native membranes. Addition of glycerol before reconstitution increases the order of lipid bilayer and shifted it to the physical state of the native membrane. These data are consistent with the hypothesis that a rise in the membrane ordering increases the accessibility of membrane-bound luteinizing hormone/human chorionic gonadotropin receptors [373].

ERR and NMR spectroscopies have been used to examine the plasma protease inhibitor pregnancy zone protein (PZP) and its complex with chymotrypsin. The results show that the pregnancy zone protein is identical to α-2-macroglobulin [374].

In the process of embryo development the human placenta regulates the transport of ions, nutrients and immunoglobulins from the maternal to the fetal circulation. An essential role is played by trophoblast organization into a syncytial layer. The structural stages by which the syncytial trophoblasts are formed are not yet understood. In order to clarify the mechanism of trophoblast development in cell cultures the EPR method using 5- and 16-nitroxystearate as spin labels has been applied to study the morphological characteristics of the syncytial trophoblast formation as well as the functional changes (transport properties and membrane microviscosity) accompanying these structural modifications. It has been found that the membrane order increases during the first 24 h of culture. This increase can be associated with the recovery of cell membranes following trypsin treatment and initial aggregation of cytotrophoblasts. [375].

GSH-transferase is involved in detoxication of electrophilic compounds via catalyzing their GSH-conjugation. Over 80% of cellular GSH-transferase are soluble dimers consisting of two either identical or different subunits. The subunits contain a binding site for GSH (G-site) and the other for electrophilic substrate (H-site). Basing on their different amino acid sequences, catalytic and immunic properties, the cytosol enzymes of mammals are subdivides into three classes, α, μ and π. The π-class possess sulfhydryl groups, that play a functional role in the catalytical area. Human placenta GSH transferase of π-class contains a highly reactive cysteine residue identified as Cyst-47 principal residue in the catalytic site. The active site of human placenta GSH glutathione transferase was characterized be means of EPR spectroscopy using spin labels. The results show that modifi-cation of the

Cyst-47 residue redirects the enzyme towards inactive conformation, thus preventing GSH-binding in the G-site [376, 377].

4-Hydroxytamoxifen (4-HTF) antiestrogen drug tamoxifen principal metabolite, is used in treating breast cancer. 4-HTF produced a blocking effect on the peroxidase/H_2O_2 system significantly faster than tamoxifen and is more efficient in damaging tumor cell DNA. The formation of glutathione thiyl radicals (a_H = 1.62 mT) has been demonstrated by EPR spectroscopy using DMPO in combination with glutathione (proton donor) [378]. The role of glutathione thiyl radicals in tamoxifen mechanism of action is not clearly understood and needs further investigation.

3.15 ONCOLOGIC DISEASES

Most investigations on the use of EPR spectroscopy in medicine are devoted to oncology. Using this method it was possible to detect decreased plasma Fe^{3+}-transferrin and increased Cu^{2+}-cerulloplasmin concentrations in oncologic patients compared to those in normal donors [379, 380]. Similar changes were found in patients with cancer of lung [381] and chronic lympholeukosis [8].

Lymphoma (hematosarcoma) is the general name of all the non-leukemic tumors of hematopoietic and lymphoid tissue. The tumoral tissues consist of lymphoid elements and derivatives thereof. Other hematopoietic tissue tumors both assigned to a cellular line (histiocytic, granulocytic) or non-differentiated, are related to the lymphoma group. Most often the lymphomas are generated in lymph nodes, spleen, Waldeier's ring lymphoid tissue, group lymph follicles. This disease can be met in any other organs or tissues having no relation either to hematopoietic or to lymphatic systems. EPR spectroscopy using spin traps has been applied for studying the active oxygen radicals produced during the respiratory burst of phorbol miristine acetate stimulated leukocytes in 17 patients with malignant lymphoma. Oxygen consumption was measured by the spin oximetry technique. The results show that the spectra produced from the PMA-stimulated patients' leukocytes are dominantly spin adducts DMPO-$HO^{\bullet}$. Besides, the patients' leukocytes are characterized by a decreased oxygen consumption during respiratory burst. Under the condition of remission, the respiratory burst and oxygen consumption become close to normal. These data are helpful for a better understanding of the mechanism of cancerous proliferation and metastasis [382].

The depletion of cellular iron can inhibit ribonucleotide reductase, preventing new DNA synthesis and hence inhibiting cell proliferation. EPR spectroscopy has been used to examine the relationship between chelation of intracellular iron and the rate of removal and regeneration of the tyrosyl radical of ribonucleotide reductase within intact human leukemia K562 cells. The base-line concentration of EPR-detectable mononuclear nonheme iron complexes was 3.15 ± 1.05 mcM. On incubation with chelators this value rises more rapidly with hydroxypyridinones than with desferrioxamine. The hydroxypyridinone-induced decrease in the intracellular iron leads to a more rapid removal of the tyrosyl radicals. Hydroxypyridinones also remove the tyrosyl radicals more rapidly, apparently as a consequence of depletion of the intracellular iron pools necessary to regenerate the active enzyme. The tyrosyl radicals are significantly faster for hydroxypyridinones than for desferrioxamine. This is consistent with their differential effects on cell cycle synchronization [383].

Gallium porphyrin (ATX-70 analog) cellular toxicity is significantly increased upon ultrasonic treatment. The mechanism of action is not well understood. The presence of short-lived ATX-70 toxic intermediates is suggested. Nevertheless, the EPR method using spin traps (POBN, DMPO DBNBS, etc.) failed to reveal any radicals [384]. This suggested needs further study to be carried out.

Some antioxidants (sodium ascorbate, gallic acid, n-propylgallate, etc.) induce human promyelocyte leukemic HL-60 cell apoptosis [385, 386]. The effect of these compounds was enhanced by $CuCl_2$ or by iron chelator, desferrioxamine, but $FeCl_3$ was decreased. The EPR data indicate that both $CuCl_2$ and $FeCl_3$ enhance ascorbic radical level, but reduce gallate radical intensity [385].

Myeloleukosis is a chronic leukosis form in which the cell-precursor of myelopoiesis is the source of tumoral process. During the last years a method for potential use in photodynamic therapy of cancer based on illumination of the tumor tissue with laser irradiation in the presence of photosensitizers has been developed. The following aminoanthroquinone derivatives were studied as photosensitizers: 1-NH_2-4,5-$(MeO)_2$-AQ (1), 1,5-$(NH_2)_2$-4,8-$(MeO)_2$-AQ (2), 1,8-$(NH_2)_2$-4,5-$(MeO)_2$-AQ (3), and 1,5-$(NHPhMe)_2$-4,8-$(MeO)_2$-AQ (4). These compounds exhibit strong absorption in the region 480-620 nm. Possible mechanisms of photosensitization were studied by measuring $^1O_2$ phosphorescence at 1270 nm, employing EPR spectroscopy (using a spin trapping technique), and measuring oxygen consumption. Anion radicals $O_2^{-\bullet}$ were detected as spin adducts DMPO in irradiated benzene solutions of compounds 1, 2 or 3 and DMPO. Aminoanthroquinones 2 and 3 sensitize photo-oxidation of NAD.H in H_2O/EtOH mixture to form the intermediate of singlet oxygen. This is indicated by the effect of sodium azide on the photo stimulated oxygen consumption. Evolution of O_2 upon addition of catalase to the illuminated solution confirms the ultimate formation of hydrogen peroxide. These findings suggest that the (di)amino-dimethoxyanthroquinones might exert photosensitization via both Type I (radical) and Type II (singlet oxygen) mechanisms. The ability to photosensitize K562 human chronic myeloid leukemic cells in culture has been tested. The cell viability was measured using the 3,4,5-diethylthiazol-2,5-diphenyl tetrazolium blue assay. DNA and possible membrane damage were assessed as well. Illumination of cells with light > 475 nm in the presence of compound 2 reduces the dark toxicity from ID50 of 23 to 250 microM, for compound 3 correspondingly from ID50 of 53 to > 300 microM, and for compound 4 from 8.85 to 14.4 microM. The mechanism of cytotoxicity of the prototype phenolic aminoanthroquinones and their methoxy counterparts is quite different. Thus, the cytotoxic action of the latter causes cell damage by the formation of a high concentration of alkali labile sites in addition to DNA strand breaks [387].

Melanoma is a malignant tumor developed from pigment-forming cells. Melanoma grows and metastasizes very quickly, and is resistant to cure. The tumor contains an enzyme ribonucleotide reductase, that controls human melanoma growth. The ribonucleotide reductase active site contains tyrosine radicals readily detected in intact cells as a doublet in the EPR spectrum. The content of tyrosine radicals is higher in young tumor tissues than in older ones. The antimelanotic drug, 4-hydroxyanisole, inhibits ribonucleotide reductase in Ehrlich ascites tumor cells. This is demonstrated by quenching of the tyrosine radicals (IC50 = 5 mcM) by the above antitumoral drug. In amelanotic melanoma tissue tyrosine radicals of the enzyme are also quenched by 4-hydroxyanisole at concentrations down to 50 mcM. These data indicate that the inactivation of ribonucleotide reductase, which provides deoxyribonucleotides for DNA synthesis, may be a hitherto unexpected mechanism for the antitumoral action of 4-hydroxyanisole [388].

Cell free extracts from metastases of human melanoma also contain a highly active ribonucleoside diphosphate reductase (RR) which uses guanosine diphosphate (GDP) as a substrate and deoxythymidine triphosphate (dTTP) as an effector. No activity could be detected in these extracts when cytidine diphosphate (CDP) was used as a substrate with adenosine triphosphate as an effector. The activity of this ribunucleotide reductase requires the presence of either Mg^{2+}or Ca^{2+} ions. There are two different free radicals in melanoma ribunucleotide reductase. One is present in the resting enzyme, and the other is used during catalytic activity. The thiolate-active site from melanoma is inhibited by the new nitrosourea antitumor drug fotemustine (IC50 = 10^{-4} M), that suppresses free radicals [389].

Neuroblastoma is a malignant tumor consisting of immature ganglion cells. Using EPR spectroscopy the mobilization of iron from intracellular ferritin by ascorbic acid has been analyzed *in situ*. This enables a distinction between ferritins and other Fe^{3+}-binding cellular components. The ordered iron core of ferritin gives rise to a resonance signal, which can be observed only at temperatures above 50°K. It has been clearly demonstrated that ascorbic acid is capable of mobilizing iron from ferritin in the cellular system by reduction of the ferric ion core in neuroblastoma SK-N-SH cells. This mechanism may open new ways in the therapy of this hardly curable tumor in the terminal stage [390].

A successful treatment of many malignant tumors is restricted by their being resistant to antitumoral drugs. The reason for this phenomenon remains poorly understood. It has been found that doxorubicin (adriamycin) (an antitumor antibiotic of the anthracyclin series) cytotoxicity may be mediated by semiquinone-free radicals derived from the drug itself and by reactive oxygen species, that form in its cells. The low concentration of reactive oxygen species and the subtoxic dose of doxorubicin are able to stimulate cell proliferation. It is suggested, therefore, that the mechanism of action of small doses of doxorubicin is connected with the generation of reactive oxygen compounds. Their nature has been investigated with the help of EPR spectroscopy using the spin trap DBNBS, which is oxidized in the presence of H_2O_2 and peroxidase enzymes and gives rise to a characteristic three-line signal in the spectrum. An identical EPR spectrum is produced by human lymphoblastic leukemic cells after exposure to doxorubicin. The EPR data have shown that subtoxic concentrations of doxorubicin are likely to induce cell proliferation via an H_2O_2 mediated mechanism [391].

Sarcoma is a malignant non-epithelial tumor that develops from embryonic mesoderm-derived cells. Sarcoma etiology and pathogenesis is unclear. The manifestation of sarcoma is favored by ionizing radiation, oncogenetic substances and viruses. Several mechanisms of drug resistance have been defined using cell lines selected for resistance in vitro. It has been found that one pretreated sarcoma line, STSAR90, is 6-fold less sensitive to doxorubicin compared to a normal fibroblast line, AG1522. The sensitivities of six other sarcoma lines are similar to that of AG1522. The cells were labeled with a photosensitive compound, a vinblastine analogue N-(p-azido-3-125I-salicyl)-N'-β-aminoethylvindesine. The results show the glutathione levels to be twice as high in STSAR90 than in STSAR11, and the glutathione peroxidase activity to be from 3.5- to 6-fold higher. This is due mostly to an increase in selenium-dependent peroxidase activity. As detected by EPR spectroscopy, after exposure to doxorubicin, STSAR90 cells form only half as many measurable hydroxyl radicals compared to STSAR11. Doxorubicin sensitivity is increased in STSAR90 cells when intracellular glutathione levels are reduced by buthionine sulfoximine. These results indicate that glutathione peroxidase-dependent detoxification of doxorubicin induced oxygen radicals may contribute to clinical doxorubicin resistance too [392].

The phenotypic expression of multidrug resistance by the doxorubicin-selected AdrR human breast tumor cell line is associated with over-expression of plasma membrane P-170 glycoprotein and increased cytosolic selenium-dependent GSH-peroxidase activity relative to the parental MCF-7 wild-type line (WT). Potential doxorubicin resistance by AdrR cells in vivo, and the possibility of biochemical differences between WT and AdrR solid tumors have been studied. For this purpose the two tumor cell lines were grown as subcutaneous xenografts in athymic hairless mice. The results show that the growth rate for the AdrR xenografts is only about half that of WT xenografts. Doxorubicin intraperitoneally injected weekly at a dose of 2-8 mg/kg, significantly diminishes the growth of WT tumors, whereas AdrR solid tumors do not respond to doxorubicin. The accumulation of ^{14}C-labeled doxorubicin is 2-fold greater in WT xenografts that in AdrR. Membrane P-170 glycoprotein mRNA is detected in the AdrR tumors, but not in WT cells. The formation of oxygen radicals in both AdrR and WT xenografts is demonstrated by EPR spectroscopy using 5,5-dimethylpyrroline-N-oxide spin trapping. The oxygen radical (superoxide and hydroxyl) concentration is found to be 2-fols greater in doxorubicin-activated WT cells compared to AdrR. On the contrary, the activities of selenium-dependent glutathione (GSH)-peroxidase, superoxide dismutase and GSH-S-aryltransferase in AdrR xenografts are elevated relative to WT. In both WT and AdrR solid tumors in vivo, the catalase, NADP.H-oxidoreductases, and glutathione di-sulfide (GSSG)-reductase activities, and the GSH and GSSG levels do not differ markedly, and are essentially the same as in cells in vitro. AdrR tumor xenografts are extremely resistant to doxorubicin and retain most of the characteristics of the altered phenotype. These results suggest that WT and AdrR breast tumor xenografts provide a useful model for the study of bio-chemical and pharmacological mechanisms of drug resistance by solid tumors *in vivo* [393].

The doxorubicin-promoted intracellular generation of free radicals has been investigated using nitroxide spin label EPR spectroscopy. The effect of 50-500 mcg/ml of doxorubicin on human tumor cells (MCF-7, breast cancer cells, and HL-60, promyelocytic leukemia, cells) was studied by measuring TEMPO absorption intensity decay (TAID) at a TEMPO concentration of 10 mcM. Doxorubicin accelerates the TAID in both cell lines with a detection limit of 50 and 500 mcg/ml doxorubicin for MCF-7 and HL-60 cells, respec-tively. Preincubation of cells with the iron chelating agent, desferri-oxamine (5 mM), partially prevents the effect of doxorubicin on the TAID. Catalase and copper, zinc-superoxide dismutase do not affect much the effect of doxorubicin on the TAID in intact cells. However, Cu,Zn-SOD completely abolished the effect of doxorubicin on the TAID in a MCF-7 cell-free system. It is suggest that doxorubicin facilitates the intracellular generation of anion radicals $O_2^{-\bullet}$, with iron being involved in this process [394].

The doxorubicin-resistant human breast tumor cell line (MCF-7), also displays resistance to mitomycin C and its two analogues, BMY 25282 and BMY 25067 isolated from *Streptomyces caespitotus*. The mitomycin C molecule contains an urethane group and the aziridine ring. Using the EPR spectroscopy it has been shown that in the presence of mitomycin C and BMY 25282 the resistant MCF-7 cells form significantly less hydroxyl radicals than the sensitive cells. In the resistant cells the activity of glutathione peroxidase and glutathione S-transferase is considerably higher (14-fold and 44-fold, respectively). The formation of free radicals is consistent with the cytotoxicity in drug-sensitive cells [395].

The absence of estrogen receptors (ER) in human breast tumor cells is considered as a bad prognosis compared to patients with ER positive breast cancer. A drug-resistant human breast tumor cell line characterized by doxorubicin-resistance exhibits a noticeably increased expression of both glutathione transferase and selenium-dependent glutathione peroxidase. The EPR method has been applied for studying the relationship between the presence of estrogen receptors in these cells and the activity of glutathione peroxidase and catalase, their capacity to generate $HO^{\bullet}$ radicals, doxorubicin redox cycling and their sensitivity to the cytotoxic effects of doxorubicin and H_2O_2 oxidants. The results have revealed an inverse relation between the expression of glutathione peroxidase (but not catalase) and the presence of ER in these cell lines. The formation of $HO^{\bullet}$ radicals induced by doxorubicin treatment of the cells is inversely proportional to the activity of glutathione peroxidase in these cells and thus directly related to the presence of estrogen receptors [396].

An enzyme-activated irreversible inhibitor of ornithine decarbocylase (ODC), α-difluoromethylornithine (DFMO) made it possible to select two highly doxorubicin-resistant cell lines, K562-DFMOr and V79-DFMOr. In both cell lines the increased expression of ornithine decarbocylase is caused by gene amplification. The K562-DFMOr cells, but not the V79-DFMOı show an elevated level of ribonucleotide reductase subunit R_2 mRNA and an increased R_2 gene copy number. Analysis of their EPR spectra indicates an increased level of the R_2 proteins in the K562-DFMOr cells compared to the wild type K562 cells. The K562-DFMOr cell line is more stable for many passages and resistant not only to DMFO, but also to hydroxyurea, an inhibitor of ribonucleotide reductase and thus DNA replication [397].

The generation of daunomycin semiquinone (an antitumor antibiotics of the anthracyclin series, produced by *Actinomyces coeruleorubidus*) in intact erythrocytes under CO atmosphere has been studied by EPR spectroscopy. The undyalized hemolyzates and the spin broadening agent chromium oxalate quench the EPR signal. This suggests the EPR-detectable radicals to externally locate and slowly diffuse inside. The reductase on the outer side of the erythrocyte membrane is thought to be the source of daunomycin radicals [398].

The sodium ascorbate-induced cytotoxicity against human glioblastoma T98G cells in an RPMI1640 medium supplemented with fetal bovine serum or human serum samples has been investigated. Human serum significantly reduces the cytotoxic activity of sodium ascorbate. This effect is independent of the sex, age or the disease of the serum donor with or without heat-inactivation of the serum. EPR spectroscopy has revealed that the above serum effect is not simply caused by the alteration of the ascorbyl radical intensity. The sodium ascorbate activity is suggested to be considerably reduced by human plasma [399].

Stomacytosis induced in human erythrocytes by antitumoral drugs vinblastine and chloropromazine has been examined by EPR spectroscopy using spun label phosphotidylcholine and sphingomyelin. A fraction of labeled lipids has been determined on the external leaflet, on the cytosol face, or trapped in endocytic vacuoles. The two drugs produce a time- and dose-dependent stomacytic shape change in parallel with an inward shift of ~ 10% to 33% of vinblastine and chloropromazine outer leaflets [400].

The *ortho*-quinone forms of 2,3- and 3,4-catechol estrogens are implicated in the carcinogenicity of these hormones. The reduction of the ortho-quinone estrogens is accompanied by a concomitant production of reactive oxygen species as shown by EPR spectroscopy. The production of hydrogen peroxide, $HO^•$ radicals, and estrone 3,4-semiquinone in estrone 3,4-quinone treated human breast cancer subcellular fractions was demon-strated even in the absence of exogenously added catalysts. Subcellular fractions of MCF-7 cells treated with estrone 3,4-quinone and NADP.H, including nuclei, mitochondria, and microsoms, provide support to H_2O_2 production. It should be mentioned that the hydrogen peroxide production and the chromosomal DNA damage are abolished by the addition of catalase. Thus, the results demonstrate that free radicals are generated during the metabolism of estrone 3,4-quinone in human cells [401].

According to the EPR spectroscopy data, the nitroxide free radicals, TEMPO, commonly used to prevent oxidative damage, increase the intracellular H_2O_2 concentration. For a better understanding of this process, various human tumor cell lines were incubated with compounds preventing the generation of active oxygen metabolites. Sodium azide, inhibitor of the respiratory chain, the iron-chelating agent desferrioxamine, superoxide dismutase and catalase have no effect on the hydrogen peroxide concentration. Metyrapone, inhibitor of the cytochrome P_{450} system, decreases, but not completely prevents, the H_2O_2 production. N-ethyl-maleimide, a sulfhydryl-bond alkylating agent, is able to completely prevent the increased hydrogen peroxide production. It is concluded that, by increasing the cellular H_2O_2 concentration, TEMPO exerts a pro-oxidant effect. This increase in hydrogen peroxide production seems to be mediated by the induction of oxidase activity in the cytochrome P_{450} system [402].

The tumor necrosis factor has been monitored by EPR spectroscopy using the spin label TEMPO. In this study the rate of TEMPO spin decay by the TNF-sensitive L929 cells served as an indicator of TNF cytolytic response. The rate of loss of the TEMPO signal is sensitive to concentration as well as to time of incubation with TNF. The TEMPO spin decay first derivative plot correlates well with the corresponding TNF cytotoxicity curve as obtained by a standard tetrazolium dye reduction procedure. DMSO (500 mM) and α-tocopherol (1 mM) inhibit the TNF-induced decrease in the rate of loss of TEMPO decay, whereas superoxide dismutase (100 U/ml), catalase (1000 U/ml), and histidine (6 mM) have little effect on the TEMPO decay rate [403].

Cancer therapy studies using proton accelerators are underway in several major medical centers in Russia, the United States, Japan and elsewhere. To facilitate dosimetry intercomparisons between these laboratories, alanine-based detectors produced at the National Institute of Standards and Technology (USA) were studied at the Institute of Theoretical and Experimental Physics (Russia) for their possible use as passive transfer dosimeters for clinical proton beams. Measurements made in a number of experiments showed average agreement between the Institute of Theoretical and Experimental Physics and National Institute of Standards and Technology dosimetry standards to 2.5% [404].

The effect of oxygen, and thus hypoxia, on tumor cell radio-sensitivity has been revealed. The tumor tissue consists of heterogeneous cells having various oxygen content. Knowledge about the oxygenation of human tumors and its importance in the response to radiotherapy is crucial for the development of improved treatment methods for radiotherapy of oncologic diseases. The oxygen content of cells is measured by numerous methods such as polarography, the comet and alkaline elution assays, nitroimidazole binding assays, hemoglobin saturation assays, magnetic resonance spectroscopy, electron spin resonance spectroscopy, phosphorescence imaging, and an assay for tumor interstitial pressure [405].

The EPR spectra of India ink are very sensitive to the partial pressure of oxygen (pO_2). That is why this oxygen-sensitive paramagnetic material is widely used to measure pO_2 in tissues of patients. This potentially provides a means to individualize and guide treatment of diseases such as cancer and peripheral vascular insufficiency [406]. pO_2 *in vivo* including human samples have measured by EPR method [407].

Using EPR oximetry with the oxygen-sensitive paramagnetic material, fusinite, the partial pressure of oxygen (pO_2) in the mouse mammary adenocarcinoma MTG-B has been measured. The results show that the average pO_2 in untreated tumors is low (about 5 mm Hg) and further decreases with the tumor growth. The EPR method and histological examination were used to localize the position of the fusinite with respect to tumor margins and vascularization. The pO_2 is generally higher in the periphery than in the center of the tumors. But there is considerable variation among tumors both during normal growth and after radiation treatment. After a single 20 Gy dose, there was an initial reduction in pO_2 (minimum occurred 6 h postirradiation). This was followed by a transient increase in pO_2 to levels higher than the preirradiation pO_2 (maximum occurred 48 h postirradiation). Thus EPR can be used to identify both potentially radioresistant tumors and tumors with slow reoxygenation [408].

Semiquinone signals (SFS constants a_N = 0.14 и a_F = 0.135 мT) were detected in the EPR spectra of human large intestine carcinoma HN-29cells upon addition of 3,6-difluoro-2,5-bis(aziridinyl)-1,4-benzoquinone. This indicates the presence of redox sites in the plasma membranes [409]. The role of NADP.H quinone oxyreductase in aziridinylbenzoquinone metabo-lism under the above pathology condition has also been studied [410].

3.16 OPHTHALMIC DISEASES

In ophthalmology the EPR spectroscopy has found much use for the lens studies. The lens of the eye is an avascular tissue, with one side facing toward the outside of the vitreous body. The research in this area is related to the development and treatment of cataract. Special attention is given to free radicals that trigger the formation of cataract. Using the EPR spin method the anion radicals ($O_2^{-\bullet}$) produced by the hypoxanthine-xanthine oxidase system and the hydroxyl radicals ($HO^\bullet$) produced by the Fenton reaction have been measured. The inhibitory effect of compounds used as anticataract drugs (pirenoxine and ascorbic acid) on radical formation has been investigated [411].

EPR Spectroscopy of the lens enabled investigation of the activity of superoxide dismutase, that eliminates anion radicals $O_2^{-\bullet}$ and glutathione (GSH), which has a defensive action against cataract. SOD activity and GSH concentrations significantly decrease in human lenses with senile cataracts as the cataract advances. The SOD activity in human lenses is localized dominantly in the lens epithelium and also in the shallow layer of the cortex [412].

The EPR method and the protein spin labels 2,2,4,4,9-pentamethyl-1,2,3,4-tetrahydro-γ-carboline-3-oxyl and 4-(N-maleimido)-2,2,6,6-tetra-methylpiperidine-1-oxyl were employed for studying the fluidity and binding ability to protein functional groups in human lens membranes. The image and the stage of cataract as well as the ultrastructural characteristics of the lens fiber cell membranes were evaluated in parallel studies.

The EPR spectra of membrane structures of transparent lenses of different ages show that the sorption parameter of the carboline label on the surface of protein-lipid components is usually weakly expressed but increases with aging. At different stages of the lens opacification the carboline analog spin label increasingly binds to the lens membrane structures. A spin label signal can be enhanced by addition to the samples of a paramagnetic probe $K_3Fe(CN)_6$. The maleimide spin label binds to the protein SH-groups in cataractous lens membranes at a slower rate, than in transparent lenses. Both the age-related and cataractogenic changes in the lens matter are accompanied by deterio-ration of the lenticular fiber plasma membranes and by formation of coalescing globules with a diameter of 220-500 nm.

At the stage of cataract with advanced opacities, there is an increase in the number of the carboline label binding sites with the surface proteins and annular lipids of mem-branes. These lipids represent a mass of amorphous aggregates contributing to significant scatter of light. It appears, therefore, that the suggested EPR spin label technique can be effectively used to monitor the aggregation process of protein membrane components, which takes place during human cataractogenesis [413].

Peroxidative reactions occurring in cataractous lenses and in the surrounding environment the lipid peroxidation increases significantly and becomes increasingly oxidized with age. In human normal lenses the fiber structure changes with age. Generation of free radicals of ascorbic acid and HO inside the cataractous lens has been revealed by the EPR method using spin labeling.

The presence of iron and copper ions in the subepithelial region of the lens is related to the increased lipid peroxidation. The lowered level of phospholipids and the higher degree of saturation of fatty acids observed in senile cataractous lenses indicate that the membrane structure is damaged by lipid peroxidation. On the epithelium of normal lenses there are receptors for low density lipoproteins. Denatured lipoproteins are incorporated into senile cataractous lenses. This suggests that the membrane barrier function deteriorates in cataractous lenses. In diabetic cataractous lenses, the levels of very low density lipoproteins show a significant increase. In diabetic patients the environment surrounding of the lenses (blood, aqueous humor, and vitreous body) has decreased levels of reduced glutathione and superoxide scavenging activity, and increased levels of lipid peroxide and glycated proteins. This may be related to a reduction in the antioxidative potential in the environment surrounding the lens due to the enhanced glycation [414, 415].

The melanin level in human melanosomes in the pigment epithelium of the eye depending on age has been determined using EPR spectroscopy [416]. Melanin is shown to be the major retinal pigment responsible for ascorbate photosensitive oxidation [417].

In the human retinal pigment epithelium a fluorescent material is generated and accumulates within the lipofuscin granules with aging. These fluorophores seem to be able to make a possible light-induced contribution to age-related macular degeneration. Some of the fluorescent components of lipofuscin granule are the products of the reaction of retinaldehyde with ethanolamine. The synthetic mixtures of this reaction can serve as a useful model for photophysical studies.

Using the EPR method it has been demon-strated that UV-irradiation of either natural or synthetic human retinal lipofuscin extracts in oxygen-free methanol leads to the formation of DMPO spin trapped carbon-centered radicals. In the presence of oxygen, superoxide anion radicals are formed. Electron or hydrogen abstraction from the substrate by these fluorophores *in vivo* and the resulting radical products may contribute to the age-related decline of the retinal pigment epithelial function and blue light damage in the retina [418].

Retinal degeneration is mainly caused by deficiency of α-tocopherol (vitamin E), the major lipid-soluble antioxidant of retinal membranes. The α-tocopherol antioxidant function is realized via scavenging peroxyl radicals, which results in the formation of phenoxyl radicals of α-tocopherol. It is suggested, that α-tocopherol phenoxyl radicals can be reduced by endogenous reductants in the retina, providing for α-tocopherol recycling.

The results of the EPR study show that: endogenous ascorbate (vitamin C) is able to reduce the α-tocopherol phenoxyl radicals, formed upon UV-irradiation of retinal homogenates and in rod outer segments. In the absence of ascorbate, either endogenous or exogenously added glutathione (GSH) are unable to protect α-tocopherol against oxidation. Exogenous dihydrolipoic acid is inefficient in direct reduction of the α-tocopherol phenoxyl radicals, but it is able to enhance the protective effect of ascorbate by regenerating it from dehydroascorbate. Thus, the vitamin E antioxidant effectiveness in the retina can be enhanced due to regeneration of α-tocopherol from its phenoxyl radicals [419].

Hydrazine derivatives involved in oxidative metabolism are used in the treatment of a number of diseases. The EPR method shows, for example, phenylhydrazine irreversibly inactivates lacrimal gland peroxidase, thus significantly affecting human eye vision function [420].

During the last years a method of vision correction using soft, drug-saturated contact lenses has found many uses in ophthalmology. This allows the action of drugs to be significantly prolonged compared to that in traditional methods of treatment. *In vitro* and *in vivo* diffusion from soft contact lenses based on atropine-, pilacorpine-, fetasnol-, and clopheline poly-2-hydroxyethalmetacrylate has been studied by the EPR method using stable nitroxyl radicals. The enthalpies of drug-sorption in soft contact lenses, diffusion rate constants and other molecular-dynamic characteristics have been determined [421].

3.17 OTORHINOLARYNGOLOGIC DISEASES

3.17.1 Nasal Secretion Free Radicals

A change in the activity of enzymes catalyzing respiration and peroxidation processes presents a substantial risk factor in upper respiratory tract (URT) pathologies [422, 423]. In particular, a distinct manifestation of the above processes is oppression of mucociliary activity of respiratory tract mucous membranes in patients with bronchial asthma. Vitamin-antioxidant deficiency plays an important role among possible reasons for lipid peroxidation in patients with chronic pharyngitis of mycotic origin [424, 425]. However, peroxide radicals are short-lived species and can be detected by EPR spectroscopy only with difficulty [425]. Therefore investigations concerning free radical complexes in otorhinolaryngology are rather scarce. The applicability of EPR spectroscopy for measurements of free radical levels in order to assess the drug-sensitivity of larynx tumor has been demonstrated.

Nasal secretion free radicals were examined in 50 patients with nasal and accessory nasal sinus lesions. Main group consisted of patients with chronic subtrophic and catarrhal rhinitis (28 and 12, respectively). Of this group 10 patients displayed vasomotor neuro-vegetative rhinitis. Diagnosis was based on symptoms, anamnesis, and the data of examinations undertaken by conventional methods accepted in clinical rhinology. The results of EPR spectroscopy of nasal secretion patients with nasal mucous membrane pathology were compared to NS free radical levels of volunteers (31 donors without pathological changes either in the nasal cavity or in the nasal sinus). NS free radical content in patients [$(0.78 \pm 0.03) \times 10^{-6}$ M] was 21% lower than that in volunteers [$(0.96 \pm 0.03) \times 10^{-6}$ M; $P < 0.001$]. In both patients with nasal mucous membrane pathology and volunteers, free radical levels in nasal secretion are reliably higher than those in blood plasma (table 3-32) [426].

Table 3-32. Blood plasma and nasal secretion free radical levels in patients with nasal and accessory nasal sinus pathology and in volunteers (M ± m)

Groups of examinees	Free radical concentration (10^{-6} M)	
	Nasal secretion	Blood plasma
Patients	0.76 ± 0.03	0.55 ± 0.03
Volunteers	0.92 ± 0.04	0.56 ± 0.03

NS free radical content does not depend on the type of upper respiratory tract pathology and is always higher than in volunteers.

A higher NS free radical level is observed in patients with vasomotor rhinitis. Thus, in chronic catarrhal, subtrophic rhinitis and rhinosinusitis the NS free radical content is (0.78 ± 0.08) x 10^{-6}, (0.75 ± 0.05) 10^{-6} и (0.72 ± 0.13) x 10^{-6} M, respectively, whereas that in vasomotor rhinitis is (0.82 ± 0.05) x 10^{-6} M [426-428].

Thus, the NS free radical level is reliably decreased in nasal and accessory nasal sinus pathologies. This indicates that there is a dependence between the free radical activity of nasal discharge and the morpho-functional state of nasal mucous membrane. The higher NS free radical level in patients with vasomotor rhinitis seems to be due to clearly expressed changes in microcirculation that occurs in the nasal mucous membrane, and intensification of transsudative and exaudative processes. These phenomena are accompanied by an emhanced trasport from the blood flow of free radicals influencing the redox reaction dynamics in the nasal mucous membrane discharge [429].

Using EPR spectroscopy the nature of nasal mucous membrane discharge free radicals in pathological changes in the nasal cavity has been studied. The g-factor values of the NS free radicals signal were determined for the patient with these pathologies (table 3-33) [426].

Table 3-33. Nasal secretion free radical g-factor values depending on the state of nasal cavity mucous membrane (NMM)

g-Factor value range	Patients with NMM pathology		Volunteers	
	n	P ±m %	n	P ± m%
2.0024 - 2.0028	5	10.0 ± 12.7	16	69.6 ± 9.6
2.0030 - 2.0040	28	56.0 ± 7.0	7	30.4 ± 9.6
2.0040 - 2.0050	14	28.0 ± 6.3	-	-
2.0050 - 2.0060	3	6.0 ± 3.3	-	-
Total	50	100.0	23	100.0

As seen from Table 3-33, the g-factor values in patients with NMM pathological changes were recorded in most cases within the 2.0030-2.0040 and 2.0040-2.0050 ranges, whereas in volunteers the g-factor values were fixed within the 2.0024-2.0028 range. The average g-factor value of free radical EPR signal of nasal patients is higher (2.0038 ± 0.0001) than that of donors (2.0025 ± 0.0002; $P < 0.001$). These data show that in NMM pathological changes, free radicals related to the class of flavo- and benzosemiquinone appear in over 80% of patients (see Table 2-5).

It should be taken into consideration that various pathological processes occurring in the nasal cavity and the respiratory system, are commonly accompanied by noticeable shifts in the cellular composition of the respiratory tract mucous membrane discharge. In particular, it is believed that an increase in the number of NS cells strictly indicates a pathological process to occur in the respiratory tracts. The peculiar cellular composition of the respiratory tract mucous membrane discharge is closely associated with the disease development and has a diagnostic and prognostic value [430].

Naturally, the increase in the amount of NS desquamated cells is accompanied by amplification of the processes of membrane stabilization, accumulation of NS intracellular structures, enzymes and semiquinones. Ultimately, this cannot but influence both the direction of free radical reactions and spectral characteristics of NS free radical EPR signal. All this allows a conclusion that EPR study of NS free radicals can provide valuable information concerning the nasal mucous membrane morpho-fuctional condition.

3.17.2 Nasal Secretion Metalloproteins

As follows from Section 3.17.1, in the assessment of functional potential for the URT local protection mechanisms it should be taken into consideration that:

1. Paramagnetic centers related to iron- and copper-containing proteins are detected in nasal secretion.

2. Nasal secretion EPR signal with $g = 4.3$ corresponds to the F^{3+} ion seemingly bound to albumin.

3. The EPR signal at $g = 2.05$ is brought about by the Cu^{2+} ion bound to a protein, that is functionally identical to ceruloplasmin and possesses oxidase activity.

Quantitative and qualitative characteristics of signals at $g = 4.3$ and $g = 2.05$ in the nasal secretion EPR spectra of patients with pathological changes of nasal mucous membrane have been studied by EPR spectroscopy. The results were compared to those obtained from the examination of plasma EPR spectra of the same patients. Comparison was also made to the level of the corresponding EPR signals for nasal cavity mucous membrane discharge and blood plasma of healthy donors.

The intensity of EPR signal at g = 4.3 was very low in almost 24% of the patients and 38% of the donors. This impeded evaluation of the Fe^{3+} and Cu^{2+} concentrations in nasal mucous membrane discharge. Therefore only the EPR spectra with distinct signals at g = 4.3 were considered in studying changes in nasal secretion Fe^{3+} concentrations [431].

The nasal secretion Fe^{3+} content was much lower in various nasal cavity pathologic changes than in the absence of lesions of this kind. The Cu^{2+} level is increased in NMM discharge of patients with subtrophic NMM changes (reduced activity of both goblet cells and glands, vascular-tissue permeability alterations, disorders in fluid and soluble species transudation, *etc.*) [429]. On the contrary, the intensity of EPR signal at g = 2.05 in patients with chronic catarrhal and vasomotor rhinites is higher in spite of normal NMM excretory function., than that in volunteers (table 3-34) [426].

Table 3-34. Nasal secretion and plasma metalloprotein PMC content in patients with nasal cavity pathology and in healthy donors

Disease	PMC	n	Nasal secretion	Blood plasma
Chronic catarrhal	Fe^{3+} (mcm/l)	7	1.17 ± 0.31	10.5 ± 1.44
rhinite	Cu^{2+} (10^{-5} M)	7	1.55 ± 0.37	2.14 ± 0.097
Chronic subtrophic	Fe^{3+} (mcm/l)	14	1.39 ± 0.13	9.6 ± 1.02
rhinite	Cu^{2+} (10^{-5} M)	14	2.45 ± 0.46	2.26 ± 0.15
Vasomotor rhinite	Fe^{3+} (mcm/l)	7	0.91 ± 0.11	12.34 ±1.83
	Cu^{2+} (10^{-5} M)	7	1.84 ± 0.38	2.14 ± 0.12
Normal	Fe^{3+} (mcm/l)	15	2.43 ± 0.27	14.98 ± 1.36
	Cu^{2+} (10^{-5} M)	15	2.04 ± 0.38	1.85 ± 0.18

With a decrease in nasal mucus immunoglobulin (IgA) concentration and suppression of the local protection potential of upper respiratory tracts [430] on the background of abrupt suppression of ciliated epithelium functional activity, the increase in Cu^{2+} concentration and Cu^{2+}/Fe^{3+} ratio (antioxidant activity index (AAI) has a compensatory character and is aimed at supporting upper respiratory tract homeostasis. The importance of Cu^{2+}/Fe^{3+} ratio is also confirmed by a tendency towards retention of the dependence between NS Fe^{3+} and Cu^{2+} levels in patients with clearly expressed dystrophic alterations of the nasal mucous membrane ($r = -0.2$, $P > 0.05$).

Since no formation of metal-containing protein complexes occurs either in nasal mucous membrane or in its discharge, the complexes are suggested to penetrate into the nasal secretion from the blood plasma. Blood microcirculation is considered to be one of the most important processes maintaining the balance of proteins, including metalloproteins. Nasal secretion blood concentration is dependent on the intensity of nasal cavity exudation processes [432]. This allows a suggestion that one of the reasons for the disbalance in nasal secretion and blood plasma metalloprotein concentration is disturbance in the transportation of metalloproteins from blood flow, that is caused by morpho-functional alterations in the nasal mucous membrane state (along with iron and copper insufficiency, reduced synthesis and decreased transportation pool). If this suggestion were true, the EPR signal at g = 4.3 must have been detected in both the patients and donors. However, the nasal secretion Fe^{3+} EPR signal was fixed only in 76% of the patients and 62% of the donors. Both groups showed no relationship between the Fe^{3+} content in nasal secretion and blood plasma. (r = -0.07, r = + 0.33, respectively). This provides evidence for the existence of other, not mentioned above, Fe^{3+} level-regulating systems in the URT mucous membrane discharge [426].

Endogenous iron is present in tissues in the Fe^{2+} form that is supported by various reducing agents [433]. Therefore it is possible that one of reliable pathways to the formation of local Fe^{3+} level in the nasal secretion involves activation of the oxidases responsible for the generation of Fe^{3+} from Fe^{2+}. To support this hypothesis we have undertaken an EPR study of nasal secretion EPR spectra derived from patients with nasal cavity pathologies and from donors prior to and after nasal secretion discharge incubation with ascorbic acid. The results show that after the addition of ascorbic acid to the nasal secretion the patients exhibit no reliable Cu^{2+} level variations (table 3-35) [426, 434].

Table 3-35. Changes in nasal secretion paramagnetic center concentration in patients with URT pathologies and donors prior to and after incubation with ascorbic acid ($M \pm m$)

Groups of examinees	PMC	Nasal secretion PMC Concentration		
		n	Prior to incubation	After incubation
Normal	Fe^{3+} (mcm/l)	13	1.09 ± 0.31	1.2 ± 0.39
	Cu^{2+} (10^{-5} M)	13	1.01 ± 0.13	0.73 ± 0.08
Patients	Fe^{3+} (mcm/l)	11	0.96 ± 0.3	1.47 ± 0.3
	Cu^{2+} (10^{-5} M)	11	1.44 ± 0.2	0.96 ± 0.2

An analogous regulation of Cu^{2+} dynamics was observed in the EPR spectra of nasal secretion incubated with divalent iron (0.01 M $FeSO_4$) (table 3-36) [434].

Thus, nasal mucous membrane pathological changes are accompanied by a clear-cut suppression of the oxidase activity of nasal secretion copper-containing protein [434].

Table 3-36. Changes in nasal secretion paramagnetic center (PMC) concentration in patients with URT pathologies and donors prior to and after incubation with 0.01 M $FeSO_4$ solution ($M \pm m$)

Groups of examinees	PMC	Nasal secretion PMC Concentration		
		n	Prior to incubation	After incubation
Normal	Fe^{3+} (mcm/l)	10	0.96 ± 0.3	12.78 ± 2.1
	Cu^{2+} (10^{-5} M)	10	0.89 ± 0.09	0.57 ± 0.099
Patients	Fe^{3+} (mcm/l)	6	1.37 ± 0.6	8.2 ± 1.3
	Cu^{2+} (10^{-5} M)	6	1.13 ± 0.15	0.78 ± 0.099

URT pathological processes are characterized by significant microflora sowing of the mucous membrane [435, 436]. Lactoferrin protects the nasal and URT mucous membrane against microbes. The lactoferrin microbial resistance is due to Fe^{3+} binding, an essential co-factor of microbial cell metabolism and to enhanced xanthinoxidase-induced generation of anion radicals $O_2^{-\bullet}$. This, in turn, intensifies lipid peroxidation and superoxidedismutase activation [437]. Evidently by the same reason, Fe^{3+} ion is revealed 14% more often in the nasal secretion of patients with nasal cavity pathologies compared to donors.

All the data obtained indicate the following reasons for changing the NS Fe- and Co-containing protein concentrations in URT pathologies:

1. Enhancement of nasal mucous membrane permeability, that intensifies metalloprotein transcapillary metabolism.

2. Imbalance of local biochemical systems maintaining the optimal Fe^{3+} and Cu^{2+} level in the URT mucous membrane.

Under the conditions of constant interaction of the nasal mucous membrane with various, sometimes hazardous environmental factors occurring in upper respiratory tracts, the adaptation mechanisms of the nasal cavity are disturbed. This may be accompanied by an increase Cu^{2+} concentration and Cu^{2+}/Fe^{3+} ratio reflecting the NS antioxidant activity level. However, in clearly-displayed disorders of the nasal mucous membrane the activity of NS ceruloplasmin oxidase is decreased. This facilitates lipid peroxidation activity and adds to upper respiratory tract pathologies.

The alveolar EPR spectra of patients with chronic tonsillitis show Cu^{2+}, Fe^{3+} and free radical signals [438].

3.18 DISEASES OF TEETH AND MOUTH CAVITY

EPR spectroscopy is widely used in stomatology to elucidate the mechanisms governing the development of teeth and mouth cavity lesions [439]. Using the EPR method the sizes of root areas after tooth extraction were measured. This technique was also applied to measure 160 tooth root surfaces. The results suggest a correlation to exist between the EPR parameters, the size of wound area and the frequency of post-extraction inflammatory disorders [440]. The EPR method was used to measure the surface of 173 of right maxillary and mandibular teeth. This technique proved to be applicable for solving some problems in tooth descriptive anatomy [441].

New approaches to the solution of a number of problems in stomatology have been developed on the basis of EPR spectroscopy data. A microwave scanning ESR microscope has been developed to study the distribution of paramagnetic ions or radicals in various materials. This device consists of a cavity with an aperture over which a sample is moved. The EPR signal intensities at all positions are composed on a computer display as an image. A new TE_{111}-mode cylindrical cavity system was designed to obtain better sensitivity and resolution. The EPR images of γ-rayed human teeth were obtained to study the effect of secondary electrons from the adjacent metal on the defect formation in the tooth tissue [442].

3.19 RADIATION-INDUCED PATHOLOGY

Ionizing radiation brings about decay of tissue resulting in the formation of free radicals. The radiation dose absorbed by the victims of accidents can be determined using the EPR method by a free radical signal of bone, dentin and tooth enamel [443-454]. Bone tissue incorporates and retains many radio nuclides. However, bone tissue hardly can be used for dosimetry of the exposed population or in clinical practice owing to the necessity of biopsy and autopsy. Tooth enamel is more preferable for this purpose. Therefore determination of the dose accumulated by human on exposure to γ- or X-rays by a method based on tooth enamel EPR spectra has been widely accepted. This technique enables using tooth enamel as an individual natural dosimeter [454]. This is of special importance for establishing the doses delivered in atomic power station accidents and other radioactive emissions.

Tooth enamel EPR dosimetry is based on the determination of $CO_3^{2-\bullet}$ ion-radical concentration of hydroxyapatite, a tooth enamel mineral component. The radical concentration and consequently, the EPR signal intensity are proportional to the absorbed dose. EPR dosimetry is restricted by the presence in the tooth enamel of a natural background signal ($g_{\perp} = 2.0018$ and $g_{\parallel} = 1.997$) of unidentified origin. At least two factors may be responsible for this signal. One of these may be related to defects in the hydroxyapatite crystal structure. The other is likely to be caused by organic radicals whose concentration can be reduced by chemical deproteinization. At a radiation dose of below 200 mGy the radiation-induced EPR signal is overlapped by background signal. By this reason a broad signal in the EPR spectrum should be neglected [455-457].

Hydroxyapatite ($Ca_{10}(PO_4)_6(OH)_2$, major constituent of tooth enamel, is located in the crystalline structure, where 2-3% of PO_4 is substituted by CO_3^{2-} anions. After exposure to radiation free electrons became trapped by CO_3^{2-} to form paramagnetic ion-radicals $CO_3^{-\bullet}$, $CO_3^{3-\bullet}$ and $CO_2^{-\bullet}$ fixed by EPR spectroscopy [458-460]. These are reactive short-lived radicals. The CO_3^{3-} ion radical is accumulated for years and does not disappear from the crystalline structure [461-463]. The amount of radicals formed and, consequently, the EPR signal intensity, are proportional to the radiation dose [464]. The EPR spectrum is brought about by free radicals with maximum g = 2.0036 and minimum g = 1.9975 [465].

Radiation-induced ion-radicals $CO_3^{2-\bullet}$ in the tooth mineral matrix are stable (life-time $\tau \sim 10^9$ y) [455]. The dependence of $CO_3^{3-\bullet}$ signal intensity on radiation dose in the low dose region from 3.1×10^{-4} C/kg to 865.6×10^{-4} C/kg was determined. For this purpose the human tooth enamel samples (particle size 350-500 micrometers) were irradiated with ^{60}Co gamma rays. The EPR spectra of samples show the appearance of ion-radical $CO_3^{3-\bullet}$ signal. There was observed a linear dependence between the true intensity of the $CO_3^{3-\bullet}$ signal and the exposure dose from 25.0×10^{-4} C/kg to 865.6×10^{-4} C/kg. Dosimetry is impeded at exposure dose of less than 25.0×10^{-4} C/kg due to the effect of organic signal and noise [466, 467].

Also the dependence of $CO_3^{3-\bullet}$ signal intensity on the tooth enamel weight was determined. There was observed a linear dependence between the $CO_3^{3-\bullet}$ signal intensity and the enamel weight up to 300 mg at the same radiation dose [466, 467].

The EPR method enables measurements of large absorbed doses (over 4 Gy). For low doses (1-2 Gy), most common in practice, EPR dosimetry is complicated by overlapping of a non-radiation (background) broad signal (g = 2.005) arisen from a tooth enamel organic component, which masks the dosimetrically important signal. In addition, the analysis is hampered by the low signal-to noise ratio due to a small quantity of tooth enamel to be examined (especially with children). To overcome these obstacles a method of selective super high frequency saturation of EPR signals has been proposed (40). The method is based on the dependence of EPR components of the tooth enamel spectrum on the microwave (mw) power. Differentiation between the background and radiation induced EPR signals is effected by subtraction of the EPR spectrum recorded at low mw power from that recorded at higher mw power. This allows determination of a radiation dose of about 100 mGy [468, 469]. However, the advantages of this method are reduced to minimum by the increasing signal-to-noise ratio and, correspondingly, the decreasing usable signal. Therefore, a search for optimal conditions for tooth enamel EPR spectra registration, which enable efficient detection of the radiation signal and improve signal-to-noise ratio seems to be of great practical value [470].

Temperature variation in recording free radicals significantly enhances the EPR method sensitivity and selectivity [471]. $CO_3^{3-\bullet}$ ion-radicals of human tooth enamel have been studied at various temperatures between 4°K and 350°K and at various microwave powers in the 9 GHz band. Signal saturation during cooling due to slowing down of the electron spin relaxation has been established. The 35 GHz (Q) band spectra of tooth enamel, irra-diated with various gamma-ray doses, are also presented. Q band ESR dosi-metry offers some advantages over the X-band dosimetry. The signal is enhanced due to the spin level population difference and especially due to a better filling factor [472].

In order to pick up a radiation-induced signal in the EPR spectrum the following is proposed:

to measure the signal intensity by its high-field, where the effect of background component is minimal;

to subtract from the experimental spectrum the EPR signal of nonirradiated enamel, namely, that of milk tooth;

to subtract from the experimental EPR spectrum the Lorenz line that approximates the background signal, thus adjusting the line position, width and amplitude by the low-field spectral region, where only the background field is present.

All this significantly complicates the EPR dosimetry technique and introduces additional errors in human absorbed dose determinations. This is clearly expressed at low doses when the radiation-induced signal is poorly seen on the background of "native" signal, that masks the radiation-induced signal at a dose of 20 Gy. EPR dosimetry based on fast scanning spectral registration is rather promising [473]. Applicability of this method has been demonstrated in tests with animals and humans [474, 475].

To estimate the individual dose of Nagasaki citizens exposed to γ-ray irradiation EPR measurements of the teeth of A-bomb survivors have been carried out. The results of EPR dosimetry show the difference in doses absorbed by individuals exposed to unshielded and shielded irradiation [476]. The EPR spectra of 100 enamel samples isolated from the extracted teeth of A-bomb survivors were compared the chromosome aberration frequency in lymphocytes of the tooth donors, and with the doses physically estimated by the Term Dosimetry System 1986 (DS86 dose) which ranged from 0 to 3.5 Gy. The results show that the ESR data closely correlate with the chromosome aberration frequency in lymphocytes [464]. Similar data were obtained during examination of Hiroshima A-bomb survivors [477, 478].

Tooth enamel samples from workers exposed to radiation at the Russian nuclear facility "Mayak" in the South Ural have been examined using EPR spectroscopy. The individual doses obtained approach 3.89 Gy [479-484]. Red bone marrow of victims exposed to radiation in the early 50ties was studied in the same manner. The individual absorbed dose was 0.24-1.6 Gy [485].

After the accident at the Chernobyl Atomic Power Station there was an abrupt increase in the possibility of negative effect of low doses on the hemopoietic system and the adaptation in children constantly living in contaminated areas. Although the role of ionizing radiation as an etiologic factor in the development of oncologic diseases is commonly accepted, there is no reliable pathogenetic concept on the effect of low doses on hemoblastosis development (acute and chronic leukemia, lymphoma) in children. Available statistical data concerning the changes in children onco-pathology structures in the postchernobyl period indicate that low doses do induce hemoblastosis growth.

Absorbed dose measurements basing on tooth enamel EPR spectra provide a reliable method comprehensive dosimetry in cases of total examination of exposed population.

EPR dosimetry of 36 enamel samples derived from children teeth (at autopsy as well) has been carried out at the Belorus Republic Scientific-Practical Center for Children Oncology and Hematology. Of these children 14 suffered from hemoblastosis. The control group consisted of 12 children with various hematological diseases having no relation to hemoblastosis. Absorbed doses were also measured in 10 healthy children living in Minsk, capital of Belorus. In the group of patients with hemoblastosis an absorbed dose ranging from 15 to 30 cGy was observed in 5 children from the Brest and Mogilev Regions (Belorus), the areas subjected to radioactive contamination. In 5 children from Minsk City and the Mogilev Region, considered to be relatively unaffected territories, the absorbed dose was from 0 to 1 cGy. In the group of children underwent γ-therapy as a prophylactic measure against neuroleukemia, the absorbed dose varied in the 30-250 cGy range. In the two control groups the absorbed dose was nearly zero (0-7 cGy for the group involving sick children and 0-5 cGy for that of healthy child-ren. The error of dose determination is 5-9 cGy depending on enamel mass. All the studied performed indicate the high sensitivity of EPR dosimetry, that provides reliable information on the level of radioactive irradiation doses in children. These doses may serve as an etiological factor of acute leukoses in children in the Belarus areas in the post-accidental period [486].

Other tooth tissues can also be the objects of dosimetry. The EPR spectra of irradiated dentin have been studied over a range of radiation energies from 50 kVp to 26 MVp. The behavior of EPR signal is similar to that of enamel signal. However, its magnitude is to 10 times smaller [448, 455].

The feasibility of using dentine from surgically extracted human teeth in EPR dosimetry *in vivo* has been investigated. The organic fraction of human dentine was removed by extraction with diethylenetriamine. The dentin granules 75 to 250 microns in size were exposed to gamma radiation at a dose ranging from 50 mGy to 8 Gy. In the post-irradiation PMR spectra the signals with g = 2.0018, line width = 0.903 mT; and g = 1.9961, line width = 0.444 mT were detected. These signals have been assigned to hydroxyapatite of bone and enamel. Several other signals were also seen but not characterized. It was concluded that doses of 500 mGy or less may be resolved with prior removal of the organic component of dentin [487].

Tooth enamel paramagnetic centers are detectable on exposure to both X- and γ-rays [474, 488] and mechanical action [489-491]. PMC are also revealed on exposure to UV light at $\lambda < 335$ nm, the wave-length range of 280-320 nm being most efficient. The PMC thermal stability and EPR spectrum are coincident with those of the ion-radicals $O_3^{3-\bullet}$ generated on exposure to γ-rays. The long wavelength tail of their creation spectrum intersects with the short wavelength tail of the solar spectrum reaching Earth's surface. All this suggests a possible effect of solar irradiation on tooth enamel EPR dosimetry [465].

TM_{110} mode and TE_{111} mode EPR microwave cavities have been developed for *in vivo* human tooth dosimetry. The EPR signal from a sample is measured by using a microwave leak from the pin hole on the cavity wall. The minimum detectable dose of 2 Gy is obtained by the signal accumulation for 100 sweeps [492, 493].

The EPR method can be applied to determine the absorbed dose not only in tooth enamel, but in bone as well. Bone is an integrating dosimeter which records the radiation history of the skeleton. During irradiation, electrons become trapped in the crystalline component of bone mineral (hydroxyapatite). The traps are very stable; at room temperature, it takes a half-life of many years to empty the traps. EPR spectra of synthetic hydroxyapatite, irradiated with ^{60}Co, were obtained at room temperature. The radiation-induced signal, with a g-value of 2.001 ± 0.001, increases linearly with the absorbed dose from 3 to 200 Gy. In contrast to pure hydroxyapatite, the EPR spectra of excised human bone show a broad "native' signal, arisen from the organic component of bone, which masks the dosimetrically important signal. This native signal is highly variable from sample to sample and precludes the use of EPR in dosimetry. However, after subtraction of the background signal from the EPR spectrum, irradiated human bone shows a linear response similar to that of synthetic hydroxyapatite. This indicates that bone can be successfully used for an *in vivo* accurate estimation of the absorbed radiation dose [494].

The feasibility of this method was checked with three victims of radiation accident at a plant in San-Salvador (South America). The absorbed dose estimated by a signal in the hydroxyapatite EPR spectrum of excised bone was about 79 Gy [495].

Thus, quantification of the absorbed dose will make it possible to establish the cause and level of radiation detriment and important biological outcomes. Most epidemiological studies of unanticipated radiation exposure fail to establish the dose for the exposed population. No one biodosimetric technique (biophysical or biological) meets all the requirements of an ideal dosimeter and thus qualify as a "gold standard". A new technique has been developed combining new results with previously published data in order to evaluate the absorbed radiation dose and first applied to the victims of three radiation accidents. The first subject was a fireman involved in the Chernobyl nuclear accident; the second was the victim of an unspecified occupational accident; and the third was exposed to a ^{60}Co sterilization source. The absorbed dose value and distribution were estimated using EPR spectroscopy, biological dosimetry and computer modeling. The use of EPR spectroscopy allows determination of the content of radiation-produced free electrons in tooth enamel and clothes of the victims. The absorbed dose was measured by a biological method employing the chromosome aberration in blood lymphocytes (cytogenetically) and by kinetics of post-radiation granulocyte and lymphocyte production. The above biodosimetric techniques show a good agreement in estimating the absorbed dose for the three accident victims [455].

The EPR method has been employed for determining the ionizing radiation absorbed by other tissues such as human nails, hair [496] and aminoacides (alanine) [497]. At initial doses between 50 and 634 Gy. The radiation center in hair is characterized by a single decay constant with $k = 4.86^{9}$ mcs^{-1} for brown and yellow hair and $k = 1.45^{10}$ mcs^{-1} for red hair. It is suggested that human hair has the potential to be used as a dosimeter for accidental exposure to ionizing radiation if measurements are carried out within initial 24 h [498].

In 1991 the International Program on the Health Effects of the Chernobyl Accident (IPHECA) was established by the World Health Organization (WHO). At present, the technical part of the Program consists of five projects related to the following areas of priority health problems: thyroid, hematology, brain damage in uterus, oral health and epidemiological registry. Important findings are:

1. a significant increase of thyroid cancer in children in Belarus and Ukraine since 1989, and in Russia since 1992. No clearly displayed relation-

ship between detected thyroid cancers and radiation exposure is has been established. Therefore the data obtained are to be refined;

2. no increase yet in the incidence of hemoblastoses in the three States;

3. no relationship established between mental retardation and radiation exposure in uterus in 4500 children investigated.

The importance of dosimetry and biological indicators of radiation damage has been recognized by IPHECA. Several methods of biological and physical dosimetry are being employed using instrumentation provided by IPHECA. Some preliminary results indicate:

1. unstable chromosomal aberrations can indicate an integral exposure;

2. a higher ratio of total aberrations in patients with hematological diseases compared to healthy persons in contaminated areas;

3. according to EPR data, the individual distribution of doses approaches a log-normal one, especially for adults, and a peak shift towards higher doses is noticeable for children [499, 500].

EPR spectroscopy, although not universally applicable, currently provides the most specific method for the detection of irradiated food [501]. Most of radiation-produced free radicals rapidly decompose, but some of them live for a long time and can be readily detected and characterized by EPR spectroscopy. A survey over recent international developments to detect the irradiation treatment of foods is given, in particular the programs on monitoring the radiation effect on food products. This is needed in order to check compliance with the existing regulations, such as the enforcement of labeling and control of prohibition, to enhance consumer confidence in the correct application of radiation processing, and to protect consumers' freedom of choice between irradiated or unirradiated food products. Some larger collaborative studies on an international scale have already taken place, e.g., ESR measurements of the absorbed doses in bones from chicken, pork, beef, frog legs and fish, vegetable and spices after γ-ray treatment [502].

After irradiation of dried fruits three different types of signals are observed in the EPR spectra. In most cases they correspond to those of sugar radicals. The spectra of irradiated sucrose-rich fruits are very similar to that of pure sucrose. The structure of the ESR spectra can change with storage. These changes are suggested to be due to radical rearrangement reactions in the samples [503].

The EPR method enables the identification of irradiated fresh fruits, seeds, grain, pips, kernels or stones. Their irradiation specific spectra were assigned to cellulose radicals. The intensity of the irradiation specific signals is linear up to doses of 11 kGy. The lifetime of the cellulose radicals (at room temperature and at 77°K) is long enough comparable to the storage time of fresh fruits. It is concluded that the EPR spectroscopy provides a routine method well suited for the identification of irradiated fresh fruits [504]. The radicals of seeds and dried spices are indistinguishable from endogenous radicals and can be destroyed over a period of days or weeks [501].

Ionizing radiation can be used to control microbial infestation of foodstuffs, inhibit sprouting, delay ripening and reduce the dangers from food-poisoning bacteria [501].

Three types of free radicals detectable by the EPR method produced by gamma-irradiation of ceftazidime at 4.2-295°K. The first radical consists of seven lines in the ESR spectrum and decays at 230°K. The signal is assigned to a $^{\bullet}C(CH_3)_2COOH$ radical. The second one shows triplet lines and decays at 293°K; it is assigned to iminoxyl radicals (>C=N-O$^{\bullet}$). The third one shows a broad singlet line and survives even at 295°K. The assignment of this spectrum is discussed. It is concluded that the O-C bond of the gamma-irradiated ceftazidime is ruptured to produce iminoxyl radicals and $^{\bullet}C(CH_3)_2COOH$ radicals. The yields of free radicals increase linearly with increasing the dose of gamma-irradiation. Half of the radicals that show a broad singlet line survive at 295°K upon storage of the irradiated ceftazidime for 159 days. It is concluded that the ESR measurement of free radicals gives a good method to distinguish irradiated antibiotics from non-irradiated ones. The EPR method also allows the estimation of the absorbed irradiation dose [505]. Analogous studies were carried out antibiotics ampicillin. Under gamma-irradiation at 295°K ampicillin shows an EPR doublet that survives over a period of 140 days [506].

Autologous transfusion of photomodified (UV-irradiated) blood provides a safe method of human adaptive-protective system stimulation. For a better understanding of the autohemotherapeutic mechanism it was necessary first to elucidate the nature of primary products formed in blood exposed to UV-irradiation. It has been found that short-lived free radicals are produced in photomodified blood. The radicals proved to be stable in the blood irradiated with UV light at a liquid nitrogen temperature (77°K). A temperature increase makes it possible to examine the behavior of radicals in time and their transformations induced by dark reactions or ultra-violet irradiation. Photoinduced paramagnetic centers in human erythrocytes, serum and plasma have been studied by EPR spectroscopy at 77°K. Human blood (volunteers of 25-40 years of age) was stabilized by sodium nitrate and irradiated with a non-filtered mercury-discharge lamp in a quartz Dewar vessel. In the course of blood irradiation, radicals were generated as a result of hydrogen abstraction from α-C atoms of the protein polypeptide chain (protein signal). Radicals of peroxide type were also generated [507].

3.20 HYPERTHERMIA IN PATHOLOGIC STATES

The EPR method extends our knowledge of regulations governing *in vivo* free radical processes conditioned by adaptive rearrangement and environmental effects. At present, the effect of hyperthermia on the organism and its cells is under extensive investigation.

The effect of heat shock on the internal microviscosity of human erythrocytes has been studied by EPR spectroscopy using nitroxyl spin probe, 4-amino-2,2,6,6-tetramethylpiperidinyloxy. Heat causes permanent damage of cells at the expense of an increase in intracellular microviscosity [508]. The effect of mitomycin-C concentration and temperature on erythrocyte membrane dynamic properties was also studied using 5- and 16-doxylstearic acid as spin labels. The results show that the temperature influences much spectral parameters, thus providing evidence for structural changes [509].

Heat-induced plasma membrane modifications such as surface blebbing, have been considered as crucial factors in cell injury and death. Different effects of free radical-induced cell damage by quinone menadione (2-methyl-1,4-naphthoquinone) and by hyperthermic shock (45°C) on the erythroleukemic cell line K562 have been investigated by the EPR method. Menadione induces the formation of surface blebs, accompanied by a rearrangement of the microfilament system and changes in the distribution of plasma membrane proteins. In contrast, heat-shocked cells show neither blebbing nor important cytoskeletal changes. Thus, an increase in membrane order not specifically related to the type of free radical-induced stress. The two different pathways induce cell injury by modifying the plasma membrane integrity and function. However, one involves cytoskeleton-dependent surface blebbing, whereas the other does not [510].

Hyperthermia is employed for malignant tumor treatment, especially in combination with chemo- or radiotherapy. For the adequate treatment and for the patients' safety the tumor tissue temperature should be maintained at 42-43°C over a period of 39-60 min, the surrounding tissue temperature being much lower. Monitoring of tumoral and surrounding tissues is a prerequisite for the method of hyperthermal therapy. The EPR method is suitable for *in vivo* temperature measurements basing on the temperature response of nitroxide stable free radicals. The response has been substantially enhanced by encapsulating the nitroxide in a medium of a fatty acid mixture inside the protein microsphere (3 mc in diameter). The mixture undergoes a phase transition in the required temperature range. The shape of the EPR spectrum is dramatically altered by the phase change, providing a highly temperature sensitive signal. The use of nitroxide dissolved in a cholesterol and a long-chain fatty acid ester, has resulted in a mixture which provides a peakheight ratio change from 3.32 to 2.11, with a standard deviation of 0.04, for a temperature change from 38°C to 48°C, that is typical in biological and medical applications. The average temperature resolution of this experimental system is 0.2°C. Therefore, this method is applicable to *in vivo* studies where the microspheres could be injected into the microvasculature having a minimum vessel diameter of the order of 8 microns. This temperature measuring method has various potential clinical applications, especially in monitoring and optimizing the treatment of cancer with hyperthermia [511].

However, not all clinicians lend credence to diagnosis of malignant hyperthermia susceptibility. The spin labeled EPR method has been employed to examine 19 donors, 10 malignant hyperthermia susceptible patients and 9 malignant hyperthermia unsusceptible ones. No evidence that this method could be used to diagnose malignant hyperthermia susceptibility has been obtained. Furthermore, there was no significant difference in the fluidity of the red blood cell membranes between the two groups. This would have been indicative of a generalized membrane abnormality in malignant hyperthermia. Thus, in this case the EPR method proved to be unsuitable to diagnosis of malignant hyperthermia susceptibility [512].

The EPR method has been used in examining the blood and blood components free radical level in workers of the Irkutsk plant of heavy machinery (Russia), occupationally exposed to high temperatures and respirable dust. The results show that the whole blood and erythrocyte free radical level was significantly higher in steel makers compared to that in machine-tool operators and donors. It has been found that complex biochemical changes leading to glutathionereductase system activation occur in foundry workers [513].

3.21 HYPERBARIC OXYGENATION

Hyperbaric oxygenation is applied in treating a number of diseases. The introduction of this method into clinical practice has initiated investigation of mechanistic aspects of hyperbaric oxygenation.

Using EPR spectroscopy the content of paramagnetic centers of rat myocardium and the dependence of this value upon hyperbaric oxygenation have been examined. In the spectrum of normal myocardium tissue there was an EPR absorption at $g = 4.3$, 2.00 and 1.94. In acute coronary insufficiency new signals at $g = 2.03$, 2.10, 2.7-2.8, 2.8-3.2 were fixed in addition to the above mentioned [367].

Hyperbaric oxygenation does not produce significant effect on the intensity of EPR signals arisen from the infarction zone tissues, which is due to the absence blood circulation in this zone [514]. It has been suggested that elevated oxygen concentrations lead to an increase in oxygen radical production, which may account for the toxic effects of excessive exposure to oxygen. Examination of blood from patients undergoing hyperbaric oxygen exposure has been performed by EPR spectroscopy. The results show a marked increase in the magnitude of a signal corresponding to a free radical (g = 2.006). The signal diminishes to zero level within 10 minutes of cessation of hyperbaric oxygenation exposure. It is also shown that the oxygen-generated free radical centers are localized in erythrocytes and give rise to a signal of ascorbic acid radicals in the EPR spectrum [515].

In biological systems most polyvalent metals are bound by specific proteins, This hinders metal-dependent catalysis of free radical reactions. Nevertheless, *in vitro* experiments show that reactive oxygen compounds release transition metals from the state of binding to proteins. In this process ascorbic acid acts as a prooxidant. Evidently, this can manifest itself in some pathologic states. However, any unfavorable effect of the addition of ascorbic acid has not been proved. The role of ascorbic acid in the process of polyvalent metal mobilization is not clearly understood yet.

Plasma and leukocyte ascorbic acid concentrations are frequently decreased in cases of hematological diseases associated with ferric overloading. This indicates an increase in ascorbic acid utilization. Ascorbic acid an unfavorably effects on ferric overloaded patients. When the blood Fe concentration exceeds the binding capacity of specific proteins, excess iron may be involved in free radical reactions, which get strengthened in the presence of ascorbic acid. It appears that, being an oxidant, ascorbic acid promotes the formation of $HO^{\bullet}$ radicals in Fenton's reaction.

Blood plasma is an important site for ascorbic acid – metal interactions. Vitamin C is an antioxidant protecting human against oxygen radicals. However, the reaction with iron enhances ascorbic acid utilization thus promoting plasma oxidative damage.

An EPR study has been made of the interaction of plasma ascorbic acid and ascorbic acid free radical (AFR) with exogenously added iron. An ascorbic acid free radical was found in the plasma of all donors and was unaffected by superoxide dismutase, catalase and a strong iron chelator desferrioxamine. These findings and the rapid decrease in the AFR content under a nitrogen atmosphere suggest that plasma AFR is probably the result of air auto-oxidation. When the iron loading of plasma exceeds the plasma latent iron-binding capacity, the intensity of the AFR signals increases to about 10 times the normal level and then decreases rapidly to undetectable levels after 15-20 min. The characteristic behavior of plasma AFR after iron loading is due to its specific iron-binding capacity and to plasma ferroxidase activity. The ferroxidase activity of plasma is important to promote the $Fe^{2+} \rightarrow$ transferrin transfer without a transient ascorbate oxidation. The EPR study using spin trap 5,5-dimethyl-1-pyrroline N-oxide and N-*t*-butyl-α-phenyl-nitrone shows that iron-overloaded plasma is unable to produce spin trap adducts even in the presence of 50-300 mcM-hydrogen peroxide or 100 mcM-azide. The formation of $HO^{\bullet}$ radicals occurs only after the addition of iron-binding EDTA. Therefore, if ascorbic acid indeed has a free radical-mediated pro-oxidant role, it is not detectable in plasma by spin trapping experiments [516].

3.22 CONDITIONS UNDER NARCOSIS

Narcosis is a pharmacologically or electrically induced deep sleep (reversible suppression of central nervous system cells). The narcotic state is associated with loss of consciousness, anesthesia, relaxation of skeletal musculature, and suppression of reflex activity. There are many theories explaining the processes occurring under narcosis. The lipid theory, for instance, points to the interaction between the anesthetic agent and intracellular lipids. However, the postulated regularities are true only for certain anesthetic agents such as acyclic hydrocarbons and inert gases. This theory is not expanding to other narcotic agents such as barbiturates, steroid anesthetics, etc. Therefore the lipid theory is far from universal. Using EPR spectroscopy it has been found that chlorpromazine and pipolfene, which potentiate general anesthesia, increase phospholipid membrane viscosity in the region of polar headgroups and in the hydrophobic area near glycerol skeleton. These results (independently of potentiating mechanism) confuse the mechanism of anesthetic action with the increased fluidity of phospholipid membranes [517].

The protein theory of narcosis is explained in terms of interaction of some anesthetic agents (chloroform, ethylene, cyclopropane) with cellular proteins. The EPR method using a spin label paramagnetic maleimide derivative (2,2,6,6-tetramethylpiperidin-1-oxyl-4-maleimide) was employed to investigate the effect of thiopental on the Na^+/K^+ ATPase activity. The results show that a Na^+/K^+-ATPase membranous preparation obtained from placental tissue is strongly inhibited by thiopental [518].

Narcotics-induced biochemical alterations occurring in the host organism have been studied by EPR spectroscopy. Using DMPO as a spin trap, plasma superoxide dismutase like activity was measured in 48 surgical patients (ranged from 10 to 74 years in age), who underwent various elective surgery. There was a marked reduction in plasma SOD like activity during surgery independently of either anesthetic agents used or the extent of surgery. Surgical duration over two hours was the most contributing factor to reduce the SOD like activity. There is no correlation between plasma SOD like activity and age or serum total protein [519, 520].

3.23 PATHOLOGIC STATES CAUSED BY INTOXICATION

Throughout the lifetime one often or constantly comes up against poisoning substances present in the environment and in food products. Heavy metals and compounds thereof rank among toxic substances of this type.

The influence of heavy metal ions on human erythrocyte membranes has been studied by EPR spectroscopy using spin labels methyl 5-doxylpalmitate or methyl 12-doxylsterate. It has been found that copper and mercury ions increase the rigidity of membrane lipid bilayers. The use of 4-maleimido-2,2,6,6-tetramethylpiperidine-1-oxyl as a spin label has revealed that the two metals also induce conformational changes of membrane proteins and modify cell internal peptides and proteins. Thus, it is concluded that copper and mercury ions decrease the internal viscosity of erythrocyte membranes [521].

Free radicals are involved in the mechanism of toxic effect of many chemicals. In some cases, the organic radicals are involved, but often oxygen radicals. Free radicals are usually very reactive, which makes them difficult to detect. The EPR techniques are frequently used to fix free radicals in a more stable form [522, 523].

To clarify the toxicological mechanism of action of paraquat and diquat, the level of ascorbate radical was examined using the ESR method. The ascorbate radical level in serum from humans acutely intoxicated with a 1:1 mixture of paraquat and diquat was increased to twice the normal level. The ascorbate radical in both intoxicated human serum and rat plasma decays much faster than that in normal samples [524].

Methoxyacetic acid, a teratogenic toxin, is the major metabolite of 2-methoxyethanol, 2-ME) which causes a wide range of toxic effects. In order to gain insight into the mechanism of methoxyacetic acid toxicity the use was made of the EPR technique in conjunction with spin labels specific for cytoskeletal proteins, bilayer lipids, cell-surface sialic acid, or cell-surface galactose and N-acetylgalactosamine residues of human erythrocyte membranes. The results show that methoxyacetic acid significantly increases the protein-protein interactions of skeletal proteins in a dose-dependent manner, while 2-ME has no effect at even a 2.5-fold higher concentration. Addition of methoxyacetic acid leads to a significant increase in the rotational motion of spin labeled terminal galactose and N-acetylgalactos-amine residues (2.0 mM of methoxyacetic acid brings about a 38% decrease in the apparent rotational correlation time). The rotational motion of spin labeled sialic acid, 70% of which is on the major transmembrane sialo-glycoprotein (glycophorin A or PAS 1), is not affected by methoxyacetic acid treatment. Methoxyacetic acid has no effect on the lipid bilayer fluidity, since no change in the motion of a lipid bilayer specific spin label (5-NS) in the erythrocyte membrane was observed. These results suggest that methoxyacetic acid may lead to teratologic toxicity by interacting not with lipid components but with specific protein components, such as transport proteins, cytoskeletal proteins or neurotransmitter receptors [525].

The azo initiator of peroxyl radicals 2,2'-azobis (2-amidinopropane) dihydrochloride (AAPH) induces oxidative hemolysis in human erythrocytes and subsequent hemoglobin oxidation. Basing on the degree of hemolysis versus time as an indication of the oxidative damage, it has been found that erythrocytes are protected against oxidative hemolysis by: (i) both reduced and oxidized α-lipoic acid; (ii) simultaneous treatment with ascorbate and dihydrolipoate or α-lipoate; (iii) glutathione in combination with dihydrolipoic acid or α-lipoic acid.

As shown by the EPR method, the spin trapping reagent DMPO forms an adduct with the peroxyl/alkoxyl radicals produced by thermal decomposition of 2,2'-azobis (2-amidinopropane) dihydrochloride in the presence of oxygen. The formation of this adduct is prevented by reduced or oxidized lipoic acid, reduced glutathione or ascorbate. This means that 2,2'-azobis (2-amidinopropane) dihydrochloride peroxyl radicals progressively damage the erythrocytes and that the released hemoglobin is subsequently oxidized to methemoglobin which might further enhance the oxidative damage. The protective effect of antioxidants is exerted outside the cells by directly scavenging AAPH-alkoxyl radicals [526].

Free radicals are formed during xenobiotics metabolism by enzymes such as cytochrome P_{450} or peroxidases. In some cases, this occurs in the redox cycle using cytochrome P_{450} reductase, which can catalyze one-electron reduction. As shown by the EPR technique, molecular oxygen can be reduced in a one electron pathway to generate anion radicals $O_2^{-\bullet}$ against which superoxide dismutase is protective. However, in the presence of iron, superoxide anion radical can generate a very reactive hydroxyl radical. Iron is therefore normally tightly controlled by transport and storage proteins. Chemicals that can release iron from these proteins can be very toxic, causing lipid, protein, and nucleic acid oxidation. The oxidation of these species is generally protected by a complex antioxidant system involving glutathione and glutathione peroxidase, vitamin E and ascorbic acid [527].

Oxidative stress is thought to be involved in the pathophysiology of several diseases and aging. Therefore reliable methods of measuring oxidative stress present one of the greatest needs in clinical diagnosis. However, standardized methods for measuring oxidative stress in humans are not yet established. The EPR technique seems to be most promising in this respect [528, 529].

Using EPR spectroscopy blood and blood components in patients with opium intoxication were examined. After hemosorption an increase in both erythrocyte free radical level ceruloplasmin/transferrin ratio is observed. This seems to be defense mechanism against opium toxic effect [530].

3.24 SEPSIS-INDUCED STATES

Sepsis is a general non-specific infectious disease of acyclic type, developed as a result of disturbed reactivity of the host organism. In the sepsis-induced state, various microorganism and toxins are brought into the blood flow from the local infectious focus.

Xanthine oxidase activity, free radical concentrations, and lipid peroxidation were examined in patients with sepsis syndrome compared with 10 non-infected critically ill patients. Xanthine oxidase activity and lipid peroxidation were increased in septic patients compared with both volunteers and non-infected patients. The highest values were in the six survived septic patients. Ascorbyl radical concentration determined by EPR spectroscopy using spin trapping was increased in septic patients compared with volunteers. Xanthine oxidase activity was lower in patients with lethal termination. The increased ascorbyl radical concentrations may be due to an increased radical generation [531].

Peculiar reactivity of the host organism, first of all, disorders in the immunic system and non-specific defense factors, are most essential in sepsis development. In this case, the initial stages of immunic reaction: leukocyte chemotaxis, opsonization, phagocytosis, leukocyte and blood serum bactericidal activity, are affected.

Endothelial oxidant generation and subsequent leukocyte chemotaxis and activation are important mechanisms of tissue damage in ischemic organs. It is still not known, if oxidant generation may be involved in triggering the subsequent leukocyte-mediated injury. Questions remain whether particular oxidants and oxygen free radicals are capable of modulating the expression of leukocyte adhesion molecules and effecting leukocyte endothelial adhesion. EPR studies were performed to determine the effect of different oxidant molecules and oxygen free radicals such as $O_2^{-\bullet}$, $HO^\bullet$, and H_2O_2 on the expression of integrin and selectin adhesion molecules on the surface of human polymorphonuclear leukocytes, and the effect of these alterations on polymorphonuclear adhesion to the endothelium. Adhesion molecule expression on the surface of human polymorphonuclear leukocytes was measured by the immunofluorescence flow cytometry. EPR spectroscopy was applied to establish exogenous free radical generation as well as that from activated polymorphonuclear leukocytes. It is found that oxygen free radicals rather than non-radical oxidants can directly trigger polymorpho-nuclear activation and adhesion to vascular endothelium [532].

3.25 PERITONITIS

Peritonitis is peritoneum inflammation, the morbidity and death rate of which remain very high. In surgery, there is no consensus of opinion on the mechanism of pathologic reactions and the cause of endogenous intoxication in peritonitis.

The EPR method was used to study paramagnetic centers in blood, urine, peritoneal exudate, intestinal contents of 61 patients with general purulent peritonitis. The vast majority of these was represented by men (71%). In these patients, peritonitis was caused by surgical invasion and traumas of peritoneal cavity organs. In 16% of the patients an intermediate, severe and extremely severe course of peritonitis was observed [533].

For EPR examination, the patients' blood was taken from the subclavian vein and centrifuged (2000 rev/min, +4°C). Peritoneal exudate was withdrawn from drainages installed during surgery in the subdiaphragmatic and subhepatic spaces and in small pelvis. The intestinal content was taken by means of stable vasointestinal probe of Miller-Ebbot type. Blood, peritoneal exudate, gastrointestinal tract contents were taken three times every 48 hours. Blood and urine nitrate levels were determined by the photoelectrocolorimetry technique [534]. In postoperational peritonic patients the samples of blood, peritoneal exudate and intestinal contents were taken every 24, 48 and 72 hours. It was found, on average: *E. coli* (26%), *Ps. aerugenosa* (23%), *Acinetobacter* (12%), *Enterococcus faecium* (14%), *Proteus mirabilis* (12%). Simultaneously, the nitrate-reducing capability of peritoneal and intestinal microorganisms was investigated by EPR spectroscopy. Most gastro-intestinal microflora are able to reduce nitrates to nitrites. This process is initiated by a nitrate reductase enzyme inhabiting these microorganisms. Nitrites, in turn, react with blood hemoglobin to form methemoglobin. The study of 9 species of the above bacteria has shown 7 of these (80%) to exhibit nitrate-reducing capability [533, 535, 536].

As shown by EPR spectroscopy, erythrocyte methemoglobin content in peritonic patients depends on the duration and severity of disease. 20 volunteers served as donors (methemoglobin concentration is 0.6 ± 0.1 g/l). The highest methemoglobin concentration is observed on the tenth to eleventh day after the onset of disease. In peritonic patients with extremely severe course of disease the methemoglobin level is increased with the very first days and further reaches the highest values (table 3-37) [533].

Table 3-37. Erythrocyte methemoglobin content in peritonic patients depending on the duration and severity of disease

State	Days					
	2-3	4-5	6-7	8-9	10-11	> 12
Intermediate	1.28 ± 0.06	1.54 ± 0.08	1.15 ± 0.13	0.83 ± 0.09		
n	4	7	4	3		
Severe	1.64 ± 0.12	1.75 ± 0.14	1.51 ± 0.12	1.18 ± 0.08		
n	5	4	7	4		
Extrem. severe	1.70 ± 0.08	1.92 ± 0.14	2.10 ± 0.18	2.18 ± 0.08	2.23 ± 0.14	2.13 ± 0.13
n	5	7	7	8	7	8

The blood and urine endogenous nitrate content in peritonic patients was determined by the EPR method and compared with that of patients undergone abdominal surgery not related to peritonitis. Since patients in the two groups were given neither water nor food, exogenous nitrate supply is excluded, so only endogenous nitrates are the case. The blood and urine nitrate content (73.9 ± 11.4 mg/l and 103.0 ± 15.2 mg/l, respectively) of peritonic patients is reliably higher compared to that in the control group (40.0 ± 5.5 mg/l и 52.8 ± 7.6 mg/l, respectively). The high nitrate level in peritonitis can be explained by amplified protein catabolism, enzymatic system activation, necrotic changes in cells. Blood methemoglobin formation is conditioned by respiratory insufficiency and endogenous nitrates [533, 537].

In donor plasma no methemoglobin is fixed. In blood plasma of peritonic patients the methemoglobin level depends on the severity of state. The presence of methemoglobin in patients plasma may be due to its absorption from the abdominal cavity by lymphatic vias. A linear correlation (r = +0.4) between methemoglobin levels in plasma and peritoneal exudate is observed [538].

The highest plasma methemoglobin concentration in peritonic patients appears on the second to fifth day and later it is decreased. This may be explained by the fact that during this time purulent exudate character, its biochemical composition, concentration and microflora virulence point out to a clearly-expressed peritonitis.

Methemoglobin and Hem-NO have been detected in peritoneal exudate by the EPR method. In the course of treatment and abdominal cavity assanation, MetHb inestinal content is decreasing independently of the severity and duration of peritonitis. The highest peritoneal exudate MetHb level is observed in peritonic patients in extremely severe state. This is explained by exudate character and contamination degree.

A Hem-NO signal, detectable in the peritoneal exudate and intestinal content EPR spectrum, corresponds to the Hb/NO complex. Hemoglobin oxidation to Hem-NO occurs in the following manner. Interacting with blood, nitrites find access into erythrocytes and get involved in the reaction with hemoglobin. In the results redox reactions desoxihemoglobin is oxidized to methemoglobin, the nitrite ions being reduced to nitric oxide.

$$Hb^{2+} + NO_2 + 2H \rightarrow MtHb + NO + H_2O$$

Nitric oxide reacts with the reduced hemoglobin to form stable Hem-NO complexes [533]. The Hem-NO signal was detected in peritoneal exudate EPR spectrum in 20% of patients. In 75% of the patients the hem-NO signal was detected in the exudate displaying a clearly-expressed hemorrhagic component. In patients with extremely severe peritonitis the Hem-NO signal appears in the spectrum late in the disease [539].

On the second to third day after the onset of disease, the intestinal methemoglobin concentration in peritonic patients does not differ much depending on the severity of illness and decreased by the fourth to fifth day. This due to intestinal probe evacuation of congested intestinal contents, detoxication measures and correcting therapy [540].

In patients with severe and extremely severe peritonitis the methemoglobin level increases proportionally with time. This is related to intestinal trophic and motor changes, which amplify fermentation, ammonia formation, intestine mucus necrotic changes accompanied by hemorrhage, growth of microflora showing nitrate reductase capability.

In peritonitis the Hem-NO signal detected in the intestinal EPR spectrum does not change significantly in the first 3 days independently of the severity of illness. On the fourth to fifth day the Hem-NO level decreases in patients with intermediate and severe peritonitis and increases in patients in the extremely heavy state. On the sixth or seventh days the patients with severe peritonitis show a tendency towards decreasing the intestinal Hem-NO level. The same is observed for next second to third days. In patients with extremely severe peritonitis, the intestinal Hem-NO concentration is increasing in the period from sixth to eleventh days since the onset of disease and the tendency towards decreasing is observed only on the twelfth day (table 3-38) [533, 538-540].

Comparison of time-dependent peritoneal exudate methemoglobin level to the intestinal one allows a conclusion that the former exhibits a tendency to decrease, whereas the latter is increasing. These suggests that the paretically affected intestine and its contents, and not the peritoneal exudate, for a long time remains the main source of intoxication [540].

In peritonic patients the paretically affected intestine provides favorable conditions for extensive growth of microorganisms and alteration of necrotic cells due to harmful agents and disorder in blood circulation.

Table 3-38. Intestinal Hem-NO concentration (10^{-5} M) in peritonic patients depen-ding on the duration and severity of disease

State	Days					
	2-3	4-5	6-7	8-9	10-11	>12
Intermediate	1.54 ± 0.44	1.36 ± 0.5				
n	5	5				
Severe	1.62 ± 0.45	1.54 ± 0.67	1.35 ± 0.41	1.25 ± 0.29		
n	4	5	6	4		
Extrem. severe	1.62 ± 0.93	2.02 ± 0.46	2.30 ± 0.61	2.50 ± 0.62	2.56 ± 0.76	2.49 ± 0.85
n	5	11	13	13	8	9

The formation of methemoglobin enhances tissue hypoxia. The iron-containing enzymes present in tissues are bound to nitric oxide and become switched out of the respiratory system. This intensifies disorder in intra-cellular metabolism and may lead to the formation of new low molecular nitrogen compounds, including nitric oxide. Nitric oxide contributes to microcirculation disorders and thrombosis followed by trophic alterations in tissues. Nitric oxide acts as a secondary mediator in intracellular signaling system owing to its effect on Ca-mobilizing cell system. This property of subcellular structures may be of special value in the development of patho-logic processes occurring on the background of ischemia and tissue hypoxia. This also results in increased NO binding by heme and non-heme iron with the formation of methemoglobin and Hem-NO. In nitrite methemoglobin-emia-induced tissue hypoxia, not only globulin transport functions are broken, but cellular iron-containing enzymes are blocked.

According to modern concepts, lipopolysaccharides of bacterial origin interact with neutrophils and macrophagues thus inducing accumulation of free radical oxygen forms. Free radical-mediated oxidative processes facili-tate the formation of nitric oxide and ammonia conversion to nitrites and nitrates. Nitric oxide is the intermediate of these processes.

Microflora of the host organism plays an important role in nitrate me-tabolism. This is mainly due to nitrate reductase activity and the ability to synthesize nitrites and nitrates from ammonia.

3.26 ACQUIRED IMMUNODEFICIENCY SYNDROME (AIDS)

AIDS is a viral disease characterized by heavy damage of the immunic system. AIDS results from a contact of the blood system with blood or saliva of an infected person and is practically incurable [541, 542]. The interactions between the amino-terminal peptide (FP-I; 23 residues 519-541) of the glycoprotein 41.000 (gp41) of Human Immunodeficiency Virus Type-1 (HIV-1) and human erythrocyte membranes has been assessed by the EPR technique using spin label.

Polypeptide aggregation with erythrocyte membranes induces hemolysis. The glycoprotein N-terminal residue is suggested to be involved in HIV-induced fusion and cytolysis [543]. The interaction between this polypeptide and human serum albumin was studied by circular dichroism spectroscopy. The results show that the peptide binds to albumin with dominant α-helical character. EPR spectra of spin labeled (glycoproteins FP-I and FP-II (N-terminal-alanine, residue 519) and FP-III (methionine, residue 537) provide evidence for peptide binding to human serum albumin.

EPR spectra of FP-II bound to HSA at 38°C indicate that the spin label at the amino-terminal residue (Ala-519) is motionally restricted. The EPR spectrum of 12-nitroxide stearate (12-NS)-labeled HSA and that obtained with FP-II are identical. This shows that the 12-NS and FP-II probes are similarly bound to albumin. Contrarily, ESR spectra of HSA labeled with FP-III indicate high mobility for the group (Met-537) at the aqueous-protein interface. This suggests that the N-terminal gp41 peptide binds as an α-helix (residues 519-536) to fatty acid sites on human serum albumin, such that Ala-519 of the peptide residues in the interior of the protein while Met-537 lies outside the protein in aqueous solution. The addition of human serum albumin to human red blood cells dramatically reduces the ability of FP-I to induce hemolysis, presumably through peptide-albumin binding that inhibits FP-I interactions with erythrocyte membranes. These results the following three conclusions to be dawn:

1. High serum levels of albumin may limit the efficacy of anti-HIV therapies using peptides based on the N-terminal gp41 domain.
2. Mechanisms of FP-I and HSA interactions with physical techniques may provide clues on the molecular features underlying viral FP-I combination with receptors on the target cell surface.
3. The affinity of albumin for the N-terminal gp41 peptide may play a subordinate role in the blocking of HIV infection *in vitro* [544].

REFERENCES

1. R. M. Nalbandyan, Denaturated agents influence on ceruloplazmin EPR spectra parameters, Biofizika. **18**:821 (1973) (Russ.).
2. E. K. Ruuge, T. M. Kerimov and A. V. Panemanglor, Lyophilic effect on animal cell free radicals, Biofizika. **21**:124 (1976) (Russ.).
3. M. V. Orlova, Kinetics of human blood ceruloplasmin properties under the effect of microcurrents at acupuncture points (letter), Biofizika. **36**:732 (1991) (Russ.).

4. M. Iu. Tarasev, E. P. Saburenkova, I. I. Dantsig, K. A. Moshkov and V. V. Ryl'kov, Effect of ceruloplasmin conformation on its activity: significance for clinical analysis, Vopr. Med. Khim. **37**:43 (1991) (Russ.).
5. R. G. Saifutdinov, Blood plasma PMC in donors and patients, Magnitnyi Rezonans v Biologii i Meditsine, Moskva (1989), pp. 82-83 (Russ.).
6. G. N. Bogdanov, V. N. Varfolomeev, E. E. Morozova, K. A. Treskunov, et al. Blood paramagnetic metallocomplexes and their relation with pathological states, Magnitnyi Rezonans v Biologii i Meditsine, Moskva (1981), pp. 191-192 (Russ.).
7. M. Bomba, A. Camagna, S. Cannistraro, P. L. Indovina and P. Samoggia, EPR study of serum ceruloplasmin and iron transferrin in myocardial infarction, Physiol. Chem. Phys. **9**:175 (1977).
8. J. A. Green, T. Pocklington, A. A. Dawson, M. Foster, ESR studies on caeruloplasmin and iron transferrin in patients with chronic lymphocytic leukaemia, Br. J. Cancer **41**:356 (1980).
9. K. R. Sedov and R. G. Saifutdinov, Electron Paramagnetic Resonance in Internal Diseases Clinics, Irkutskii Dom Pechati, Irkutsk (1993) (Russ.).
10. L. I. Idel'son, Hipochromic Anemia, Meditsina, Moskva (1981).
11. V. N. Petrov, Physiology and Pathology of Iron Metabolism, Nauka, Leningrad (1982) (Russ.).
12. T. V. Golosova, E. M. Vlasikhina and I. N. Pobedinskaya, Serologic markers of viral hepatitis B in hematological patients, Gematologiya i transfuziologiya. N1:7 (1987) (Russ.).
13. P. D. Howotle, G. M. Wood, S. G. A. Romaskile and M. S. Losowsky, Chronic hepatitis "B" carriers found at blood donation. Do they need regular follow-up? Vox. Sang. **52**:200 (1987).
14. M. S. Kushakovskii, Clinical Forms of Hemoglobin Damage, Meditsina, Leningrad (1968) (Russ.).
15. M. S. Moxness, L. S. Brunauer and W. H. Huestis, Hemoglobin oxidation products extract phospholipids from the membrane of human erythrocytes, Biochemistry. **35**:7181 (1996).
16. D. A. Svistunenko, R. P. Patel and M. T. Wilson, An EPR investigation of human methaemoglobin oxidation by hydrogen peroxide: methods to quantify all paramagnetic species observed in the reaction, Free Radic. Res. **24**:269 (1996).
17. Neto L. Martin, M. Tabak and O. R. Nascimento, Effect of hydration in metHb: reversible changes monitored by ESR of iron, J. Inorg. Biochem. **40**:309 (1990).
18. N. Nedjar-Arroume, A. Castellano, J. M. Piot and D. Guillochon, Stabilizing effect of water/alcohol solvents towards autoxidation of human hemoglobin, Biotechnol. Appl. Biochem. **18**(Pt 1):25 (1993).
19. N. Kosover and E. Kosover, Glutathione disulfate systems, in: Free Radicals in Biology, Mir, Moskva (1979), pp. 65-95 (Russ.).
20. N. P. Skakun and V. I. Shendevitskii, Pharmacokinetics of MetHb-forming agents, Terapevt. Arkhiv. N1:100 (1980) (Russ.).
21. M. Nagai, K. Mawatari, Y. Nagai, S. Horita, Y. Yoneyama and H. Hori, Studies of the oxidation states of hemoglobin M Boston and hemoglobin M Saskatoon in blood by EPR spectroscopy, Biochem. Biochem. Biophys. Res. Commun. **210**:483 (1995).
22. I. B. Afanas'ev, Free radical inhibitors and promotors of biological processes, in: Oxygen Radicals in Chemistry, Biology and Medicine, Zinatne, Riga (1988), pp. 9-25 (Russ.).

23. G. A. Hamilton and R. D. Libby, The valence of copper and the role of supperoxide in the D-galactoae oxidase catalyzed reaction, Biochem. Biophys. Res. Commun. **55**:333 (1973).
24. A. G. Shanturov, E. V. Nosulya and R. G. Saifutdinov, EPR study of the nasal secretion metalloproteins, Vestnik otorinolaringologii. N6:13 (1991) (Russ.).
25. E. V. Nosulya, A. G. Shanturov and R. G. Saifutdinov, EPR study of the nasal secretion in normal and upper respiratory tract pathology, Magnitnyi Rezonans v Khimii i Biologii, May 26 - June 2, Zvenigorod (1996), p.61 (Russ.).
26. R. G. Saifutdinov, E. V. Nosulia and A. G. Shanturov, Investigation by ESR of the nose secret received from human, YI International Symposium on Magnetic Resonance in Colloid and Interface science, Ference (1992), p. 177.
27. M. Machnicki, Biological properties of lactotransferrin, Folia Biol. (Praha). **37**:65 (1991).
28. C. L. Day, K. M. Stowell, E. N. Baker and J. W. Tweedie, Studies of the N-terminal half of human lactoferrin produced from the cloned cDNA demonstrate that interlobe interactions modulate iron release, J. Biol. Chem. **267**:13857 (1992).
29. D. Legrand, J. Mazurier, D. Colavizza, J. Montreuil and G. Spik, Properties of the iron-binding site of the N-terminal lobe of human and bovine lactotransferrins. Importance of the glycan moiety and of the non-covalent interactions between the N- and C-terminal lobes in the stability of the iron-binding site, Biochem. J. **266**:575 (1990).
30. R. G. Saifutdinov, Paramagnetic Centers in Human Biological Fluids and Their Diagnostic and Pathogenetic Role in Some Internal Diseases, Thesis Dr.Sci., Medical Institute, Tomsk (1989) (Russ.).
31. T. M. Lachocki, D. F. Church and W. A. Pryor, Persistent free radicals in the smoke of common household materials: biological and clinical implications, Environ. Res. **45**:127 (1988).
32. O. Zschornig, H. Machill, O. M. Panasenko, T. V. Volnova, K. Arnold and O. A. Azizova, Influence of polar polymers on the apoprotein region of human serum lipoproteins: an electron paramagnetic resonance (EPR) study, Gen. Physiol. Biophys. **12**:113 (1993).
33. M. Kveder, G. Pifat, S. Pecar and M. Schara, The ESR characterization of molecular mobility in the lipid surface layer of human serum lipoproteins, Chem. Phys. Lipids. **70**:101 (1994).
34. M. Kveder, G. Pifat, S. Pecar, M. Schara, P. Ramos and H. Esterbauer, Nitroxide reduction with ascorbic acid in spin labeled human plasma LDL and VLDL, Chem. Phys. Lipids. **85**:1(1997).
35. H. Lottin, C. Motta and G. Simard, Differential effects of glycero- and sphingo-phospholipolysis on human high-density lipoprotein fluidity, Biochim. Biophys. Acta. **1301**:127 (1996).
36. O. V. Sviridov, M. N. Ermolenko, E. S. Pyshko, S. V. Khalimondhik, N. A. Gurinovich, S. L. Romanov, P. A. Kiselev and O. A. Strel'chenok, Isolation, identification and properties of a new thyroxine-binding protein-apolipoprotein A-1 from human blood serum, Biokhimiia. **56**:2281 (1991) (Russ.).
37. O. M. Panasenko, T. V. Volnova, O. A. Azizova, Y. A. Vladimirov, Free radical modification of lipoproteins and cholesterol accumulation in cells upon atherosclerosis, Free Radic. Biol. Med. **10**:137 (1991).

38. C. E Thomas, G. Ku and B. Kalyanaraman, Nitrone spin trap lipophilicity as a determinant for inhibition of low density lipoprotein oxidation and activation of interleukin-1 β release from human monocytes, J. Lipid. Res. **35**:610 (1994).
39. P. K. Witting, J. M. Upston and R. Stocker, Role of α-tocopheroxyl radical in the initiation of lipid peroxidation in human Loe-density lipoprotein exposed to horse radish peroxidase, Biochemistry. **36**:1251 (1997).
40. S. R. Thomas, P. K. Witting and R. Stocker, 3-Hydroxyanthranilic acid is an efficient, cell-derived co-antioxidant for alpha-tocopherol, inhibiting human low density lipoprotein and plasma lipid peroxidation, J. Biol. Chem. **271**:32714 (1996).
41. J. M. Upston, J. Neuzil and R. Stocker, Oxidation of LDL by recombinant human 15-lipoxygenase: evidence for α-tocopherol-dependent oxidation of esterified core and surface lipids, J. Lipid Res. **37**:2650 (1996).
42. P. K. Witting, J. M. Upston and R. Stocker, Role of α-tocopheroxyl radical in the initiation of lipid peroxidation in human Loe-density lipoprotein exposed to horse radish peroxidase, Biochemistry. **36**:1251 (1997).
43. V. E. Kagan, E. A. Serbinova, T. Forte, G. Scita and L. Packer, Recycling of vitamin E in human low density lipoproteins, J. Lipid. Res. **33**:385 (1992).
44. M. F. Muldoon, S. B. Kritchevsky, R. W. Evans and V. E. Kagan, Serum total antioxidant activity in relative hypo- and hypercholesterolemia, Free Radic. Res. **25**:239 (1996).
45. O. M Panasenko, S. A. Evgina and V. I. Sergienko, A spin-probe study of the structural change in human blood lipoproteins under the action of sodium hypochlorite, Biull. Eksp. Biol. Med. **116**:153 (1993) (Russ.).
46. O. M. Panasenko, S. A. Evgina, S. V. Nikitin and V. I. Sergienko, The effect of hypochlorite on electrical surface properties of human blood lipoproteins, Biofizika. **40**:545 (1995) (Russ.).
47. C. G. Taylor, M. K. Dame, H. S. Murphy, P. A. Ward and J. Varani, Spontaneous injury to human umbilical vein endothelial cells increases during in vitro culture and is blocked by protein kinase activation, Lab. Invest. **70**:822 (1994).
48. S. L. Hempel, G. R. Buettner, D. A. Wessels, G. M. Galvan and Y. Q. O'Malley, Extracellular iron (II) can protect cells from hydrogen peroxide, Arch. Biochem. Biophys. **330**:401 (1996).
49. V. O'Donnell and M. J. Burkitt, Mitochondrial metabolism of a hydroperoxide to free radicals in human endothelial cells: an electron spin resonance spin-trapping investigation, Biochem. J. **304** (Pt 3):707 (1994).
50. J. Van der Zee, Formation of peroxide- and globin-derived radicals from the reaction of methaemoglobin and metmyoglobin with t-butyl hydroperoxide: an ESR spin-trapping investigation, Biochem. J. **322**(Pt 2):633 (1997).
51. L. Pronai, K. Hiramatsu, Y. Saigusa and H. Nakazawa, Low superoxide scavenging activity associated with enhanced superoxide generation by monocytes from male hypertriglyceridemia with and without diabetes, Atherosclerosis. **90**:39 (1991).
52. K. R. Sedov and R. G. Saifutdinov, Blood and blood components free radicals and their significance in IHD diagnosis, IV Vsecoyuznyi S'ezd Kardiologov, Moskva, (1986), pp. 40-41 (Russ.).
53 K. R. Sedov and R. G. Saifutdinov, Blood free radicals and their value in IHD diagnosis, Bulleten' Sibirskogo Otdeleniya AMN SSSR, Novosibirsk, N6:73 (1985) (Russ.).
54. R. G. Saifutdinov, A study of blood free radicals in myocardial infarction patients, Novye Napravleniya v Kardiologii. Infarkt Miokarda, Nedostatochnost' Krovo-obrashcheniya, Novokuznetsk (1987), pp. 46-47 (Russ.).

55. R. G. Saifutdinov and K. R. Sedov, Differential diagnosis of stenocardia from cardiology, Pat SSSR N 1061800 (1982), Bull. Izobr. N47:21 (1983) (Russ.).
56. I. K. Shvatsabaya, Ischaemic Heart Disease, Meditsina, Moskva (1975) (Russ.).
57. M. E. Raiskina, The role of catecholamines in cardiac metabolism, in: Biogene Amines, Meditsina, Moskva (1967), pp. 86-102 (Russ.).
58. R. G. Saifutdinov, Adrenalin effect on free radical levels in humen plasma and erythrocytes, Bulleten' Eksperimental'noi Biologii i Meditsiny. N10:78 (1982) (Russ.).
59. A. M. Shaposhnikova and E. F. Derkachev, Ceruloplasmin interaction with cell electron-transporting systems, Biokhimiya. **36**:3 (1971) (Russ.).
60. L. Broman, Chromatographic and magnetic studies on human ceruloplasmin, Acta Soc. Med. Upsal. **69**:1 (1964).
61. T. Kakuda, H. Tanaka, E. Kimoto and F. Morishige, EPR spectrum of human serum copper, Appl. Spectrosc. **34**:276 (1980).
62. A. Mouithys-Mickalad, P. Hans, G. Deby-Dupont, M. Hoebeke, C. Deby and M. Lamy, Propofol reacts with peroxynitrite to form a phenoxyl radical: demonstration by electron spin resonance, Biochem. Biophys. Res. Commun. **249**:833 (1998).
63. J. G. Coghlan, W. D. Flitter, A. E. Holley, M. Norell, A. G. Mitchell, C. D. Ilsley and T. F. Slater, Detection of free radicals and cholesterol hydroperoxides in blood taken from the coronary sinus of man during percutaneous transluminal coronary angioplasty, Free. Radic. Res. Commun. **14**:409 (1991).
64. A. J. Tortolani, S. R. Powell, V. Misik, W. B. Weglicki, G. J. Pogo and J. H. Kramer, Detection of alkoxyl and carbon-centered free radicals in coronary sinus blood from patients undergoing elective cardioplegia, Free Radic. Biol. Med. **14**:421 (1993).
65. E. D. Grech, N. J. Dodd, M. J. Jackson, W. L. Morrison, E. B. Faragher, D. R. Ramsdale, Evidence for free radical generation after primary percutaneous transluminal coronary angioplasty recanalization in acute myocardial infarction, Am. J. Cardiol. **77**:122 (1996).
66. J. L. Zweier, R. Broderick and P. Kuppusamy, Determination of the mechanism of free radical generation in human aortic endothelial cells exposed to anoxia and reoxygenation, J. Biol. Chem. **269**:24156 (1994).
67. S. Pietri, J. R. Seguin, P. D. d'Arbigny and M. Culcasi, Ascorbyl free radical: a noninvasive marker of oxidative stress in human open-heart surgery, Free Radic. Biol. Med. **16**:523 (1994).
68. J. L. Zweier, P. Wang, A. Samouilov and P. Kuppusamy, Enzyme-independent formation of nitric oxide in biological tissues [see comments], Nat. Med. **1**:804 (1995).
69. M. Pissarek, F. Janichen, I. E. Blasig, R. Haseloff, T. Keller, E. Tapp and EG. Krause, Cardioprotective potency of the radical scavenger S-2-(3 aminopropylamino) ethylphosphorothioic acid in the post-ischaemic rat heart, Mol. Cell. Biochem. **145**:121 (1995).
70. A. Movahed, K. G. Nair, T. F. Ashavaid and P. Kumar, Free radical generation and the role of allopurinol as a cardioprotective agent during coronary artery bypass grafting surgery, Can. J. Cardiol. **12**:138 (1996).
71. K. B. Shumaev, V. Ya. Bykhovskii, V. V. Kukharchuk and E. K. Ruuge, Hydrophobic antioxidant radical forms: possible role in POLL protection, Fizika i Khimiya Elementarnykh Khimicheskikh Protsessov; V Vserossiiskaya Konferentsiya, Posvyashchennaya 80-letiyu Akademika V.V. Voevodskogo, September 29 - October 3, Chernogolovka (1997), pp. 296-298 (Russ.).

72. V. E. Kagan, H. J. Freisleben, M. Tsuchiya, T. Forte and L. Packer, Generation of probucol radicals and their reduction by ascorbate and dihydrolipoic acid in human low density lipoproteins, Free Radic. Res. Commun. **15**:265 (1991).
73. P. Ascenzi, M. Coletta, A. Desideri, R. Petruzzelli, F. Polizio, M. Bolognesi, S. G. Condo and B. Giardina, Spectroscopic properties of the nitric oxide derivative of ferrous man, horse, and ruminant hemoglobins: a comparative study, J. Inorg. Biochem. **45**:31 (1992).
74. P. Ascenzi, M. Coletta, A. Desideri, F. Polizio, A. Bertollini, R. Santucci and G. Amiconi, Effect of bezafibrate and clofibric acid on the spectroscopic properties of the nitric oxide derivative of ferrous human hemoglobin, J. Inorg. Biochem. **48**:47 (1992).
75. T. G. Gantchev and M. B. Shopova, Characterization of spin-labelled fatty acids and hematoporphyrin binding sites interactions in serum albumin, Biochim. Biophys. Acta. **1037**:422 (1990).
76. E. P. Sidorik, Yu. B. Bogaev and AP. Burlaka, Instrumental problems in EPR diagnosis of tumors lung diseases, Problemy tekhniki v meditsine, Tomsk (1983), pp. 58-59 (Russ.).
77. E. I. Chazov, Coronary Insufficiency (Progress in Theoretical and Clinical Cardiology), Meditsina, Moskva, (1977) (Russ.).
78. N. M. Mukharlyamov, Early Stages of Circulation Deficiency and its Compensatory Mechanism, Meditsina, Moskva, (1978) (Russ.).
79. L. I. Ol'binskaya and P. F. Litvitskii, Coronary and Myocardic Insufficiency. Meditsina, Moskva (1986) (Russ.).
80. V. Z. Gorkin, Modern view on contribution of E and K vitamins and related compounds to oxidative phosphorylation, Ukrainskii biokhimicheskii zhurnal. **31**:270 (1959) (Russ.).
81. R. G. Saifutdinov, Effect of α-tocopherole on concentration of α-tocopherylquinone human blood lipids, Bulleten' Eksperimental'noi Biologii i Meditsiny. N9:296 (1985) (Russ.).
82. L. R. Cantilena Jr, R. P. Smith, S. Frasur, H. Kruszyna, R. Kruszyna and D. E. Wilcox, Nitric oxide hemoglobin in patients receiving nitroglycerin as detected by electron paramagnetic resonance spectroscopy [see comments], J. Lab. Clin. Med. **120**:902 (1992).
83. H. Kosaka, S. Tanaka, T. Yoshii, E. Kumura, A. Seiyama and T. Shiga, Direct proof of nitric oxide formation from a nitrovasodilator metabolised by erythrocytes, Biochem. Biophys. Res. Commun. **204**:1055 (1994).
84. D. Sh. Burbaev, Paramagnetic Iron-Serum Center in Electron Transport Chain of Membrane Biological Structure, Thesis Dr.Sci., Moskva (1986) (Russ.).
85. M. M. Muratov, Formaldehyde Effect on the Electron Transport Chain of Heart Conserved Mitochondria. Thesis Ph.D., Kazan (1981) (Russ.).
86. K. R. Sedov and R. G. Saifutdinov, EPR posibility in circular deficience patients, Novye Napravleniya v Kardiologii. Infarkt Miokarda, Nedostatochnost' Krovoobrashcheniya, Novokuznetsk (1987), pp. 11-12 (Russ.).
87. A. M Komarov and C. S. Lai, Detection of nitric oxide production in mice by spin-trapping electron paramagnetic resonance spectroscopy, Biochim. Biophys. Acta. **1272**:29 (1995).
88. V. Quaresima, H. Takehara, K. Tsushima, M. Ferrari and H. Utsumi, In vivo detection of mouse liver nitric oxide generation by spin trapping electron paramagnetic resonance spectroscopy, Biochem. Biophys. Res. Commun. **221**:729 (1996).

89. S. Al-Dabbagh and R. Dolt, N-nitroso compounds: occurence, biological effects and relevance to human cancer, (IARC Sci. Publ. N57), Lyon (1983), pp. 901-909.
90. R. G. Saifutdinov and A. V. Sukhanov, Study of nitrate-reducing ability of saliva. Magnetic resonance and related phenomena, XXVII-th Congress Ampere, August 21-28, Kazan (1994), pp. 910-911.
91. R. G. Saifutdinov and A. V. Sukhanov, The nitrate-reducing ability of the saliva, XYI International Conference on Magnetic Resonance in Biological Systems (ISMRBS), The Netherlands, Veldhoven, (1994).
92. A. V. Sukhanov, Biological Fluid Nitrate-Reducing Capability in Some Internal Diseases, Thesis Ph.D., Institut Usovershenstvovaniya Vrachei, Irkutsk (1999) (Russ.).
93. R. G. Saifutdinov and A. V. Sukhanov, Nitrate-reducing ability of saliva, International Conference EPR-spectroscopy of nitric oxide in biological systems, December 11-15, Suzdal (1996), pp. 27-28.
94. A. E. Kalmanson, On the Nature and Role of Free Radicals in Biological Processes, Thesis Ph.D., Institute of Chem. Phys, Moskva (1977) (Russ.).
95. A. E. Kalmanson, A. V. Belozerova and N. S. Motavkina, EPR spectra of intestinal and coccal group pathogenes, IV Mezhdunarodnyi Biofizicheskii Kongress, Moskva (1972), v. 4, pp. 220-221 (Russ.).
96. R. G. Saifutdinov, K. V. Protasov and V. S. Kustos, Intestinal microflora PMCs in IHD patients taking dairy products, Magnitnyi Rezonans v Khimii i Biologii, May 26 - June 2, Zvenigorod (1996), pp. 69-70 (Russ.).
97. R. G. Saifutdinov, V. S. Kustos and K. V. Protasov, Investigation of faeces by EPR/ESR at patients with coronary heart deasease, taking sour milk products, 2eme Symposium Satellite "Vitamins and Biofactors", September 27-28, France, Nancy (1996).
98. G. G. Kuznetsova, Intestinal biocenosis disorders in chronic colites, Sov. Meditsina. N11:62 (1972) (Russ.).
99. A. I. Oparin, E. F. Kharat'yan and N. S. Gel'man, Dehydrogenase localisation and relation to lactobacterium pentoaceticum cellular oxygen, Dokl. Akad. Nauk SSSR. **64**:211 (1964) (Russ.).
100. E. A. Boichenko and A. I. Oparin, Metal compounds interaction in hydrogenase evolution, Zhurnal evolyutsionnoi biokhimii. **16**:3 (1980).
101. E. W. Ainscough and A. M. Brodie, ESR spectrometry and its application to biochemistry, Chem. N. Z. **47**:86 (1983).
102. K.V. Protasov, Holesterin-Transfering Ability of IHD Patient Intestal Microflora, Institut Usovershenstvovaniya Vrachei, Irkutsk (1998) (Russ.).
103. M. Persson, P. Hammarstrum, M. Lindgren, B. H. Jonsson, M. Svensson and U. Carlsson, EPR mapping of interactions between spin-labeled variants of human carbonic anhydrase II and GroEL: evidence for increased flexibility of the hydrophobic core by the interaction, Biochemistry. **38**(1):432 (1999).
104. L. A. Wheeler, F. B. Soderberg and P. Goldman, The relationship between nitro group reduction and the intestinal microflora, J. Pharmacol. Exp. Ther. **194**:135 (1975).
105. Ya. I. Azhipa, Medical-Biological Aspects of Electron Paramagnetic Resonance Applications, Nauka, Moskva (1983) (Russ.).
106. S. L. Gorbach and E. Bengt, Gustafsson memorial lecture. Function of the normal human microflora, Scand. J. Infect. Dis. Suppl. **49**:17 (1986).
107. A. L. Myasnikov, Hypertension Disease and Atherosclerosis, Meditsina, Moskva (1965) (Russ.).

108. B. M. Braginskii, M. A. Lis and V. P. Vodoevich, Thiamine sufficiency in patients with hypertensive diseases and atherosclerosis, Zdravookhranenie Belorussii. N8:26 (1972) (Russ.).
109. Yu. V. Zobnin, Blood PMCs and Lipid Free Radicals Oxidation in the Baikal-Amur Railway Road Builders with Arterial Hypertension, Thesis Ph.D., Cardiology Institute, Tomsk (1986) (Russ.).
110. A. L. Kleshchev, Nitric Oxide as a General Metabolite of Nitrogen-Containing Vasodilators and Guanylacyclase Activators. Hypertensive properties of free and iron-bonding NO, Thesis Ph.D., 1st Medical Institute, Moskva (1986) (Russ.).
111. J. R. Stone, R. H. Sands, W. R. Dunham and M. A. Marletta, Electron paramagnetic resonance spectral evidence for the formation of a pentacoordinate nitrosyl-heme complex on soluble guanylate cyclase, Biochem. Biophys. Res. Commun. **207**:572 (1995).
112. P. F. Chen, A. L. Tsai, V. Berka and K. K. Wu, Mutation of Glu-361 in human endothe-lial nitric-oxide synthase selectively abolishes L-arginine binding without perturbing the behavior of heme and other redox centers, J. Biol. Chem. **272**:6114 (1997).
113. A. K. Nussler, D. A. Geller, M. A. Sweetland, M. Di Silvio, T. R. Billiar, J. B. Madariaga, R. L. Simmons and J. R. Lancaster Jr., Induction of nitric oxide synthesis and its reactions in cultured human and rat hepatocytes stimulated with cytokines plus LPS, Biochem. Biophys. Res. Commun. **194**:826 (1993).
114. J. Vasquez-Vivar, A. M. Santos, V. B. Junqueira and O. Augusto, Peroxynitrite-mediated formation of free radicals in human plasma: EPR detection of ascorbyl, albumin-thiyl and uric acid-derived free radicals, Biochem. J. **314**(Pt 3):869 (1996).
115. R. B. Clarkson, S. W. Norby, A. Smirnov, S. Boyer, N. Vahidi, R. W. Nims and D. A. Wink, Direct measurement of the accumulation and mitochondrial conversion of nitric oxide within Chinese hamster ovary cells using an intracellular electron paramagnetic resonance technique, Biochim. Biophys. Acta. **1243**:496 (1995).
116. A Wennmalm, G Benthin, A Edlund, N Kieler-Jensen, S Lundin, AS Petersson and F. Waagstein, Nitric oxide synthesis and metabolism in man, Ann. N.-Y. Acad. Sci. **714**:158 (1994).
117. T. Yamamura, Techniques for measurement of nitric oxide in biological systems: principles and practice, Nippon Yakurigaku Zasshi. **107**:173 (1996).
118. A. Wennmalm, B. Lanne and A. S. Petersson, Detection of endothelial-derived relaxing factor in human plasma in the basal state and following ischemia using ESR spectrometry, Anal. Biochem. **187**:359 (1990).
119. G. G. Borisenko, A. N. Osipov, K. D. Kazarinov and Yu. A. Vladimirov, Photo-chemical reactions of nitrosyl hemoglobin during exposure to low-power laser irradiation, Biokhimiay. **62**:661 (1997) (Russ.).
120. H. Kruszyna, R. Kruszyna, L. G. Rochelle, R. P. Smith and D. E. Wilcox, Effects of temperature, oxygen, heme ligands and sulfhydryl alkylation on the reactions of nitroprusside and nitroglycerin with hemoglobin, Biochem. Pharmacol. **46**:95 (1993).
121. D. N. Rao, S. Elguindi and P. J. O'Brien, Reductive metabolism of nitroprusside in rat hepatocytes and human erythrocytes, Arch. Biochem. Biophys. **286**:30 (1991).
122. D. Pietraforte and M. Minetti, One-electron oxidation pathway of peroxynitrite decomposition in human blood plasma: evidence for the formation of protein tryptophan-centred radicals, Biochem. J. **321** (Pt 3):743 (1997).
123. M. R. Gunther, L. C. Hsi, J. F. Curtis, J. K. Gieras, L. J. Marnett, T. E. Eling and R. P. Mason, Nitric oxide trapping of the tyrosyl radical of prostaglandin H synthase-2

leads to tyrosine iminoxyl radical and nitrotyrosine formation, J. Biol. Chem. **272**:17086 (1997).

124. M. Minetti, G. Scorza and D. Pietraforte, Peroxynitrite induces long-lived tyrosyl radical(s) in oxyhemoglobin of red blood cells through a reaction involving CO_2 and a ferryl species, Biochemistry. **38**:2078 (1999).

125. H. Kosaka and A. Seiyama, Physiological role of nitric oxide as an enhancer of oxygen transfer from erythrocytes to tissues, Biochem. Biophys. Res. Commun. **218**:749 (1996).

126. M. B. Johnson, J. G. Adamson and A. G. Mauk, Functional comparison of specifically cross-linked hemoglobins biased toward the R and T states, Biophys. J. **75**:3078 (1998).

127. M. F. R. Martin, K. E. Surrall, F. McKenna, J. S. Dixon, H. A. Bird and V. Wright., Captopril: a new treatment for rheumatoid arthritis, Lancet. **1**:1325 (1984).

128. W. H. van Gilst, P. A. de Graeff, H. Wesseling and C. D. J. de Langen, Reduction of reperfusion arrhythmia in the ischemic isolated rat heart by angiotensin cinverting enzyme inhibitors: A comparison of captopril, enalapril and HOE 498, J. Cardiovascular Pharmacol. **8**:722 (1986).

129. V. Misik, I. T. Mak, R. E. Stafford and W. B. Weglicki, Reactions of captopril and epicaptopril with trandition metal ions and OH radicals - an EPR spectroscopy study, Free Radic. Biol. Med. **15**:611 (1993).

130. D. Bagchi, R. Prasad and D. K. Das, Direct scavenging of free radicals by captopril an angiotensin converting enzyme inhibitor, Biochem. Biophys. Res. Comm. **158**:52 (1989).

131. O. I. Aruoma, D. Akanmu, R. Cecchini and B. Halliwell, Evaluation of the ability of the angiotensin-converting enzyme inhibitor captopril to scavenge reactive oxygen species, Chem.-Biol. Interact. **77**:303 (1991).

132. D. Jay, A. Cuellar, E. G. Jay, C. Garcia, R. Gleason and E. Munoz, Study of a Fenton type reaction: effect of captopril and chelating reagents, Arch. Biochem. Biophys. **298**:740 (1992).

133. J. Reguli and V. Misik, Superoxide scavenging by thiol/copper complex of captopril-an EPR spectroscopy study, Free Radic. Res. **22**:123 (1995).

134. D. Jay, A. Cuellar, R. Zamorano, E. Munoz and R. Gleason, Captopril does not scavenge superoxide: captopril prevents O^{2-}.production by chelating copper, Arch. Biochem. Biophys. **290**:463 (1991).

135. A. Stasko, V. Brezova, S. Biskupic, K. Ondrias and V. Misik, Reactive radical intermediates formed from illuminated nifedipine, Free Radic. Biol. Med. **17**:545 (1994).

136. C. Dragu, L. Copaescu and V. V. Morariu, Manganese transport through human erythrocyte membranes. An EPR study, Biochim. Biophys. Acta. **1328**:90 (1997).

137. J. N. Cohn, S. M. Ziesche, L. E. Loss and G. T. Anderson, Effect of felodipine on short-term exercise and neurohormones and long-term mortality in heart failure: Results of V-HeFT III, Circulation. **92**(Suppl 1):1 (1995).

138. C. M. O'Connor, R. N. Belkin, P. E. Carson, et al., For the PRAISW investigators. Effect ofamlodipine on mode of death in severe chronic heart failure, The PRAISW trial (Abstr), Circulation. **92**(Suppl 1):1 (1995).

139. K. Tsuda, Y. Minatogawa, H. Iwahashi, I. Nishio, R. Kido and H. Masuyama, Spin-labelling study of biomembranes in spontaneously hypertensive rats: calcium-and calmodulin-dependent regulation, Clin. Exp. Pharmacol. Physiol. **22**(Suppl 1):S234 (1995).

140. R. G. Saifutdinov and O. N. Koshchina, Blood PMCs in iron-deficient anemic patients before and after treatment by iron-containing medicines, Environment and Human Health (Regional Meeting), Irkutsk Medical Institute, Irkutsk (1990), v. 3, pp. 406-411 (Russ.).
141. Y. Chen-Barrett, P. M. Harrison, A. Treffry, M. A. Quail, P. Arosio, P. Santambrogio and N. D. Chasteen, Tyrosyl radical formation during the oxidative deposition of iron in human apoferritin, Biochemistry. **34**:7847 (1995).
142. B. R. Crouse, V. M. Sellers, M. G. Finnegan, H. A. Dailey and M. K. Johnson, Site-directed mutagenesis and spectroscopic characterization of human ferrochelatase: identification of residues coordinating the [2Fe-2S] cluster, Biochemistry. **35**:16222 (1996).
143. R. G. Saifutdinov, The Possibility of ESR method in clinical medicine, XXVII-th Congress Ampere, August 21-28, Kazan (1994), pp. 908-909.
144. P. L. Indovina, ESR studies on different pathological states, Ann. Ist. Super. Sanita. **14**:863 (1978).
145. O. B. Priima, Non-enzymatic cation proteins of peripheral blood leukocytes as a factor of nonspecific response of living organism to injury, Klinicheskaya meditsina. N2:4 (1997) (Russ.).
146. V. E. Pigarevskii, Clinical Morphology of Neutrophilic Granulocytes, Meditsina, Moskva (1988) (Russ.).
147. B. S. Nagoev, Neutrophilic Granulocytosis, Nal'chik (1986) (Russ.).
148. W. Chamulitrat, M. S. Cohen and R. P. Mason, Free radical formation from organic hydroperoxides in isolated human polymorphonuclear neutrophils, Free Radic. Biol. Med. **11**:439 (1991).
149. C. E. Cooper and E. Odell, Interaction of human myeloperoxidase with nitrite, FEBS Lett. **314**:58 (1992).
150. C. E. Cooper and E. Odell, EPR differences between human myeloperoxidase isoenzymes, Biochem. Soc. Trans. **20**:108S (1992).
151. B. E. Britigan, T. J. Coffman and G. R. Buettner, Spin trapping evidence for the lack of significant hydroxyl radical production during the respiration burst of human phagocytes using a spin adduct resistant to superoxide-mediated destruction, J. Biol. Chem. **265**:2650 (1990).
152. C. L. Ramos, S. Pou, B. E. Britigan, M. S. Cohen and G. M. Rosen, Spin trapping evidence for myeloperoxidase-dependent hydroxyl radical formation by human neutrophils and monocytes, J. Biol. Chem. **267**:8307 (1992).
153. V. R. Muzykantov, T. A. Kushnareva, M. D. Smirnov and E. K. Ruuge, Avidin attachment to biotinylated human neutrophils induces generation of superoxide anion, Biochim. Biophys. Acta. **1177**:229 (1993).
154. M. L. McCormick, T. L. Roeder, M. A. Railsback and B. E. Britigan, Eosinophil peroxidase-dependent hydroxyl radical generation by human eosinophils, J. Biol. Chem. **269**:27914 (1994).
155. T. R. Green, J. H. Fellman, A. L. Eicher and K. L. Pratt, Antioxidant role and subcellular location of hypotaurine and taurine in human neutrophils, Biochim. Biophys. Acta. **1073**:91 (1991).
156. D. Constantin, K. Mehrotra, B. Jernstrom, A. Tomasi and P. Moldeus, Alternative pathways of sulfite oxidation in human polymorphonuclear leukocytes, Pharmacol. Toxicol. **74**:136 (1994).

157. D. Pietraforte, E. Tritarelli, U. Testa and M. Minetti, gp120 HIV envelope glycoprotein increases the production of nitric oxide in human monocyte-derived macrophages, J. Leukoc. Biol. **55**:175 (1994).
158. K. Hajela, R.. Kayestha and J. Sumati, Carbohydrate induced modulation of cell membrane. IV: Interaction with mucin and fucoidan totally immobilizes the human platelet membrane, Indian. J. Biochem. Biophys. **33**:308 (1996).
159. K. Hajela, R. Kayestha and J. Sumati, Carbohydrate induced modulation of cell membrane: II. Spin label study of fluidity changes in peripheral blood lymphocyte membrane, FEBS Lett. **380**:165 (1996).
160. K. Hajela, R. Kayestha and J. Sumati, Carbohydrate induced modulation of cell membrane. V : ESR characterization of molecular mobility in lipid bilayer of human neutrophil plasma membrane, Indian. J. Exp. Biol. **34**:513 (1996).
161. B. Meier, H. H. Radeke, S. Selle, G. G. Habermehl, K. Resch and H. Sies, Human fibroblasts release low amounts of reactive oxygen species in response to the potent phagocyte stimulants, serum-treated zymosan, N-formyl-methionyl-leucyl-phenylalanine, leukotriene B4 or 12-O-tetradecanoylphorbol 13-acetate, Biol. Chem. Hoppe Seyler. **371**:1021 (1990).
162. C. D. Black, A. Samuni, J. A. Cook, C. M. Krishna, D. C. Kaufman, H. L. Malech and A. Russo, Kinetics of superoxide production by stimulated neutrophils, Arch. Biochem. Biophys. **286**:126 (1991).
163. D. K. Perry, W. L. Hand, D. E. Edmondson and J. D. Lambeth, Role of phospholipase D-derived diradylglycerol in the activation of the human neutrophil respiratory burst oxidase. Inhibition by phosphatidic acid phosphohydrolase inhibitors, J. Immunol. **149**:2749 (1992).
164. B. M. Babior, Oxidants from phagocytes: agents of defense and destruction, Blood. **64**:959 (1984).
165. B. O. Anderson, J. M. Brown, A. H. Harken, Mechanisms of neutrophil-mediates tissue injury, J. Surgical. Res. **51**:170 (1991).
166. J. Reguli, Z. Durackova, J. Pogady, D. Martisova and A. Stasko, EPR spin trapping of reactive oxygen products of the respiratory burst of phagocytes (use of EPR spectroscopy in biology and medicine II), Bratisl. Lek. Listy. **93**:557 (1992).
167. L. Seawright, M. Tanigawa, T. Tanigawa, Y. Kotake and E. G. Janzen, Can spin trapping compounds like PBN protect against self-inflicted damage in polymorphonuclear leukocytes?, Free Radic. Res. **23**:73 (1995).
168. Y. Mao, L. Zang and X. Shi, Singlet oxygen generation in the superoxide reaction, Biochem. Mol. Biol. Int. **36**:227 (1995).
169. T. Tanigawa, Y. Kotake, M. Tanigawa and L. A. Reinke, Mutual contact of adherent polymorphonuclear leukocytes inhibits their generation of superoxide, Free Radic. Res. **22**:361 (1995).
170. C. Opper, C. Fett, B. Capito, S. Raha and W. Wesemann, Plasma membrane properties in heterogeneous human blood platelet subfractions modulate the cellular response at the second messenger level, Thromb. Res. **72**:39 (1993).
171. P. Gaffet, F. Basse and A. Bienvenue, Loss of phospholipid asymmetry in human platelet plasma membrane after 1-12 days of storage. An ESR study, Eur. J. Biochem. **222**:1033 (1994).
172. Y. Bayon, M. Croset, D. Daveloose, F. Guerbette, V. Chirouze, J. Viret, J. C. Kader and M. Lagarde, Effect of specific phospholipid molecular species incorporated in human platelet membranes on thromboxane A2/prostaglandin H2 receptors, J. Lipid. Res. **36**:47 (1995).

173. Y. Bayon, M. Croset, F. Guerbette, D. Daveloose, V. Chirouze, J. Viret, J. C. Kader and M. Lagarde, Selective modifications of the phospholipid fatty acid composition in human platelet membranes using nonspecific and specific lipid transfer proteins, Anal. Biochem. **230**:75 (1995).
174. L. Pronai, K. Ichimori, H. Nozaki, H. Nakazawa, H. Okino, A. J. Carmichael and C. M. Arroyo, Investigation of the existence and biological role of L-arginine/nitric oxide pathway in human platelets by spin-trapping/EPR studies, Eur. J. Biochem. **202**:923 (1991).
175. Z. Y. Wen, Z. Y. Yan, T. Gao, H. Dou, J. Lu, D. Sun and Z. Lu, A study of effects of WGA and ConA on RBC membrane receptors using a new ektacytometric method, Clin. Hemorheol. Microcirc. **17**:467 (1997).
176. L. Iuliano, F. Violi, J. Z. Pedersen, D. Pratico, G. Rotilio and F. Balsano, Free radical-mediated platelet activation by hemoglobin released from red blood cells, Arch. Biochem. Biophys. **299**:220 (1992).
177. L. Iuliano, J. Z. Pedersen, D. Pratico, G. Rotilio and F. Violi, Role of hydroxyl radicals in the activation of human platelets, Eur. J. Biochem. **221**:695 (1994).
178. C. Watala, T. Pietrucha, K. Gwozdzinski, U. Kralisz and C. S. Cierniewski, Microenviron-ment changes in human blood platelet membranes associated with binding of tissue-type plasminogen activator, Eur. J. Biochem. **215**:867 (1993).
179. T. Pietrucha, W. J. Stec, A. Okruszek, B. Uznanski, M. KozioEkiewicz, A. Wilk, A. Peucienniczak, M. Swiatkowska and C. S. Cierniewski, The epidermal growth factor-like domain from tissue plasminogen activator. Cloning in E. coli, purification and ESR studies of its interaction with human blood platelets, Acta Biochim. Pol. **41**:25 (1994).
180. V. L. Nienaber and L. J. Berliner, Subtle differences in active site structure between bovine and human thrombins: ESR and fluorescence studies, Thromb. Haemost. **65**:40 (1991).
181. L. J. Berliner and J. K. Woodford, Biophysical studies of interactions of hirudin analogs with bovine and human thrombin by ESR and fluorescence labelling studies, Adv. Exp. Med. Biol. **340**:51 (1993).
182. R. O. Hynes, Fibronectins, Springer-Verlag, New York (1990).
183. C. Narasimhan and C. S. Lai, Differential behavior of the two free sulfhydryl groups of human plasma fibronectin: effects of environmental factors, Biopolymers. **31**:1159 (1991).
184. S. Suzuki, K. Sugai, H. Sato, M. Sakatume and M. Arakawa, Inhibition of active oxygen generation by dipyridamole in human polymorphonuclear leukocytes, Eur. J. Pharmacol. **227**:395 (1992).
185. C. Watala and K. Gwozdzinski, Effect of aspirin on conformation and dynamics of membrane proteins in platelets and erythrocytes, Biochem. Pharmacol. **45**:1343 (1993).
186. Y. Sato, T. Miura and Y. Suzuki, Interaction of pentoxifylline with human erythrocytes. II. Effects of pentoxifylline on the erythrocyte membrane, Chem. Pharm. Bull. (Tokyo). **38**:555 (1990).
187. S. Suzuki, K. Sugai, H. Sato and M. Arakawa, Effect of a platelet-activating factor antagonist, WEB 2086, on inhibition of active oxygen generation in human polymorphonuclear leukocytes, Pharmacology. **48**:111 (1994).
188. G. I. Dorofeev and V. M. Uspenskaya, Gastroduodenal Diseases in Young Age, Meditsina, Moskva (1984) (Russ.).

189. V. G. Podoprigorova, LPO disorder and antioxidant system state as a partial mechanism in ulceric disease, XIX Vsesoyuznyi S'ezd Terapevtov, Tashkent (1987), v. 2, pp. 349-360 (Russ.).
190. V. Kh. Vasilenko, A. L Grebnev and A. A. Sheptulin, Ulcer Disease, Meditsina, Moskva (1987) (Russ.).
191. D. J. Pountney, J. K. Shergill, R. J. Simpson and J. M. Wrigglesworth, EPR and visible spectroscopic analysis of a duodenal b-type cytochrome implicated in the process of intestinal iron absorption, Biochem. Soc. Trans. **23**:140S (1995).
192. R. G. Saifutdinov, I. L. Petrunjko, M. B. Shapochnik and S. P. Chicoteev, Investigation of gastric Mucosa by EPR, YII International Symposium on Magnetic Resonance in Colloid and Interface (ISMRCIS YII), Spain, Madrid (1995).
193. K. R. Sedov and R. G. Saifutdinov, Method EPR in clinical medicine, The II-nd International Symposium of Japan-Russia Medical Exchange Foundation and the Hear Region, Vladivostok (1994), p. 289.
194. J. B. Hibbs, B. R. Taintor, Z. Vavrin, D. L. Granger, J. C. Drapier, I. J. Amber and J. R. Lancaster, Synthesis of nitric oxide from a terminal guanidino nitrogen atom of L-arginine: a molecular mechanism regulating cellular proliferation that targets itracellular iron, in: Nitric Oxide from L-Arginine: a Bioregulatory System, S. Moncada, E. A. Higgs, eds., Elsevier Science Publishers B.V. (Biochemical Division), (1990), pp. 159-223.
195. A. Wennmalm, B. Lanne and A. S. Petersson, Detection of endothelial-derived relaxing factor in human plasma in the basal state and following ischemia using electron paramagnetic resonance spectrometry, Analytical Biochem. **187**:359 (1990).
196. V. P. Kryshen' and M. F. Nesterova, Food channel microbial flora in gastric and duodenal ulcer, Vracheb delo. N9:16 (1982) (Russ.).
197. H. M. Swartz and T. Walczak, In vivo EPR: prospects for the '90s., Phys. Med. **9**:41 (1993).
198. G. R. Eaton, S. S. Eaton and K. Ohno, eds., EPR Imaging and In-vitro EPR, CRC Press, FL, Boca Raton (1991).
199. V. V. Khramtsov, L. M. Vainer, V. A. Rar, T. A. Berezina, V. V. Martin and L. B. Volodarskii, Macromolecular spin pH-probes on the basis of human serum albumin, Biokhimiya. **55**:1014 (1990) (Russ.).
200. B. Gallez, K. Mader and H. M. Swartz, Noninvasive measurement of the pH inside the gut by using pH-sensitive nitroxides. An in vivo EPR study, Magn. Reson. Med. **36**:694 (1996).
201. T. Ohara, R. Sasaki, D. Shibuya, S. Asaki and T. Toyota, Effect of omeprazole on ascorbate free radical formation, Tohoku J. Exp. Med. **167**:185 (1992).
202. T. Yoshikawa, Y. Naito, T. Tanigawa and M. Kondo, Free radical scavenging activity of the novel anti-ulcer agent rebamipide studied by electron spin resonance, Arzneimittelforsch. **43**:363 (1993).
203. N. Yoshida, T. Yoshikawa, S. Iinuma, M. Arai, S. Takenaka, K. Sakamoto, T. Miyajima, Y. Nakamura, N. Yagi, Y. Naito, F. Mukai and M. Kondo, Rebamipide protects against activation of neutrophils by Helicobacter pylori, Dig. Dis. Sci. **41**:1139 (1996).
204. I. M. Drake, M. J. Davies, N. P. Mapstone, M. F. Dixon, C. J. Schorah, K. L. White, D. M. Chalmers and A. T. Axon, Ascorbic acid may protect against human gastric cancer by scavenging mucosal oxygen radicals, Carcinogenesis. **7**:559 (1996).

205. L. Pronai, I. Yukinobu, I. Lang and J. Feher , The oxygen-centered radicals scavenging activity of sulfasalazine and its metabolites. A direct protection of the bowel, Acta Physiol. Hung. **80**:317 (1992).
206. J. Hargreaves, Y. Popineau, M. Le Mestc and M. A. Hemminga, Molecular flexibility in wheat gluten proteins submitted to heating, FEBS Lett. **372**:103 (1995).
207. J. Hargreaves, Y. Popineau, M. Cornec and J. Lefebvre, Relations between aggregative, viscoelastic and molecular properties in gluten from genetic variants of bread wheat, Int. J. Biol. Macromol. **18**:69 (1996).
208. H. M. Swartz, K. Chen and J. A. Roth, Further evidence that the pigment in the Dubin-Johnston syndrome is not melanin, Pigment Cell Res. **1**:69 (1987).
209. K. Keddad, P. Therond, C. Motta, C. Baussan and A. Legrand, Alterations in erythrocyte membrane fluidity and fatty acid composition in glycogen storage disease, Biochim. Biophys. Acta. **1315**:61 (1996).
210. G. P. Butcher, A. Raqabah, M. J. Jackson, J. Hoffman, J. M. Rhodes and M. C. Symons, Failure of electron paramagnetic resonance spectroscopy studies to detect elevated free radical signals in liver biopsy specimens from patients with alcoholic liver disease, Free Radic. Res. **22**:99 (1995).
211. P. Clot, G. Bellomo, M. Tabone, et al., Detection of antibodies against proteins modified by hydroxyethyl free radicals in patients with alcoholic cirrhosis, Gastroent. **108**:201 (1995).
212. D. N. Rao, M. X. Yang, J. M. Lasker and A. I. Cederbaum, 1-Hydroxyethyl radical formation during NADPH- and NADH-dependent oxidation of ethanol by human liver microsomes, Mol. Pharmacol. **49**:814 (1996).
213. A. Icen, Glutathione reductase of human erythroeytes, Scand. J. Clin. Invest. **96**:1 (1967).
214. L. Chruscinski, M. Dyba, M. Jezowska-Bojczuk, Henryk-Kozwski, G. Kupryszewski, Z. Mackiewicz and A. Majewska, Specific interactions of Cu2+ ions with fragments of envelope protein of hepatitis B virus, J. Inorg. Biochem. **63**:49 (1996).
215. T. Matsuo, H. Shinzawa, H. Togashi, M. Aoki, K. Sugahara, K. Saito, T. Saito, T. Takahashi, I. Yamaguchi, M. Aoyama and H. Kamada, Free Radic. Biol. Med. **25**:929 (1998.)
216. J. Sun, T. M. Loehr, A. Wilks and P. R. Ortiz de Montellano, Identification of histidine 25 as the heme ligand in human liver heme oxygenase, Biochemistry. **33**:13734 (1994).
217. L. H. Patterson, S. Jones and A. Gescher, Generation of a free radical from calphostin C by microsomal cytochrome P450 reductase in rat and human liver, Biochem. Pharmacol. **51**:599 (1996).
218. Y. Dai, J. Rashba-Step and A. I. Cederbaum, Stable expression of human cytochrome P4502E1 in HepG2 cells: characterization of catalytic activities and production of reactive oxygen intermediates, Biochemistry. **32**:6928 (1993).
219. A. Berson, C. Wolf, C. Chachaty, C. Fisch, D. Fau, D. Eugene, J. Loeper, J. C. Gauthier, P. Beaune, D. Pompon, P. Maurel and D. Pessayre, Metabolic activation of the nitroaromatic antiandrogen flutamide by rat and human cytochromes P-450, including forms belonging to the 3A and 1A subfamilies, J. Pharmacol. Exp. Ther. **265**:366 (1993).
220. S. Baron, D. Coppenhaver, F. Dianzani, et al., Interferon principles and medical applications, Gavelstone (1992), pp. 47-64.

221. L. M. Pfeffer and F. R. Landsberger, Interferon-alpha modulates the plasma membrane-cytoskeletal complex of human lymphoblastoid cells sensitive to the antiproliferative action of interferon-alpha, J. Interferon. Res. **10**:91 (1990).
222. A. Aszalos, P. M. Grimley, E. Balint, K. C. Chadha and J. L. Ambrus, On the mechanism of action of interferons: interaction with nonsteroidal anti-inflammatory agents, pentoxifylline (Trental) and cGMP inducers, J. Med. **22**:255 (1991).
223. A. E. Medlock and H. A. Dailey, Human coproporphyrinogen oxidase is not a metalloprotein, J. Biol. Chem. **271**:32507 (1996).
224. Z. H. Yang, K. Wang and X. T. Liu, Studies of electron spin resonance on bilirubin free radicals, Sci. China. **35B**:1093 (1992).
225. L. X. Chen, J. F. Lu and K. Wang, The influence of bilirubin on fluidity and rotational correlation times of human erythrocyte membrane, Cell Biol. Int. Rep. **16**:567 (1992).
226. B. Commoner, J. L .Townsend and G. Pake, Free radicals in biological materials, Nature. **174**:689 (1954).
227. R. G. Saifutdinov and A. V. Shcherbakova, EPR study of bile cholecystitis patients, Magnitnyi Rezonans v Khimii i Biologii; May 26 - June 2, Zvenigorod (1996), p. 72 (Russ.).
228. A. V. Shcherbakova, Biological Media PMCs in Cholecystitic Patients, Thesis Ph.D., Medical University, Irkutsk (1997) (Russ.).
229. C. A. Hubel, A. V. Kozlov, V. E. Kagan, R. W. Evans, S. T. Davidge, M. K. McLaughlin and J. M. Roberts, Decreased transferrin and increased transferrin saturation in sera of women with preeclampsia: implications for oxidative stress, Am. J. Obstet. Gynecol. **175** (Pt 1):692 (1996).
230. B. Kalyanarman, C. C. Felix and R. C. Sealy, Semiquinone anion radicals of catechol (amine)s, catechol estrogens, and their metal ion complexes, Environ. Health. Perspectives. **64**:185 (1985).
231. B. M. Mullock and L. J. Shaw, Sources of proteins in human bile, Gut. **26**:500 (1985).
232. A. V. Sherbakova and R. G. Saifutdinov, Human bile study by EPR method, YII International Symposium on Magnetic Resonance in Colloid and Interface (ISMRCIS YII), Spain, Madrid (1995), ref. 188.
233. G. H. Reed, EPR Studies of Mn (II) Complexes with enzymes and substrates, Biochem. Soc. Trans. **13**:567 (1985).
234. R. G. Saifutdinov and A. V. Sherbakova, The problem of the human protein containing cuprum, 2nd Winter Research Conferences, January 22-27, France, Les Deux Alpes (1995), poster n. 4.
235. R. G. Saifutdinov and A. V. Sherbakova, Examination of Cu2+ and Fe3+ in blood and bile by EPR method (EPR/ESR) in diagnostics of the cholecystitis, 5th International Meeting on Trace Elements in Medicine and Biology, February 4-7, France, Meribel (1996), p. 163.
236. A. V. Shcherbakova, R. G. Saifutdinov, Blood EPR study of cholecystitis patients, Magnitnyi Rezonans v Khimii i Biologii, May 26 - June 2, Zvenigorod (1996), p. 71 (Russ.).
237. I. T. Lott, R. Dipaolo, S. S. Raghavan, P. Clopath, Y. A. Milunsky, W. C. Robertson and J. N. Kanfer, Abnormal copper metabolism in Menke's steely-hair syndrome, Pediat. Res. **13**:845 (1979).
238. M. Lejoyeux, M. P. Bouvard, J. Viret, D. Daveloose, J. Ades and M. Dugas, Modifications of erythrocyte membrane fluidity from patients with anorexia nervosa before and after refeeding, Psychiatry Res. **59**:255 (1996).

239. K. R. Sedov, A. F. Kolpakova and N. G. Maksimov, Blood PMC of nonspecific chronic lung disease patients, 2 Vsesoyuznyi Kongress po Boleznyam Organov Dykhaniya, September 16-19, Chelyabinsk (1991), p. 85 (Russ.).
240. S. A. Baglushkin, R. G. Saifutdinov and T. P. Sizykh, Plasma PMC of bronchial asthma patients, Diagnostika i Lechenie Tuberkuleza v Novykh Epidemiologicheskikh Usloviyakh, Ulan-Ude (1992), pp. 10-14 (Russ.).
241. V. Vallyathan and X. Shi, The role of oxygen free radicals in occupational and environmental lung diseases, Environ Health Perspectives. **105**(Suppl 1):165 (1997).
242. S. Pinamonti, M. Leis, A. Barbieri, D. Leoni, M. Muzzoli, S. Sostero, M. C. Chicca, A. Carrieri, F. Ravenna, L. M. Fabbri and A. Ciaccia, Detection of xanthine oxidase activity products by EPR and HPLC in bronchoalveolar lavage fluid from patients with chronic obstructive pulmonary disease, Free Radic. Biol. Med. **25**:771 (1998).
243. N. S. Dalal, J. Newman, D. Pack, S. Leonard and V. Vallyathan, Hydroxyl radical generation by coal mine dust: possible implication to coal workers' pneumoconiosis (CWP), Free Radic. Biol. Med. **18**:11 (1995).
244. N. S. Dalal, B. Jafari, M. Petersen, F. H. Green and V. Vallyathan, Presence of stable coal radicals in autopsied coal miners' lungs and its possible correlation to coal workers' pneumoconiosis, Arch. Environ. Health. **46**:366 (1991).
245. H. J. H. Fenton, Oxidation of tartaric acid in presence of iron, J. Chem. Soc. **65**:899 (1994).
246. X. Shi, Y. Mao, L. N. Daniel, U. Saffiotti, N. S. Dalal and V. Vallyathan, Silica radical-induced DNA damage and lipid peroxidation, Environ. Health Perspectives. **102**(Suppl 10):149 (1994).
247. V. Vallyathan, J. F. Mega, X. Shi and N. S. Dalal, Enhanced generation of free radicals from phagocytes induced by mineral dusts, Am. J. Respir. Cell. Mol. Biol. **6**:404 (1992).
248. A. J. Ghio, M. B. Kadiiska, Q. H. Xiang and R. P. Mason, In vivo evidence of free radical formation after asbestos instillation: an ESR spin trapping investigation, Free Radic. Biol. Med. **24**:11 (1998).
249. T. Ishizaki, E. Yano, N. Urano and P. H. Evans, Crocidolite-induced reactive oxygen metabolites generation from human polymorphonuclear leukocytes, Environ. Res. **66**:208 (1994).
250. T. Takeuchi, K. Morimoto, H. Kosaka and T. Shiga, Spin trapping of superoxide released by opsonized asbestos from human promyelocytic leukemia cell line, HL60, Biochem. Biophys. Res. Commun. **194**:57 (1993).
251. Y. Rojanasakul, X. Shi, D. Deshpande, W. W. Liang and L. Y. Wang, Protection against oxidative injury and permeability alteration in cultured alveolar epithelium by transferrin-catalase conjugate, Biochim. Biophys. Acta. **1315**:21 (1996).
252. W. A. Pryor, Biological effects of cigarette smoke, wood smoke, and the smoke from plastics: the use of electron spin resonance, Free Radic. Biol. Med. **13**:659 (1992).
253. K. K. Stone, E. Bermudez and W. A. Pryor, Aqueous extracts of cigarette tar containing the tar free radical cause DNA nicks in mammalian cells, Environ. Health Perspectives. **102**(Suppl 10):173 (1994).
254. M. Tsuchiya, D. F. Thompson, Y. J. Suzuki, C. E. Cross and L. Packer, Superoxide formed from cigarette smoke impairs polymorphonuclear leukocyte active oxygen generation activity, Arch. Biochem. Biophys. **299**:30 (1992).
255. K. R. Maples, T. Sandstrom, Y. F. Su and R. F. Henderson, The nitric oxide/heme protein complex as a biologic marker of exposure to nitrogen dioxide in humans, rats, and in vitro models, Am. J. Respir. Cell. Mol. Biol. **4**:538 (1991).

256. I. Ueno, M. Hoshino, T. Miura and N. Shinriki, Ozone exposure generates free radicals in the blood samples in vitro. Detection by the ESR spin-trapping technique, Free Radic. Res. **29**:127 (1998).
257. B. Gonet, Free radical aspects of the ozone influence on blood, Physiol. Chem. Phys. Med. NMR. **26**:273 (1994).
258. W. A. Pryor, Mechanisms of radical formation from reactions of ozone with target molecules in the lung, Free Radic. Biol. Med. **17**:451 (1994).
259. M. K. Pulatova, V. L. Sharygin and Iu. P. Reva, Detection and nature of melanin-containing centers in bronchoalveolar smears and blood of persons responsible for cleaning up after the accident at the Chernobyl Atomic Energy Station, Dokl. Akad. Nauk. **345**:130 (1995) (Russ.).
260. S. F. Piao, Studies of free radical metabolism in patients with cor pulmonale, Chung. Hua. Chieh. Ho Ho Hu Hsi Tsa Chih. **15**:214, 254 (1992).
261. G. Supinski, Free radical induced respiratory muscle dysfunction, Mol. Cell. Biochem. **179**:99 (1998).
262. R. F. Baikeev, K. M. Ziiatdinov, I. R. Akhmadeev, G. T. Rakhmaeva and N. G. Demidova, The diagnosis of the destructive changes in pulmonary and osteoarticular tuberculosis, Problemy Tuberkuleza. N11-12:28 (1992) (Russ.).
263. K. Takeda, M. Haida, S. Shioya, T. Matsumori, Y. Kimura and K. Hasumi, Estimation of the malignancy of pleural effusions by electron spin resonance, Anal. Biochem. **241**:47 (1996).
264. P. E. James, G. Bacic, O. Y. Grinberg, F. Goda, J. F. Dunn, S. K. Jackson and H. M. Swartz, Endotoxin-induced changes in intrarenal pO2, measured by in vivo electron paramagnetic resonance oximetry and magnetic resonance imaging, Free Radic. Biol. Med. **21**:25 (1996).
265. K. Gwozdzinski and M. Janicka, Oxygen free radicals and red blood cell damage in acute renal failure, Biochem. Soc. Trans. **23**:635S (1995).
266. J. Pincemail, O. Detry, C. Philippart, J. O. Defraigne, C. Franssen, K. Burhop, C. Deby, M. Meurisse and M. Lamy, Diaspirin crosslinked hemoglobin (DCLHb): absence of increased free radical generation following administration in a rabbit model of renal ischemia and reperfusion, Free Radic. Biol. Med. **19**:1 (1995).
267. Z. Zdrojewski, J. Nowak, A. Raszeja-Specht, A. Skibowska, J. Kustosz and B. Rutkowski, Evaluation of platelet membrane structure in patients with chronic glomerulonephritis (CGN), Przegl. Lek. **53**:431 (1996).
268. D. Roccatello, G. Mengozzi, M. Ferro, G. Cesano, R. Polloni, R. Mosso, G. Bonetti, T. Inconis, L. Paradisi, L. M. Sena and G. Piccoli, Isosorbide 5 mononitrate administration increases nitric oxide blood levels and reduces proteinuria in IGA glomerulonephritis patients with abnormal urinary endothelin cyclic GMP ratio, Clin. Nephrol. **44**:163 (1995).
269. A. Kashem, M. Endoh, F. Yamauchi, N. Yano, Y. Nomoto, H. Sakai, L. Pronai, M. Tanaka and H. Nakazawa, Superoxide dismutase activity in human glomerulonephritis, Am. J. Kidney Dis. **28**:14 (1996).
270. K. Yokoyama, Y. Tomino, Y. Yaguchi, H. Koide, D. Ohmori and F. Yamakura, Serum superoxide dismutase (SOD) activity in patients with renal disease by a spintrap method using electron spin resonance (ESR), Nippon Jinzo Gakkai Shi. **35**:809 (1993).
271. S. E. Roselaar, N. B. Nazhat, P. G. Winyard, P. Jones, J. Cunningham and D. R. Blake, Detection of oxidants in uremic plasma by electron spin resonance spectroscopy, Kidney Int. **48**:199 (1995).

272. H. Trachtman, D. Wilson and P. S. Rao, The role of oxygen free radicals in the development of chronic renal failure, Life Sci. **50**:1877 (1992).
273. G. M. Bogdan, Blood PMC of Chronic Renal Failure Patients, Thesis Ph.D., Krasnoyarsk (1993) (Russ.).
274. G. M. Bogdan and R.G. Saifutdinov, Paramagnetic blood centres by patients with chronic renal failure, XXVII-th Congress Ampere, August 21-28, Kazan (1994), pp. 879-880.
275. V. S. Finin, V. I. Soklakov, N. E. Savchenko and V. A. Deshko, Spin probe study of blood serum on uremia, Fizika i Khimiya Elementarnykh Khimicheskikh Protsessov; V Vserossiiskaya Konferentsiya, Posvyashchennaya 80-letiyu Akademika V.V. Voevodskogo, September 29 - October 3, Chernogolovka (1997), pp. 291-293 (Russ.).
276. W. A. Pettinger, F. G. Soyanco and J. A. Oates, Inhibition of monoamine oxidase in man by furazolidone, Clin. Pharmacol. Ther. **2**:442 (1968).
277. B. H. Ali, Pharmacology and toxicity of furazolidone in man and animals: some recent research, Gen. Pharmacol. **20**:557 (1989).
278. L. Rossi, I. De Angelis, J. Z. Pedersen, E. Marchese, A. Stammati, G. Rotilio and F. Zucco, N-[5-nitro-2-furfurylidene]-3-amino-2-oxazolidinone activation by the human intestinal cell line Caco-2 monitored through noninvasive electron spin resonance spectroscopy, Mol. Pharmacol. **49**:547 (1996).
279. T. Ozawa and Y. Nakano, The radical scavenging abilities of a new anti-nephritic drug, OPC-15161, Biochem. Mol. Biol. Int. **38**:231 (1996).
280. I. A. Menon, S. D. Persad, H. F. Haberman, P. K. Basu, J. F. Norfray, C. C. Felix and B. Kalyanaraman, Characterization of the pigment from homogentisic acid and urine and tissue from an alkaptonuria patient, Biochem. Cell. Biol. **69**:269 (1991).
281. J. Pincemail, J. O. Defraigne, C. Franssen, P. Bonnet, G. Deby-Dupont, J. Pirenne, C. Deby, M. Lamy, M. Limet and M. Meurisse, Evidence for free radical formation during human kidney transplantation, Free Radic. Biol. Med. **15**:343 (1993).
282. M. J. Payne, A. Chapman and R. Cammack, Evidence for an [Fe]-type hydrogenase in the parasitic protozoan Trichomonas vaginalis, FEBS Lett. **317**:101 (1993).
283. M. S. Vidakovic, G. Fraczkiewicz and J. P. Germanas, Expression and spectroscopic characterization of the hydrogenosomal [2Fe-2S] ferredoxin from the protozoan Trichomonas vaginalis, J. Biol. Chem. **271**:14734 (1996).
284. J. M. Richard, E. E. Creppy, J. L. Benoit-Guyod and G. Dirheimer, Orellanine inhibits protein synthesis in Madin-Darby canine kidney cells, in rat liver mitochondria and in vitro: indication for its activation prior to in vitro inhibition, Toxicology. **67**:53 (1991).
285. J. M. Richard, D. Cantin-Esnault and A. Jeunet, First electron spin resonance evidence for the production of semiquinone and oxygen free radicals from orellanine, a mushroom nephrotoxin, Free Radic. Biol. Med. **19**:417 (1995).
286. H. Oubrahim, J. M. Richard, D. Cantin Esnault, F. Seigle Murandi and F. Trecourt, Novel methods for identification and quantification of the mushroom nephrotoxin orellanine. Thin-layer chromatography and electrophoresis screening of mushrooms with electron spin resonance determination of the toxin, J. Chromatogr. **A758**:145 (1997).
287. A. G. Maksina, N. P. Mikaelian, Iu. A. Kniazev and B. A. Dainiak, Structural changes in erythrocyte membranes in diabetes mellitus using spin labelled fatty acids, Biofizika. **37**:306 (1992) (Russ.).

288. C. Watala, Hyperglycaemia alters the physico-chemical properties of proteins in erythrocyte membranes of diabetic patients, Int. J. Biochem. **24**:1755 (1992).
289. E. A. Preoteasa, C. Ionescu-Tirgoviste, V. V. Grecu and R. Georgescu, Electron spin resonance study of the erythrocyte membrane fluidity changes in diabetes, Rom. J. Intern. Med. **31**:271 (1993).
290. S. R. Wente, M. Villalba, V. L. Schramm and O. M. Rosen, Mn2(+)-binding properties of a recombinant protein-tyrosine kinase derived from the human insulin receptor, Proc. Natl. Acad. Sci. USA. **87**:2805 (1990).
291. M. T. Santini, P. L. Indovina, J. R. Simmons and S. W. Peterson, Human erythrocyte insulin receptor down-regulation is accompanied by a transient decrease in membrane order, Biochim. Biophys. Acta. **1054**:333 (1990).
292. M. T. Santini, R. Masella, A. Cantafora and S. W. Peterson, Changes in erythrocyte membrane lipid composition affect the transient decrease in membrane order which accompanies insulin receptor down-regulation, Experientia. **48**:36 (1992).
293. W. Malorni, R. Masella, M. T. Santini, F. Iosi, P. Samoggia, A. Cantafora, D. Merrell and S. W. Peterson, Human erythrocyte insulin receptor processing is affected by the oxidizing agent menadione, Exp. Cell. Res. **206**:195 (1993).
294. C. J. Mullarkey, D. Edelstein and M. Brownlee, Free radical generation by early glycation products: a mechanism for accelerated atherogenesis in diabetes, Biochem. Biophys. Res. Commun. **173**:932 (1990).
295. M. C. Delmas-Beauvieux, E. Peuchant, M. J. Thomas, L. Dubourg, A. P. Pinto, M. Clerc and H. Gin, The place of electron spin resonance methods in the detection of oxidative stress in type 2 diabetes with poor glycemic control, Clin. Biochem. **31**:221 (1998).
296. R. G. Saifutdinov and K. A. Apartsin, Spleen tissue paramagnetic centers prior to spleen autotransplantation in experiment and clinics, Aktual'nye Voprosy Rekonstruktivnoi i Vosstanovitel'noi Khirurgii, Irkutsk (1993), pp. 197-198 (Russ.).
297. J. A. Corbett, M. A. Sweetland, J. L. Wang, J. R. Lancaster Jr and M. L. McDaniel, Nitric oxide mediates cytokine-induced inhibition of insulin secretion by human islets of Langerhans, Proc. Natl. Acad.. Sci. USA. **90**:1731 (1993).
298. M. L. Brader, D. Borchardt and M. F. Dunn, The T to R transition in the copper(II)-substituted insulin hexamer. Anion complexes of the R-state species exhibiting type 1 and type 2 spectral characteristics, Biochemistry. **31**:4691 (1992).
299. M. Brank, M. Sentjurc, A. Stalc and Z. Grubic, The influence of soman simulator on reactivation by HI-6 of soman-inhibited acetylcholinesterase in preparations of rat and human skeletal muscle, Biochem. Pharmacol. **45**:499 (1993).
300. D. E. Bredesen, M. Wiedau-Pazos, J. J. Goto, S. Rabizadeh, J. A. Roe, E. B. Gralla, L. M. Ellerby and J. S. Valentine, Cell death mechanisms in ALS, Neurology. **47**(Suppl 2):S36, (discussion S38) (1996).
301. J. Liu, M. K. Shigenaga, L. J. Yan, A. Mori and B. N. Ames, Antioxidant activity of diethyldithiocarbamate, Free Radic. Res. **24**:461 (1996).
302. E. G. Janzen, Spin trapping, Accounts. Chem. Res. **4**:31 (1971).
303. J. R. Harbour, V. Chow and J. Bolton, An electron spin resonance study of the spin adduct of OH and HO_2 radicals with nitrons in the ultra violet photolysis of aqueous hydrogen peroxide solution, Can. J. Chem. **52**:3549 (1974).
304. E. G. Janzen, D. E. Nutter and E. D. Davis, On spin trapping hydroxyl and hydroperoxyl radicals, Can. J. Chem. **56**:2237 (1978).
305. M. Hiramatsu, M. Kohno, R. Edamatsu, K. Mitsuta and A. Mori, Increased superoxide dismutase activity in aged human cerebrospinal fluid and rat brain

determined by electron spin resonance spectrometry using the spin trap method, J. Neurochem. **58**:1160 (1992).

306. B. Halliwell and J. M. C. Gutterige, Free Radicals in Biology and Medicine, Clarendon Press, Oxford (1989), pp. 324-332.
307. R. Nakagawa and N. Tsuchihasi, A free radical in human cerebrospinal fluid investigated by EPR, Chem. Letters. 551 (1997).
308. T. Tamada, H. Inoue and A. Mori, Superoxide dismutase activity in cerebrospinal fluid and its relation to compression of the lumbosacral nerve root, Acta Med. Okayama. **50**:197 (1996).
309. S. Kaakkola, P. H. Rosenberg, A. Alila, T. Erkinjuntti, R. Sulkava and J. Palo, Platelet membrane fluidity in Alzheimer's disease and multi-infarct dementia: a spin label study, Acta Neurol. Scand. **84**:18 (1991).
310. M. Wender, J. Szczech, S. Hoffmann and W. Hilczer, Electron paramagnetic resonance analysis of heavy metals in the aging human brain, Neuropatol. Pol. **30**:65 (1992).
311. K. M. Lundberg, C. J. Stenland, F. E. Cohen, S. B. Prusiner and G. L. Millhauser, Chem. Biol. **4**:345 (1997).
312. D. A. Butterfield, K. Hensley, M. Harris, M. Mattson and J. Carney, β-Amyloid peptide free radical fragments initiate synaptosomal lipoperoxidation in a sequence-specific fashion: implications to Alzheimer's disease, Biochem. Biophys. Res. Commun. **200**:710 (1994).
313. D. A. Butterfield, K. Hensley, P. Cole, R. Subramanian, M. Aksenov, M. Aksenova, P. M. Bummer, B. E. Haley and J. M. Carney, Oxidatively induced structural alteration of glutamine synthetase assessed by analysis of spin label incorporation kinetics: relevance to Alzheimer's disease, J. Neurochem. **68**:2451 (1997).
314. D. A. Butterfield, β-Amyloid-Associated free radical oxidative stress and neurotoxicity: implications for Alzheimer's disease, Chem. Res. Toxicol. **10**:495 (1997).
315. M. E. Harris, J. M. Carney, P. S. Cole, K. Hensley, B. J. Howard, L. Martin, P. Bummer, Y. Wang, N. W. Pedigo Jr and D. A. Butterfield, beta-Amyloid peptide-derived, oxygen-dependent free radicals inhibit glutamate uptake in cultured astrocytes: implications for Alzheimer's disease, Neuroreport. **6**:1875 (1995).
316. B. Gonthier, H. Eysseric, A. Soubeyran, D. Daveloose, R. Saxod and L. Barret, Free Radic. Res. **27**:645 (1997).
317. Y. Zhou, J. S. Richardson, M. J. Mombourquette and J. A. Weil, Free radical formation in autopsy samples of Alzheimer and control cortex, Neurosci. Lett. **195**:89 (1995).
318. G. Nattero, G. Mengozzi, T. Inconis and L .Paradisi, Nitric oxide, endothelin-1, and transcranial Doppler in migraine. Findings in interictal conditions and during migraine attack, Headache. **36**:307 (1996).
319. S. W. Norby, J. A. Weyhenmeyer and R. B. Clarkson, Stimulation and inhibition of nitric oxide production in macrophages and neural cells as observed by spin trapping, Free Radic. Biol. Med. **22**:1 (1997).
320. L. Zecca, T. Shima, A. Stroppolo, C. Goj, G. A. Battiston, R. Gerbasi, T. Sarna and H. M. Swartz, Interaction of neuromelanin and iron in substantia nigra and other areas of human brain, Neuroscience. **73**:407 (1996).
321. S. Aime, B. Bergamasco, D. Biglino, G. Digilio, M. Fasano, E. Giamello and L. Lopiano, Biochim. Biophys. Acta. **1361**:49 (1997).

322. W. S. Enochs, M. J. Nilges and H. M. Swartz, Purified human neuromelanin, synthetic dopamine melanin as a potential model pigment, and the normal human substantia nigra: characterization by electron paramagnetic resonance spectroscopy, J. Neurochem. **61**:68 (1993).
323. L. Mosca, C. Blarzino, R. Coccia, C. Foppoli and M. A. Rosei, Melanins from tetrahydroisoquinolines: spectroscopic characteristics, scavenging activity and redox transfer properties, Free Radic. Biol. Med. **24**:161 (1998).
324. L. Zecca and H. M. Swartz, Total and paramagnetic metals in human substantia nigra and its neuromelanin, J. Neural. Transm. Park. Dis. Dement. Sect. **5**:203 (1993).
325. S. Nishibayashi, M. Asanuma, M. Kohno, M. Gomez-Vargas and N. Ogawa, Scavenging effects of dopamine agonists on nitric oxide radicals, J. Neurochem. **67**:2208 (1996).
326. J. K. Shergill, R. Cammack, C. E. Cooper, J. M. Cooper, V. M. Mann and A. H. Schapira, Detection of nitrosyl complexes in human substantia nigra, in relation to Parkinson's disease, Biochem. Biophys. Res. Commun. **228**:298 (1996).
327. D. Constantin, A. Bini, E. Meletti, P. Moldeus, D. Monti and A. Tomasi, Age-related differences in the metabolism of sulphite to sulphate and in the identification of sulphur trioxide radical in human polymorphonuclear leukocytes, Mech. Ageing Dev. **88**:95 (1996).
328. M. T. Carri, A. Battistoni, F. Polizio, A. Desideri and G. Rotilio, Impaired copper binding by the H46R mutant of human Cu,Zn superoxide dismutase, involved in amyotrophic lateral sclerosis, FEBS Lett. **356**:314 (1994).
329. D. Rosen, T. Siddique, D. Patterson, D. A. Figlewicz, P. Sapp, A. Hentati, D. Donaldson, J. Goto, J. P. Oregan, H. X. Deng, Z. Rahmani, A. Krizus, D. McKennayasek, A. Cayabyab, S. M. Gaston, R. Berger, R. E. Tanzi, J. J. Halperin, W. Y. Hung, T. Bird, G. Deng, D. W. Mulder, C. Smyth, ... J. S. Gusella, H. R. Horvitz and R. H. Brown, Mutations in Cu/Zn superoxide dismutase gene are associated with familial amyotrophic lateral sclerosis, Nature. **362**(N6415):59 (1993).
330. D. E. Bredesen, L. M. Ellerby, P. J. Hart, M. Wiedau-Pazos and J. S. Valentine, Do posttranlational modifications of CuZnSOD lead to sporadic amyotrophic lateral sclerosis?, Annals of Neurology. **42**:135 (1997).
331. G. Ferretti, A. Tangorra and G. Curatola, Effects of intramembrane particle aggregation on erythrocyte membrane fluidity: an electron spin resonance study in normal and in dystrophic subjects, Exp. Cell. Res. **191**:14 (1990).
332. Y. Ihara, A. Mori, T. Hayabara, R. Namba, K. Nobukuni, K. Sato, S. Miyata, R. Edamatsu, J. Liu and M. Kawai, Free radicals, lipid peroxides and antioxidants in blood of patients with myotonic dystrophy, J. Neurol. **242**:119 (1995).
333. M. Rybczynska, A. L. Pawlak, S. K. Hoffmann and R. Ignatowicz, Carriers of ataxia-telangiectasia gene display additional protein fraction and changes in the environment of SH groups in erythrocyte membrane, Biochim. Biophys. Acta. **1022**:260 (1990).
334. A. V. Musaev, S. G. Guseinova and A. P. Mamedov, The combined use of electromagnetic decimeter waves and deresinated naphthalan in patients with vertebrogenic humeroscapular periarthrosis (its experimental and clinical validation), Voprosy Kurortol. Fizioter. Lech. Fiz. Kult. **4**:9 (1994) (Russ.).
335. P. J. Bacon, S. A. Love, A. K. Gupta, P. J. Kirkpatrick and D. K. Menon, Plasma antioxidant consumption associated with ischemia reperfusion during carotid endarterectomy, Stroke. **27**:1808 (1996).

336. E. V. Onuchina, Clinical and Diagnostic Value of Paramagnetic Antioxydants in Biological Fluids of Patients with Rheumatoid Arthritis and Osteoarthrosis, Thesis Ph.D., Institute of Rheumatology, Moskva (1993) (Russ.).
337. E. V. Popova, Yu. A. Goryaev and R. G. Saifutdinov, Differential diagnosis of rheumatoid arthritis and osteoarthrosis deformans, Pat SSSR N 1810829 (1989), Bull Izobr. N15:126 (1993) (Russ.).
338. E. V. Popova, Yu. A. Goryaev and R. G. Saifutdinov, Differential diagnosis of rheumatoid arthritis activity, Pat SSSR N 1686357 (1989), Bull Izobr. N39:163 (1991) (Russ.).
339. C. L. Hawkins and M. J. Davies, Direct detection and identification of radicals generated during the hydroxyl radical-induced degradation of hyaluronic acid and related materials, Free Radic. Biol. Med. **21**:275 (1996).
340. A. V. Suchkov and A. I. Sinopal'nikov, Cyanosis, Klinicheskaya Meditsina. **2**:63 (1997) (Russ.).
341. K. Stolze, A. Dadak, Y. Liu and H. Nohl, Hydroxylamine and phenol-induced formation of methemoglobin and free radical intermediates in erythrocytes, Biochem. Pharmacol. **52**:1821 (1996).
342. H. Nohl and K. Stolze, The effects of xenobiotics on erythrocytes, Gen. Pharmacol. **31**:343 (1998).
343. R. P. Bonomo, A. De Flora, E. Rizzarelli, A. M. Santoro, G. Tabbi and M. Tonetti, Copper(II) complexes encapsulated in human red blood cells, J. Inorg. Biochem. **59**:773 (1995).
344. D. Singh, N. B. Nazhat, K. Fairburn, T. Sahinoglu, D. R. Blake and P. Jones, Electron spin resonance spectroscopic demonstration of the generation of reactive oxygen species by diseased human synovial tissue following ex vivo hypoxia-reoxygenation, Ann. Rheum. Dis. **54**:94 (1995).
345. L. Pronai, Y. Ichikawa, H. Nakazawa and S. Arimori, Superoxide scavenging activity of leukocytes in rheumatoid arthritis and Behcet's diseases, Tokai J. Exp. Clin. Med. **15**:93 (1990).
346. R. G. Saifutdinov and E. V. Onuchina, The activity of glutathione reductase of the red cells of patients with rheumatoid arthritis, XXVII-th Congress Ampere, August 21-28, Kazan (1994), pp. 909-910.
347. V. P. Timofeev, V. A. Lapuk and B. A. Samarianov, Differentiation of the spatial mobility of lysine residues in immunoglobulin G using spin labels, Mol. Biol. (Moskva), **28**:1315 (1994) (Russ.).
348. R. Chatterjee, U. Bandyopadhyay, A. Mazumdar and R. K. Banerjee, Lactoperoxidase-catalysed oxidation of indomethacin, a nonsteroidal antiinflammatory drug, through the formation of a free radical, Biochem. Pharmacol. **52**:1169 (1996).
349. S. Sankarapandi and J. L. Zweier, Bicarbonate is required for the peroxidase function of Cu, Zn-superoxide dismutase at physiological pH, J. Biol. Chem. **274**:1226 (1999).
350. C. C. Lin, J. M. Lu, J. J. Yang, S. C. Chuang and T. Ujiie, Anti-inflammatory and radical scavenge effects of Arctium lappa, Am. J. Clin. Med. **24**:127 (1996).
351. R. L. Webber, Oral imaging as a diagnostic tool for assessing osseous changes, J. Bone Miner. Res. Suppl. 2:S543 (1993).
352. M. C. Symons, Radicals generated by bone cutting and fracture, Free Radic. Biol. Med. **20**:831 (1996).
353. E. S. Green, C. Cooper, J. Wrigglesworth and C. Rice-Evans, A novel hydrogen-donating drug suppresses haem damage from myoglobin mediated by oxidised low density lipoproteins, Biochem. Soc. Trans. **20**:330S (1992).

354. T. Kitagawa, M. R. Ondrias, D. L. Rousseau, M. Ikeda-Saito and T. Yonetani, Evidence for hydrogen bonding of bound dioxygen to the distal histidine of oxycobalt myoglobin and hemoglobin, Nature. **298**(N5877):869 (1982).
355. M. Ikeda-Saito, R. S. Lutz, D. A. Shelley, E. J. McKelvey, R. Mattera and H. Hori, Free radicals, lipid peroxides and antioxidants in blood of patients with myotonic dystrophy, J. Biol. Chem. **266**:23641 (1991).
356. K. Yamaguchi, Y. Watanabe and I. Morishima, Push effect on the heterolytic O-O bond clevage of peroxoiron (III) porphyrin adducts, Inorg. Chem. **31**:156 (1992).
357. S. Adachi, S. Nagano, K. Ishimori, Y. Watanabe, I. Morishima, T. Egawa, T. Kitagawa and R. Makino, Roles of proximal ligand in heme proteins: replacement of proximal histidine of human myoglobin with cysteine and tyrosine by site-directed mutagenesis as models for P-450, chloroperoxidase, and catalase, Biochemistry. **32**:241 (1993).
358. X. Chen, G . Guo and S. Li, A study of human keloid with electron spin resonance, Chung Hua Cheng Hsing Shao Shang Wai Ko Tsa Chih. **12**:9 (1996).
359. Z. V. Kuropteva, S. G. Mikaelyan and T. V. Sanikadze, EPR study of skin and liver changes in experimental and clinical burns, Magnitnyi Rezonans v Biologii i Meditsine, Moskva (1981), pp. 221-222 (Russ.).
360. J. Fuchs, H. J. Freisleben, M. Podda, G. Zimmer, R. Milbradt and L. Packer, Nitroxide radical biostability in skin, Free Radic. Biol. Med. **15**:415 (1993).
361. Y. Kawasaki, D. Quan, K. Sakamoto and H. I. Maibach, Dermatology. **194**:238 (1997).
362. A. Iannone, A. Marconi, G. Zambruno, A. Giannetti, V. Vannini and A. Tomasi, Free radical production during metabolism of organic hydroperoxides by normal human keratinocytes, J. Invest. Dermatol. **101**:59 (1993).
363. K. Mader, G. Bacic and H. M. Swartz, In vivo detection of anthralin-derived free radicals in the skin of hairless mice by low-frequency electron paramagnetic resonance spectroscopy, J. Invest. Dermatol. **104**:514 (1995).
364. H. Zhang and R. L. Zheng, Promotion of human sperm capacitation by superoxide anion, Free Radic. Res. **24**:261 (1996).
365. F. W. Kleinhans, V. S. Travis, J. Du, P. M. Villines, K. E. Colvin and J. K. Critser, Measurement of human sperm intracellular water volume by electron spin resonance, J. Androl. **13**:498 (1992).
366. J. Du, F. W. Kleinhans, P. Mazur and J. K. Critser, Human spermatozoa glycerol permeability and activation energy determined by electron paramagnetic resonance, Biochim. Biophys. Acta. **1194**:1 (1994).
367. E. N. Burgova, Human and Animal Tissue PMCs in Some Pathological States, Thesis Ph.D., Institute of Chem. Phys. RAS, Moskva (1986) (Russ.).
368. C. Benedetto, A. Bocci, M. U. Dianzani, B. Ghiringhello, T. F. Slater, A. Tomasi and V. Vannini, Electron Spin Resonance studies on normal human uterus and cervix and on benign and malignant uterine tumors, Cancer Res. **41**:2936 (1981).
369. A. Tomasi, C. Benedetto, M. Nilges, F. Slater, H. M. Swartz and M. C. R. Symons, Studies on human uterine cervix and rat uterus using S-band, X-band and Q-band electron spin resonance spectroscopy, Biochem. J. **224**:431 (**1984**).
370. D. R. Orazvalieva, S. V. Stepankov, E. N. Burgova, N. I. Zorina and A. F. Vanin, Ribonucleatide reductase paramagnetic form in human tissue, Biofizika. **34**:139 (1989) (Russ.).
371. N. Madan, S. Kapoor, U. Rusia, S. Sharma, V. L. Nayyar, K. R. Sundaram and S. K. Sood, ESR and iron status in pregnancy, Indian J. Pathol. Microbiol. **40**:521 (1997).

372. K. Eguchi, T. Sawai, Y. Mizutani and M. Yonezawa, Comparative study of erythrocyte deformability in maternal and cord blood, Am. J. Perinatol. **12**:39 (1995).
373. J Kolena, K Matejcikova and G. Srenkelova, Osmolytes improve the reconstitution of luteinizing hormone/human chorionic gonadotropin receptors into proteoliposomes, Mol. Cell. Endocrinol. **83**:201 (1992).
374. P. Gettins and L. Sottrup-Jensen, NMR and ESR studies on human pregnancy zone protein. Comparison with human alpha 2-macroglobulin, J. Biol. Chem. **265**:7268 (1990).
375. L. Mazzanti, R. A. Rabini, G. Biagini, A. Pugnaloni, R. de Pirro, E. Faloia, V. Mancini, C. Romanini and N. Cester, Changes in membrane fluidity and Na+/K+-ATPase activity during human trophoblast cell culture, Eur. J. Biochem. **206**:881 (1992).
376. A. M. Caccuri, F. Polizio, F. Piemonte, P. Tagliatesta, G. Federici and A. Desideri, Investigation of the active site of human placenta glutathione transferase pi by means of a spin-labelled glutathione analogue, Biochim. Biophys. Acta. **1122**:265 (1992).
377. A. Desideri, A. M. Caccuri, F. Polizio, R. Bastoni and G. Federici, Electron paramagnetic resonance identification of a highly reactive thiol group in the proximity of the catalytic site of human placenta glutathione transferase, J. Biol. Chem. **266**:2063 (1991).
378. M. E. Malone, E. A. Martin, R. M. Jones, R. Jukes, C. K. Lim, L. L. Smith and I. N. White, Peroxidase activation of 4-hydroxytamoxifen to free radicals detected by EPR spectroscopy, Free Radic. Biol. Med. **22**:423 (1997).
379. M. A. Foster, ESR in haematological research, Br. J. Haematol. **47**:1 (1981).
380. C. Mailer, H. M. Swartz, M. Konieczny, et al., Identify of the paramagnetic element found in increased concentrations in plasma of cancer patients and its relationship to other pathological processes, Cancer Res. **34**:637 (1974).
381. E. P. Sidorik, L. M. Korchevaya and A. P. Burlaka, Bile EPR study in tumor process diagnosis, Onkologiya (Kiev). **14**:68 (1979) (Russ.).
382. X. Li, Active oxygen radicals produced by the leukocytes in patients with malignant lymphoma, Chung Hua I Hsueh Tsa Chih. (Taipei). **70**:334, 24 (1990).
383. C. E. Cooper, G. R. Lynagh, K. P. Hoyes, R. C. Hider, R. Cammack and J. B. Porter, The relationship of intracellular iron chelation to the inhibition and regeneration of human ribonucleotide reductase, J. Biol. Chem. **271**:20291 (1996).
384. N. Miyoshi, V. Misнk and P. Riesz, Sonodynamic toxicity of gallium-porphyrin analogue ATX- 70 in human leukemia cells, Radiat. Res. **148**:43 (1997).
385. K. Satoh and H. Sakagami, Effect of copper and iron ions on cytotoxicity induced by ascorbate, gallate and caffeate, Anticancer Res. **17**:2181 (1997).
386. A. Nogaki, K. Satoh, K. Iwasaka, H. Takano, M. Takahama, Y. Ida and H. Sakagami, Radical intensity and cytotoxic activity of curcumin and gallic acid, Anticancer Res. **18**:3487 (1998).
387. K. J. Reszka, P. Bilski, C. F. Chignell, J. A. Hartley, N. Khan, R. L. Souhami, A. J. Mendonca and J. W. Lown, Photosensitization by anticancer agents. 11. Mechanisms of photosensitization of human leukemic cells by diaminoanthraquinones: singlet oxygen and radical reactions, J. Photochem. Photobiol. **B15**:317 (1992).
388. G. Lassmanm, B. Liermann, W. Arnold and K. Schwabe, Ribonucleotide reductase in melanoma tissue. EPR detection in human amelanotic melanoma and quenching of the tyrosine radical by 4-hydroxyanisole, J. Cancer Res. Clin. Oncol. **117**:91 (1991).

389. K. U. Schallreuter, T. E. Elgren, L. S. Nelson Jr, S. MacFarlan, I. Yan-Sze and H.P. Hogenkamp, Ribonucleotide diphosphate reductase from human metastatic melanoma, Melanoma Res. **2**:393 (1992).

390. S. L. Baader, E. Bill, A. X. Trautwein, G. Bruchelt and T. Matzanke, Mobilization of iron from cellular ferritin by ascorbic acid in neuroblastoma SK-N-SH cells: an EPR study, FEBS Lett. **381**:131 (1996).

391. M. Yang, N. B. Nazhat, X. Jiang, S. M. Kelsey, D. R. Blake, A. C. Newland and C. J. Morris, Adriamycin stimulates proliferation of human lymphoblastic leukaemic cells via a mechanism of hydrogen peroxide (H2O2) production, Br. J. Haematol. **95**:339 (1996).

392. B. L. Samuels, J. L. Murray, M. B. Cohen, A. R. Safa, B. K. Sinha, A. J. Townsend, M. A. Beckett and R. R. Weichselbaum, Increased glutathione peroxidase activity in a human sarcoma cell line with inherent doxorubicin resistance, Cancer Res. **51**:521 (1991).

393. E. G. Mimnaugh, C. R. Fairchild, J. P. Fruehauf and B. K. Sinha, Biochemical and pharmacological characterization of MCF-7 drug-sensitive and AdrR multidrug-resistant human breast tumor xenografts in athymic nude mice, Biochem. Pharmacol. **42**:391 (1991).

394. E. E. Voest, E. Van Faassen, J. P. Neijt, J. J. Marx and B. S. Van Asbeck, Doxorubicin-mediated free radical generation in intact human tumor cells enhances nitroxide electron paramagnetic resonance absorption intensity decay, Magn. Reson. Med. **30**:283 (1993).

395. L. Dusre, S. Rajagopalan, H. M. Eliot, J. M. Covey and B. K. Sinha, DNA interstrand cross-link and free radical formation in a human multidrug-resistant cell line from mitomycin C and its analogues, Cancer Res. **50**:648 (1990).

396. A. J. Townsend, C. S. Morrow, B. K. Sinha and K. H. Cowan, Selenium-dependent glutathione peroxidase expression is inversely related to estrogen receptor content of human breast cancer cells, Cancer Commun. **3**:265 (1991).

397. A. Ask, L. Persson, A. Rehnholm, L. Frostesjo, I. Holm and O. Heby, Development of resistance to hydroxyurea during treatment of human myelogenous leukemia K562 cells with alpha-difluoromethylornithine as a result of coamplification of genes for ornithine decarboxylase and ribonucleotide reductase R2 subunit, Cancer Res. **53**:5262 (1993).

398. J. Z. Pedersen, L. Marcocci, L. Rossi, I. Mavelli and G. Rotilio, Generation of daunomycin radicals on the outer side of the erythrocyte membrane, Biochem. Biophys. Res. Commun. **168**:240 (1990).

399. H. Sakagami, K. Satoh, K. Sugaya, M. Iida, N. Hirota, K. Matsumoto, S. Kimura, K. Gomi, S. Taguchi, S. Kato and M. Takeda, Effect of the type of serum in the medium on sodium ascorbate-induced cytotoxicity, Anticancer Res. **16**:1937 (1996).

400. S. L. Schrier, A. Zachowski and P. F. Devaux, Mechanisms of amphipath-induced stomatocytosis in human erythrocytes, Blood. **79**:782 (1992).

401. L. M. Nutter, Y. Y. Wu, E. O. Ngo, E. E. Sierra, P. L. Gutierrez and Y. J. Abul-Hajj, An o-quinone form of estrogen produces free radicals in human breast cancer cells: correlation with DNA damage, Chem. Res. Toxicol. **7**:23 (1994).

402. E. E. Voest, E. van Faassen, B. S. van Asbeck, J. P. Neijt and J. J. Marx, Increased hydrogen peroxide concentration in human tumor cells due to a nitroxide free radical, Biochim. Biophys. Acta. **1136**:113 (1992).

403. S. Mishra, A spin decay assay for tumor necrosis factor cytotoxicity, Indian J. Biochem. Biophys. **32**:254 (1995).

404. D. Nichiporov, V. Kostjuchenko, J. M. Puhl, D. L. Bensen, M. F. Desrosiers, C. E. Dick, W. L. McLaughlin, T. Kojima, B. M. Coursey and S. Zink, Investigation of applicability of alanine and radiochromic detectors to dosimetry of proton clinical beams, Appl. Radiat. Isot. **46**:1355 (1995).

405. H. B. Stone, J. M. Brown, T. L. Phillips and R. M. Sutherland, Oxygen in human tumors: correlations between methods of measurement and response to therapy, Summary of a Workshop, Bethesda, Maryland, November 19-20 (1992), Radiat. Res. **136**:422 (1993).

406. H. M. Swartz, K. J. Liu, F. Goda and T. Walczak, India ink: a potential clinically applicable EPR oximetry probe, Magn. Reson. Med. **31**:229 (1994).

407. H. M. Swartz, G. Bacic, B. Friedman, F. Goda, O. Grinberg, P. J. Hoopes, J. Jiang, K. J. Liu, T. Nakashima, O'Hara J, et al., Measurements of pO2 in vivo, including human subjects, by electron paramagnetic resonance, Adv. Exp. Med. Biol. **361**:119 (1994).

408. J. A. O'Hara, F. Goda, K. J. Liu, G. Bacic, P. J. Hoopes and H. M. Swartz, The pO2 in a murine tumor after irradiation: an in vivo electron paramagnetic resonance oximetry study, Radiat. Res. **144**:222 (1995).

409. C. Giulivi and E. Cadenas, Extracellular activation of fluorinated aziridinylbenzoquinone in HT29 cells EPR studies, Chem. Biol. Interact. **113**:191 (1998).

410. X. B. Qiu and E. Cadenas, The role of NAD(P)H:quinone oxidoreductase in quinone-mediated p21 induction in human colon carcinoma cells, Arch. Biochem. Biophys. **346**:241 (1997).

411. H. Fujiwara, N. Tanaka and T. Suzuki, The effects of anti-cataract drugs on free radicals formation in lenses, Nippon Ganka Gakkai Zasshi. **95**:1071 (1991).

412. H. Fujiwara, Y. Takigawa, T. Suzuki and K. Nakata, Superoxide dismutase activity in cataractous lenses, Jpn. J. Ophthalmol. **36**:273 (1992).

413. M. A. Babizhayev, B. A. Dainyak and A. H. Maxina, ESR spin label and ultrastructural monitoring of protein-lipid interactions in the lens fiber-cell plasma membranes in relation to human ageing and cataractogenesis, Mech. Ageing Dev. **64**:133 (1992).

414. Y. Obara, The oxidative stress in the cataract formation, Nippon Ganka Gakkai Zasshi. **99**:1303 (1995).

415. N. H. Ansari, L. Wang, A. A. Erwin and D. F. Church, Glucose-dependent formation of free radical species in lens homogenate, Biochem. Mol. Med. **59**:68 (1996).

416. A. B. Guliaev, A. E. Dontsov, V. B. Il'iasova and M. A. Ostrovskii, Determination of the melanin level in human melanosomes in the pigment epithelium of the eye depending on age, Dokl. Akad. Nauk. **333**:257 (1993) (Russ.).

417. M. Rybanowska, A. Bober, J. M. Burke and T. Sarna, The role of retinal pigment epithelium melanin in photoinduced oxidation of ascorbate, Photochem. Photobiol. **65**:472 (1997).

418. K. Reszka, G. E. Eldred, R. H .Wang, C. Chignell and J. Dillon, The photochemistry of human retinal lipofuscin as studied by EPR, Photochem. Photobiol. **62**:1005 (1995).

419. D. A. Stoyanovsky, R. Goldman, R. M. Darrow, D. T. Organisciak and V. E. Kagan, Endogenous ascorbate regenerates vitamin E in the retina directly and in combination with exogenous dihydrolipoic acid, Curr. Eye Res. **14**:181 (1995).

420. A. Mazumdar, S. Adak, R. Chatterjee and R. K. Banerjee, Mechanism-based inactivation of lacrimal-gland peroxidase by phenylhydrazine: a suicidal substrate to probe the active site, Biochem. J. **324**:713 (1997).

421. A. I. Mikhailov, I. A. Shilova, M. G. Keshelava and A. A. Kivaev, An application of spin-labeling medicine analogs for investigation of the diffusion processes in mild contact lenses, Fizika i Khimiya Elementarnykh Khimicheskikh Protsessov, V Vserossiiskaya Konferentsiya, Posvyashchennaya 80-letiyu Akademika V.V. Voevodskogo, September 29 - October 3, Chernogolovka (1997), pp. 298-299 (Russ.).
422. V. D. Dragomiretskii, Yu. I. Bazhora and A. G. Zazhivilov, A comprehensive study of URT mucous membrane immunity, VII S'ezd Otorinolaringologov Ukrainskoi SSR, Kiev (1989), p. 302 (Russ.).
423. H. H. Naumann, The mechanism of the respiratory mucosa to wards infections, Acta otolaryng. **89**:165 (1980).
424. L. G. Tsirul'nikov, V. Ya. Kunel'skaya and B. V. Starosvetskii, Lipid peroxidation in patients with chronic pharyngitisis of fungal etiology, Zhurnal Ushnykh, Nosovykh i Gorlovykh Boleznei. N6:45 (1990) (Russ.).
425. O. A. Azizova, A. I. Islomov and D. I. Roshchupkin, Free radicals of lipid biological membranes after exposure to UV-irradiation, Biofizika. **24**:396 (1979) (Russ.).
426. A. G. Shanturov, R. G. Saifutdinov and E. V. Nosulya, EPR Spectroscopy in Clinical Rhinology, Vostochno-Sibirskoe Knizhnoe Izdatelstvo, Irkutsk (1992) (Russ.).
427. E. V. Nosulya and R. G. Saifutdinov, Diagnostic importance of nasal secretion free radical EPR study, Fizika i Khimiya Elementarnykh Khimicheskikh Protsessov, V Vserossiiskaya Konferentsiya, Posvyashchennaya 80-letiyu Akademika V.V. Voevodskogo, September 29 - October 03, Chernogolovka (1997), pp. 275-276 (Russ.).
428. A. G. Shanturov, E. V. Nosulia and R. G. Saifutdinov, Free Radicals of nasal secretion: the nature and diagnostic importance in chronic non-specific diseases of upper respiratory tract, 27th Congress Ampere, August 21-28, Kazan (1994), pp. 912-913.
429. S. Z. Piskunov, The Functional Diagnosis and Treatment of Different Rhinitis Forms, Thesis Dr.Sci., Nauchno-Issledovatel'skii Institut Ukha, Gorla, Nosa i Rechi., Moskva, (1986) (Russ.).
430. G. V. Lavrenova, Nasal Cavity Mucociliar System on Chronic Dust Action, Thesis Dr.Sci., Nauchno-Issledovatel'skii Institut Ukha, Gorla, Nosa i Rechi, Leningrad, (1986) (Russ.).
431. E. V. Nosulya, A. G. Shanturov and R. G. Saifutdinov, Diagnosis of maxillary sius cysts, Pat SSSR N 1826165, (1991) (Russ.).
432. D. Passali and L. Bellussi, Circadian changes in the secretory sctivity of nasal mucosa, Acta otolaryngol. **106**:281 (1998).
433. A. V. Kozlov, D. Yu. Egorov, Yu. A. Vladimirov and S. A. Azizova, The formation of iron-ascorbic acid complexes in tissues, Biofizika. **35**:513 (1990) (Russ.).
434. E. G. Nosulya, High Respiratory Chronic Disease: Diagnostic and Pathogenetic Role of Nasal Secret PMC, Thesis Dr.Sci., Nauchno-Issledovatel'skii Institut Ucha, Gorla, Nosa i Rechi, Sankt-Peterburg (1994) (Russ.).
435. A. Yu. Krotov, Vital problems of otorhinolaryngology, Aktual'nye Voprosy Otorinolaringologii, Alma-Ata (1989), pp. 109-111 (Russ.).
436. B. Majumbar and P. D. Bull, The incidence and bacterilogy of maxillary sinusitis in nasal polyposis, J. Laryng. **96**:937 (1982).
437. J.V. Bannister, W. H. Bannister, H. A. O. Hill and P. J. Thornally, Enhanced production of hydroxyl radicals by the xanthine-xanthine oxidase reaction in the presence of lactoferrin, Biochim. Acta. **715**:116 (1982).

438. A. G. Shanturov, E. V. Nosulya, R. G. Saifutdinov and E. A. Shpakova, Faucial tonsilla tissues PMCs in patientes with chronic tonsilitis, in: Aktual'nye Voprosy Klinicheskoi Meditsiny, Medical Universuty, Irkutsk (1997), pp. 366-367 (Russ.).
439. M. F. Ottaviani, A. Fiorini, P. N. Mason and C. Corvaja, Electron spin resonance studies of dental composites: effects of irradiation time, decay over time, pulverization, and temperature variations, Dent. Mater. **8**:118 (1992).
440. H. G. Goetzke and J. Klammt, Size of wound area and frequency of alveolitis after tooth extraction, Dtsch. Z. Mund. Kiefer. Gesichtschir. **15**:306 (1991).
441. H. G. Goetzke and B. Ebert, Experimental studies of the size of the root surface of permanent human teeth, Anat. Anz. **172**:117 (1991).
442. A. Hochi, M. Furusawa and M. Ikeya, Applications of microwave scanning ESR microscope: human tooth with metal, Appl. Radiat. Isot. **44**:401 (1993).
443. S. Mascarenhas, A. Hasegawa and K. Takeshita, ESR dosimetry of bones from Hiroshima A-bomb site, Bull. Amer. Phys. Soc. **18**:579 (1973).
444. B. Pass and J. E. Aldrich, Dental enamel as in vivo dosimetr, Med. Phys. **12**:305 (1985).
445. J. E. Aldrich and B. Pass, Dental Enamel as an in vivo radiation dosemeter: separation of the diagnostic X ray dose from the dose due to natural sources, Radiation. Protection Dosimetry. **17**:175 (1986).
446. I. Caracelli, M. C. Terrile and S. Mascarenhas, ESR and dosimetric properties of bones, Health Phys. **50**:259 (1986).
447. Y Nishiwaki and T. Shimano, Uncertainties in dose estimation under emergency conditions and ESR dosimetry with human teeth, Radiat. Protect. Dosim. **34**:295 (1990).
448. B. Pass, J. E. Aldrich and P. L. Scallion, An analysis of paramagnetic centers in irradiated dentin using electron spin resonance, Calcif. Tissue. Int. **46**:166 (1990).
449. E. J. Rhodes and R. Grun, ESR behavior of the paramagnetic centre at g = 2.0018 in tooth enamel, Third International Symposium on ESR Dosimetry and Applications, October 14-18, Japan (1991), p. 62.
450. E. J. Rhodes and R. Grun, ESR behavior of the paramagnetic centre at g = 2.0018 in tooth enamel, Ancient. **9**:14 (1991).
451. M. F. Desrosiers, M. J. Avila, D. A. Schauer, B. M. Coursey and N. J. Parks, Experimental validation of radiopharmaceutical absorbed dose to mineralized bone tissue, Appl. Radiat. Isot. **44**:459 (1993).
452. T. Straume, L. R. Anspaugh, E. H. Haskell, J. N. Lucas, A. A. Marchetti, I. A. Likhtarev, V. V. Chumak, A. A. Romanyukha, V. T. Khrouch, Y. I. Gavrilin and V. F. Minenko, Emerging technological bases for retrospective dosimetry. Radiation injury and the chernobyl catastrophe, Stem. Cells (AlphaMed Press), **15**(Supll 2):183 (1997).
453. A. I. Ivannikov, V. G. Skvortzov, V. F. Stepanenko, D. D. Tikunov, I. M. Fedosov, A. A. Romanyukha and A. Wieser, Wide-scale EPR retrospective dosimetry: results and problems, Radiat. Protect. Dosimetry. **71**:175 (1997).
454. E. H. Haskell, R. B. Hayes and G. H. Kenner, An EPR dosimetry method for rapid scanning of children following a radiation accident using deciduous teeth, Health Phys. **76**:137 (1999).
455. B. Pass, A. E. Baranov, E. D. Kleshchenko, J. E. Aldrich, P. L. Scallion and R. P. Gale, Collective biodosimetry as a dosimetric "gold standard": a study of three radiation accidents, Health Phys. **72**:390 (1997).

456. A. A. Romanyukha, D. Regulla, E. Vasilenko and A. Wieser, South Ural nuclear workers: comparison of individual doses from retrospective EPR dosimetry and operational personal monitoring, Appl. Radiat. Isot. **45**:1195 (1994).
457. A. Wieser, E. Haskell, G. Kenner and F. Bruenger, EPR dosimetry of bone gains accuracy by isolation of calcified tissue, Appl. Radiat. Isot. **45**:525 (1994).
458. F. Callens, P. Moens and R. Verbeeck, An EPR study of intact and powdered human tooth enamel dried at 400 degrees C, Calcif. Tissue Int. **56**:543 (1995).
459. F. Callens, M. Verbeeck, P. Matthys, L. Martens and E. Boesman, The contribution of CO_3(3-) and CO_2(-) to the ESR spectrum near g = 2 of powdered human tooth enamel, Calcif. Tiss Int. **41**:124 (1987).
460. A. Rossi and G. Poupeau, Radiation damage in bioapatites: the ESR spectrum of irradiated dental enamel revisited, Nucl. Tracks. Radiat. Meas. **17**:537 (1990).
461. M. Ikeya, J. Mijajima and S. Okajima, ESR dosimetry for atomic bomb survivors using shell buttons and tooth enamel, Jap. J. Appl. Phys. **23**:L697 (1984).
462. M. Ikeya, New Applications of ESR. Dating, Dosimetry and Microscopy, World Scientific, Singapore (1993).
463. H. Schwarcz, ESR studies of tooth enamel, Nucl. Tracks. **10**:865 (1985).
464. N. Nakamura, C. Miyazawa, M. Akiyama, S. Sawada and A. A. Awa, Biodosimetry: chromosome aberration in lymphocytes and electron paramagnetic resonance in tooth enamel from atomic bomb survivors, World Health Stat. Q. **49**:67 (1996).
465. G. Liidja, J. Past, J. Puskar and E. Lippmaa, Paramagnetic resonance in tooth enamel created by ultraviolet light, Appl. Radiat. Isot. **47**:785 (1996).
466. M. Iwasaki, C. Miyazawa and T. Shimano, Relation between the weight of human tooth enamel and the CO_3(3-) signal intensity on the ESR dosimetry, Ohu Daigaku Shigakushi. **17**:95 (1990).
467. M. Iwasaki, C. Miyazawa and T. Shimano, A problem with ESR dosimetry utilizing human tooth enamel in low dose region, Ohu Daigaku Shigakushi. **17**:303 (1990).
468. E. A. Ignatiev, A. A. Romanyukha, A. A. Koshta and A. Wieser, Selective saturation method for EPR dosimetry with tooth enamel, Appl. Radiat. Isot. **47**:333 (1996).
469. E. A. Ignatiev, EPR Spectroscopy as a Dosimetry Method in Carbonated Hydroxy Apatite, Thesis Ph.D., Ural University, Ekaterinburg (1998) (Russ.).
470. V. Skvortzov, A. Ivannikov and U. Eichhoff, Assessment of individual accumulated irradiation dose using EPR spectroscopy of tooth enamel, J. Molec. Struct. **347**:321 (1995).
471. J. E. Aldrich, B. Pass and C. Mailer, Changes in the paramagnetic centres in irradiated and heated dental enamel studied using EPR, Int. J. Radiat. Biol. **61**:433 (1992).
472. T. Room, G. Liidja and E. Lippmaa, Temperature and frequency effects in tooth enamel electron spin resonance dosimetry, Appl. Radiat. Isot. **45**:1061 (1994).
473. V. E. Galtsev, E. V. Galtseva, O. Ia. Grinberg and Ia.S. Lebedev, Increase in sensitivity of ESR-dosimetry from tooth enamel, Dokl. Akad. Nauk. **334**:649 (1994) (Russ.).
474. J. Aldrich and B. Passm, Determining radiation exposure from nuclear accidents and atomic tests using dental enamel, Health Physics. **54**:469 (1988).
475. G. A. Klevezal, V. A. Serezhenkov and V. N. Kalyakin, Radiation dosage accumulated by reindeer from Novaya Zemlya, Appl. Radiat. Isot. **46**:1077 (1995).
476. J. Tatsumi-Miyajima and S. Okajima, Physical dosimetry at Nagasaki-Europium-152 of stone embankment and electron spin resonance of teeth from atomic bomb survivors, J. Radiat. Res. (Tokyo). (Suppl 32):83 (1991).

477. N. Nakamura, C. Miyazawa, S. Sawada, M. Akiyama and A. A. Awa, A close correlation between electron spin resonance (ESR) dosimetry from tooth enamel and cytogenetic dosimetry from lymphocytes of Hiroshima atomic-bomb survivors, Int. J. Radiat. Biol. **73**:619 (1998).
478. N. Nakamura, J. F. Katanic and C. J. Miyazawa, Contamination from possible solar light exposures in ESR dosimetry using human tooth enamel, Radiat. Res. (Tokyo). **39**:185 (1998).
479. A. A. Romanyukha, M. O. Degteva, V. P. Kozheurov, A. Wieser, P. Jacob, E. A. Ignatiev and M. I. Vorobiova, Pilot study of the Urals population by tooth EPR dosimetry, Radiat. Environ. Biophys. **35**:305 (1996).
480. A. A. Romanyukha and D. Regulla, Aspects of retrospective ESR dosimetry (Invited Paper), Appl. Radiat. Isot. **47**:1293 (1996).
481. A. A. Romanyukha, D. Regulla, E. Vasilenko, A. Wieser, E. G. Drozhko, A. F. Lyzlov, N. A. Koshurnikova, N. S. Shilnikova and A. P. Panfilov, Verification of occupational doses at the first nuclear plant in the Former Soviet Union, Appl. Radiat. Isot. **47**:1277 (1996).
482. A. A. Romanyukha, E. A. Ignatiev, M. O. Degteva, V. P. Kozheurov, A. Wieser and P. Jacob, Radiation doses from Ural region [letter], Nature. **381**(N6579):199 (1996).
483. A. A. Romanyukha, A. Wieser and D. Regulla, EPR dosimetry with different biological and synthetic carbonated materials, Radiation Protect. Dosimetry. **65**:389 (1996).
484. A. Wieser, A. A. Romanyukha, M. O. Degteva, V. P. Kozheurov and G. Petzoldt, Tooth enamel as a natural veta dosimeter for bone seeking radionuclides, Radiat. Protect. Dosimetry. **65**:413 (1996).
485. M. Bauchinger, K. Salassidis, H. Braselmann, A. Vozilova, S. Pressl, G. Stephan, G. Snigiryova, V. P. Kozheurov and A. Akleyev, FISH-based analysis of stable translocations in a Techa River population, Int. J Radiat. Biol. **73**:605 (1998).
486. O. V. Aleinikov, V. S. Finin, V. A. Deshko and A. V. Alekseichik, EPR dosimetry measuring of radioactive irradiation doses in oncohematological children, Fizika i Khimiya Elementarnykh Khimicheskikh Protsessov, V Vserossiiskaya Konferentsiya, Posvyashchennaya 80-letiyu Akademika V.V. Voevodskogo, September 29 - October 3, Chernogolovka (1997), pp. 190-191 (Russ.).
487. E. H. Haskell, G. H. Kenner and R. B. Hayes, Electron paramagnetic resonance dosimetry of dentine following removal of organic material, Health Phys. **68**:579 (1995).
488. T. Shimano, M. Iwasaki, C. Miyazawa, T. Miki, A. Kai and M. Ikeya, Human tooth dosimetry for gamma-rays and dental X-rays using ESR, Appl. Radiat Isot. **40**:1035 (1989).
489. M. Desrosiers, M. Simic, F. Eichmiller, F. Johnston and R. Bowen, Mechanically induced generation of radicals in tooth enamel, Appl. Radiat. Isot. **40**:1195 (1989).
490. M. F. Desrosiers and A. A. Romanyukha, Technical aspects of the EPR method for tooth enamel dosimetry, in: Biomarkers: Medical and Workplace Application, M. L. Mendelson, L. C. Mohr and J. P. Peters, eds., DC: Josef Henry Press, Washington (1998).
491. V. Polyakov, E. Haskell, G. Kenner, G. Huett and R. Hayes, Effect of mechanically induced background signal on EPR dosimetry of tooth enamel, Radiat. Meas. **24**:249 (1995).

492. C. Yamanaka, M. Ikeya and H. Hara, In vivo ESR dosimeter for human tooth enamel, Third International Symposium on ESR Dosimetry and Applications, October 14-18, Japan (1991), p.23.
493. C. Yamanaka, M. Ikeya and H. Hara, ESR cavities for in vivo dosimetry of tooth enamel, Appl. Radiat. Isot. **44**:77 (1993).
494. S. L. Breen and J. J. Battista, Radiation dosimetry in human bone using electron paramagnetic resonance, Phys. Med. Biol. **40**:2065 (1995).
495. M. F. Desrosiers, EPR dosimetry following the San Salvador radiation accident, Third International Symposium on ESR Dosimetry and Applications, October 14-18, Japan (1991), p. 24.
496. L. A. Blumenfel'd and A. E. Kalmanson, EPR spectra of biological samples, Biofizika. **2**:552 (1957) (Russ.).
497. S. Onori, F. dErrico, C. De Angelis, E. Egger, P. Fattibene and I. Janovsky, Alanine dosimetry of proton therapy beams, Med. Phys. **24**:447 (1997).
498. R. Kudynski, J. Kudynska and H. A. Buckmaster, The decay constant for irradiated human hair using EPR spectroscopy, Appl. Radiat. Isot. **45**:645 (1994).
499. V. G. Bebeshko, V. I. Klimenko, L. N. Iukhimuk and I. M. Pogontseva, The paramagnetic centers of the blood in persons with a history of acute radiation sickness as a result of the accident at the Chernobyl Atomic Electric Power Station, Lik. Sprava. **4**:7 (1992).
500. W. Kreisel, International program on the health effects of the Chernobyl accident, Stem. Cells (Dayt). Suppl 1:33 (1995).
501. N. J. Dodd, Free radicals and food irradiation, Biochem. Soc. Symp. **61**:247 (1995).
502. H. Delincee, International cooperation on the detection of irradiated food, Z. Lebensm. Unters. Forsch. **197**:217 (1993).
503. J. Mischke, N. Helle, B. Linke, G. A. Schreiber and K. W. Bogl, Electron spin resonance measurements on dried fruit, Carbohydrate composition and ESR signal structure of irradiated fruit, Z. Ernahrungswiss. **33**:258 (1994).
504. N. Helle, A. Wolbert, B. Linke, D. Ehlers and K. E. Kruger, Electron spin resonance spectroscopy investigations of fresh fruit. Evidence of treatment with ionizing radiation, Z. Lebensm. Unters. Forsch. **201**:355 (1995).
505. T. Miyazaki, T. Kaneko, T. Yoshimura, A. S. Crucq and B. Tilquin, Electron spin resonance study of radiosterilization of antibiotics: ceftazidime, J. Pharm. Sci. **83**:68 (1994).
506. T. Miyazaki, J. Arai, T. Kaneko, K. Yamamoto, M. Gibella and B. Tilquin, Electron spin resonance study of radiosterilization of antibiotics: ceftazidime, J. Pharm. Sci. **83**:1643 (1994).
507. V. E. Kholmogorov, Photoinduced paramagnetic centers in human blood, Biofizika. **39**:888 (1994) (Russ.).
508. M. Koter, Effect of hyperthermia on the internal microviscosity of erythrocytes and lymphocytes: a spin-label study, Int. J. Radiat. Biol. **58**:157 (1990).
509. O. Korkmaz and M. Korkmaz, The effects of mitomycin-C and temperature on dynamical properties of human erythrocyte membrane, Biophys. Chem. **69**:167 (1997).
510. W. Malorni, S. Paradisi, F. Iosi and M. T. Santini, Two different pathways for necrotic cell death induced by free radicals, Cell. Biol. Toxicol. **9**:119 (1993).
511. J. J. Eckburg, J. C. Chato, K. J. Liu, M. W. Grinstaff, H. M. Swartz, K. S. Suslick and F. P. Auteri, The measurement of temperature with electron paramagnetic resonance spectroscopy, J. Biomech. Eng. **118**:193 (1996).

512. P. J. Halsall, F. R. Ellis and P. F. Knowles, Evaluation of spin resonance spectroscopy of red blood cell membranes to detect malignant hyperthermia susceptibility [see comments], Br. J. Anaesth. **69**:471 (1992).
513. S. K. Sedov, The Influence of the Sanitary-hygienic Condition of Labour on the Healthy of the Foundry Workers, Thesis Ph.D., Institute Profpatologii, Moskva (1983) (Russ.).
514. A. F. Vanin, L. V. Vakhnina and A. G. Chetverikov, On nature EPR signal in cancer tissues, Biofizika. **15**:1044 (1970) (Russ.).
515. C. K. Narkowicz, J. H. Vial and P. W. McCartney, Hyperbaric oxygen therapy increases free radical levels in the blood of humans, Free Radic. Res. Commun. **19**:71 (1993).
516. M. Minetti, T. Forte, M. Soriani, V. Quaresima, A. Menditto and M. Ferrari, Iron-induced ascorbate oxidation in plasma as monitored by ascorbate free radical formation. No spin-trapping evidence for the hydroxyl radical in iron-overloaded plasma, Biochem. J. **282** (Pt 2):459 (1992).
517. A. Burt, I. B. Abramov and V. I. Sachkov, Is fluidization of the phospholipid bilayer by general anesthetics a reason for general anesthesia? Bull. Eksp. Biol. Med. **112**:387 (1991) (Russ.).
518. L. Mazzanti, R. A. Rabini, R. Staffolani, G. Benedetti, N. Cester and G. Lenaz, Modifications induced by general anesthetics on Na+/K+ ATPase obtained from human placenta, Biochem. Biophys. Res. Commun. **173**:1248 (1990).
519. M. Kudou, T. Kudou and A. Matsuki, Changes in plasma superoxide dismutase like activity during general anesthesia and surgery in man, Masui. **39**:1172 (1990).
520. S. Shirasaki, M. Kudo, H. Ishihara, A. Matsuki and T. Takazawa, The effect of enflurane anesthesia on plasma superoxide dismutase (SOD)-like activity in elderly patients, Masui. **40**:1079 (1991).
521. K. Gwozdzinski, A spin label study of the action of cupric and mercuric ions on human red blood cells, Toxicology. **65**:315 (1991).
522. R. P. Mason, In vitro and in vivo detection of free radical metabolites with electron spin resonance, in: Free Radicals. A Practical Approach, N. A, Punchard, F. J. Kelly, eds., Oxford University Press, New York (1996), pp. 12-24.
523. R. P. Mason, Electron spin resonance investigation of free radical toxicology, in: Free Radicals in Biology and Environment, F. Minisci., ed., Kluwer Academic Publishers, The Netherlands (1997), pp. 1-27.
524. K. Minakata, O. Suzuki, S. Saito and N. Harada, Ascorbate radical levels in human sera and rat plasma intoxicated with paraquat and diquat, Arch. Toxicol. **67**:126 (1993).
525. J. Lee, C. H. Trad and D. A. Butterfield, Electron paramagnetic resonance studies of the effects of methoxyacetic acid, a teratologic toxin, on human erythrocyte membranes, Toxicology. **83**:131 (1993).
526. A. Constantinescu, H. Tritschler and L. Packer, α-Lipoic acid protects against hemolysis of human erythrocytes induced by peroxyl radicals, Biochem. Mol. Biol. Int. **33**:669 (1994).
527. S. D. Aust, C. F. Chignell, T. M. Bray, B. Kalyanaraman and R. P. Mason, Free radicals in toxicology, Toxicol. Appl. Pharmacol. **120**:168 (1993).
528. G. F. Weber, The measurement of oxygen-derived free radicals and related substances in medicine, J. Clin. Chem. Clin. Biochem. **28**:569 (1990).
529. W. A. Pryor and S. S. Godber, Noninvasive measures of oxidative stress status in humans, Free Radic. Biol. Med. **10**:177 (1991).

530. Yu. V. Zobnin, R. G. Saifutdinov and I. P. Provado, The influence of hemosorption on the contens of the paramagnetic centres in the patients with opium abstinence, XXVII-th Congress Ampere, August 21-28, Kazan (1994), pp. 927-928.

531. H. F. Galley, M. J. Davies and N. R. Webster, Xanthine oxidase activity and free radical generation in patients with sepsis syndrome, Crit. Care Med. **24**:1649 (1996).

532. A. Fraticelli, C. V. Serrano Jr, B. S. Bochner, M. C. Capogrossi and J. L. Zweier, Hydrogen peroxide and superoxide modulate leukocyte adhesion molecule expression and leukocyte endothelial adhesion, Biochim. Biophys. Acta. **1310**:251 (1996).

533. L. A. Sadokhina, A Study of NO Endogenous Synthesis Markers in Diffuse Abscess Peritonitis, Thesis Ph.D., Medical University, Irkutsk (1998) (Russ.).

534. R. G. Saifutdinov, L. A. Sadochina and E. G. Grigorjev, EPR study of MetHb in the peritonitis patient intestinal content, Rossiiskii Zhurnal Gastroenterologii, Gepatologii, Koloproktologii. **6**:60 (1996) (Russ.).

535. R. G. Saifutdinov, L. A. Sadochina and E. G. Grigorjev, Investigated of paramagnetic centres of the intestinal content by the method of ESR at the patients with the peritonitis, The Ninth Annual Conference of the International Society for Environmental Epidemiology, Academia Sinica International Center, Taiwan, Republic of China Taipei (1997), p. 242.

536. R. G. Saifutdinov, L. A. Sadochina and E. G. Grigorjev, Paramagnetic centres of the intestinal content investigated by the method of the ESR at the patients with the peritonitis, Eighth International Conference on Bioinorganic Chemistry, 27 July - 1 August, Japan, Yokohama, (1997), J. Inorganic. Biochem. **67**:132 (1997).

537. R. G. Saifutdinov, L. A. Sadokhina and E. G. Grigoriev, An EPR study of MetHb intestinal-content in peritonitic patients, International Workshop Fal'ka, June 20, Sankt-Peterburg (1996), p. 72.

538. R. G. Saifutdinov and L. A. Sadokhina, New aspects of endogenous intoxication in peritonitis, Rossiiskii Gastroenterologicheskii Zhurnal. N4:131 (1997) (Russ.).

539. R. G. Saifutdinov, L. A. Sadochina and. G. Grigoriev, Application of EPR/ESR at prognosis of the course of peritonitis judging by the contents of methaemoglobin in the intestinal contents, The 4th Russia-Japan International Medical Symposium, September 3-8, Irkutsk (1996), pp. 28-29.

540. R. G. Saifutdinov, L. A. Sadochina and E. G. Grigoriev, Application of EPR/ESR at prognosis of the course of peritonitis judging by the contents of methaemoglobin in the intestinal contents, International Conference EPR-Spectroscopy of Nitric Oxide in Biological Systems, December 11-15, Suzdal (1996), pp. 28-29.

541. A. Ya. Lysenko and M. V. Pavlovskaya, SPID-Assotsiiruemye Infektsii i Invazii, Medicine, Moskva (1992) (Russ.).

542. A. E. Ades, C. F. Davison, F. J. Holland, D. M. Gibb, C. N. Hudson, A. Nicholl, G. D. Goldberg and C. S. Peckham, Vertically transmitted HIV infection in the British Isles, Brit. Med. J. **306**:1296 (1993).

543. L. M. Gordon, C. C. Curtain, Y. C. Zhong, A. Kirkpatrick, P. W. Mobley and A. J. Waring, The amino-terminal peptide of HIV-1 glycoprotein 41 interacts with human erythrocyte membranes: peptide conformation, orientation and aggregation, Biochim. Biophys. Acta. **1139**:257 (1992).

544. L. M. Gordon, C. C. Curtain, V. McCloyn, A. Kirkpatrick, P. W. Mobley and A. J. Waring, The amino-terminal peptide of HIV-1 gp41 interacts with human serum albumin, AIDS Res. Hum. Retroviruses. **9**:1145 (1993).

Chapter 4

OTHER ASPECTS OF EPR INVESTIGATION OF BIOLOGICAL SPECIMENS

The development of new techniques, particularly with the use of spin probes and spin labels, has extended the capabilities of studying the biological specimens using EPR. This resulted in the increase in the number of investigations of both the materials obtained from humans in various pathological states, and food products, medicines and other biological objects.

4.1 FOOD PRODUCTS

Phenolics in red wine inhibit lipid oxidation [1]. By using EPR spectroscopy red (Brown Bros., 12L, cask) and white (Tisdall, 12L, cask) wines have been studied. EPR signals are not detected in the spectra of regular wines. But after the wines are evaporated in vacuum in the cold down to one tenth of the initial volume, an Mn^{2+} ion signal is detected in the spectrum of the syrup. Beside this signal, the ESR spectrum of the evaporated red wine (but not white wine) shows a free radical signal (half width of 0.2 ± 0.01mT and g = 2.0038 ± 0.0001). This was thought to be due to the fact, that red wine had been fermented on oak, and the white wine had not. Nevertheless, in the spectrum of the "Cab Mac" red wine, not subjected to oak-fermentation, both the Mn^{2+} and free radical signals were detected. These two signals were observed in the spectrum of the oak-fermented Semillon (Yaldara) white wine. These data suggest that all the wines contain Mn^{2+} playing an important role in the grape physiology. EPR spectroscopy was also used to study the concentrated extracts and juices of red (shiraz) and white (semillon) grapes. Free radicals were detected in the extracts of red grapes, but not in the extracts of white grapes. Thus, free radicals are present in the natural red grape juice and absent in the white grape juice. These radicals are associated with phenols, because the intensity of EPR signal corresponding to the red wine extracts treated with polyvinylpyrrolidone is reduced by approximately 80%. That is why the red wine *in vitro* inhibits lipid oxidation [2].

Carbohydrates (sugars) are identified as the source of free radicals in beers, stouts, coffees and ersatz coffees, a number of other food flavoring agents, *etc*. These radicals are not derived from phenolic constituents, in contrast to those in wine [3].

The effect of 8 different flavonoids on erythrocyte deformability, osmotic stability and aggregation capability has been shown by EPR spectroscopy [4].

A number of food products are able to trap active radicals, which prevents tissue degradation. The green tea contains a lot of catechins – epicatechin, picatechingallate, epigallocatechin and epigallocatechingallate, which possess an antioxidant effect [5, 6]. The EPR technique was applied to study free radical "scavengers" such as green tea extract, sunflower seeds. Vitamins A, C and E, fruit and vegetable flavonoids [7-10]. A 0.05% solution of medicine "β-catechine" completely scavenges 1,1,-diphenyl-2-picrylhydrazyl (6.1 x 10^{15} sp/ml). A 10% solution of β-catechin completely scavenged superoxide radicals (4.2 x 10^{15} spins/ml) generated by the hypoxanthine-xanthine oxidase system, and an undiluted solution of this preparation scavenged about 90% of hydroxyl radicals (3.5 x 10^{15} spins/ml) generated by the Fenton reaction. "β-Catechin" is recommended as an oxidant having a prophylactic effect against aging and neurological diseases related to free radical mechanisms [8-10].

The free radical scavenging activity of the Japanese herbal medicine, Toki-Shakuyaku-San (TJ-23; TSUMURA & Co., Tokyo, Japan), was examined using EPR spectrometry. This medicine scavenges 1,1-diphenyl-2-picrylhydrazyl radicals (DPPH), superoxide, and hydroxyl radicals in a dose-dependent manner [1].

BG-104, a compound of Chinese herbs, inactivates superoxide ion-radical in cell free systems. The EPR technique was used to study the effect of BG-104 treatment in patients with Behcet's disease, Sjogren's syndrome or hematological malignancy, and compared with that of vitamin E. The plasma superoxide scavenging activity was significantly lower in these patients as compared to that in healthy controls. The treatment with BG-104 and/or vitamin E significantly enhanced the plasma superoxide scavenging activity and lowered the erythrocyte sedimentation rates, the absolute number of neutrophils, as well as C-reactive protein levels. These results indicate that the BG-104 has an anti-inflammatory effect [12].

EPR spectroscopy provides a very convenient precision instrument to determine stable free radical concentrations on radiation-exposed food (meat and fish bones, marine products, fruits, vegetables [13].

4.2 DRUGS

EPR spectroscopy and fluorescence polarization of fluorescence quenching of membrane tryptophan residues have been used to investigate the interaction of bee venom (melittin) with erythrocyte membrane ghosts. Melittin induces disorders in the lipid-protein matrix both in the hydrophobic core of bilayer and at the polar/non-polar interface of melittin complexes with erythrocyte membranes. The peptide acts most efficiently at a concentration of the order of 10^{-10} mol/mg of membrane protein. The effect of crude honey bee venom is of similar magnitude as the effect of pure melittin at the same concentration. The EPR data using two lipophilic spin labels (methyl 5-doxylpalmitate and 16-doxylstearic acid) show that melittin at an increasing concentration induces a well marked rigidization in the deeper regions of lipid bilayer. The effect of rigidization near the membrane surface maximizes at melittin concentration of 10^{-10} mol/mg (10^{-4} M of melittin per mole of membrane phospholipid). Melittin also effectively immobilizes membrane proteins. Thus, melittin induces rigidization of the lipid bilayer and the rearrangements in membrane protein pattern and, consequently, alters the lipid-protein interactions [14].

The scavenging effect of berbamine on active oxygen radicals was studied using EPR spectroscopy (with the use of a spin trapping technique) and chemiluminescence methods in phorbol myristate acetate stimulated polymorphonuclear leukocytes and in cell-free superoxide or hydroxyl radical generating systems. Berbamine (0.1 to 0.3 mM) effectively reduces active oxygen radicals in PMA-stimulated polymorphonuclear leukocytes. EPR oxymetry shows the absence of obvious effect on oxygen consumption during the respiratory burst of these cells. Berbamine (0.3 mM) quenched $O_2^{-\bullet}$ in xanthine/xanthine oxidase and irradiation riboflavin systems, and radicals $HO^\bullet$ in the Fenton reaction. Radicals $HO^\bullet$ were trapped as Vitamin E. Berbamine scavenging action was stronger than that of Vitamin E in the xanthine/xanthine oxidase system but weaker than that of Vitamin E in the riboflavin system [15].

EPR showed that natural Kampo medicines possess $HO^\bullet$ radical scavenging activity [16] whereas 1-O-hexyl-2.3.5-trimethylhydroquinone quenches superoxide anion radical, hydroxyl and *tert*-butyl peroxyl radicals and singlet oxygen [17].

Using EPR spin trapping technique, it has been found that baicalin (B) can scavenge hydroxyl radicals generated from Fenton reaction, as well as superoxide radicals generated from the reaction system containing xanthine and xanthine oxidase. Kinetic studies on the competition between baicalin and a spin trap (DMPO) in trapping $HO^{\bullet}$ showed that baicalin had a kinetic reactive rate constant of the order of 7.7×10^{11} $M^{-1}s^{-1}$. The studies on the competition between baicalin and ferricytochrome C in trapping $O_2^{-\bullet}$ gave a kinetic reactive rate constant of 3.2×10^{6} $M^{-1}s^{-1}$. The baicalin could prevent hydrogen peroxide-induced erythrocyte hemolysis and protect the conformation of sulfhydryl groups (SH) on membrane proteins. It also reduced the membrane fluidity of erythrocytes incubated with hydrogen peroxide. The baicalin could protect the membranes of erythrocytes from free radical injuries more effectively than α-tocopherol [18].

The effect of different flavonoids on the lifetime of ascorbic acid radicals has studied by EPR spectroscopy. The results show that the introduction of flavonoids into the ascorbic acid/ascorbate oxidase leads to in increase in both the initial concentration and lifetime of ascorbate acid free radicals [19]

EPR and fluorescence techniques were used to study the effects of L-carnitine [$(CH_3)_3N^+CH_2CH(OH)CH_2COO^-$, Vitamin B_T] and its short-chain esters (acetyl-L-carnitine and propionyl-L-carnitine) on the natural fluidity gradient and molecular packing of phospholipid headgroups of erythrocyte membrane in intact cells. For this purpose, the purified erythrocyte suspensions (labeled with different stearic acid derivatives containing a stable doxyl radical ring at the C-5, C-7, C-12 and C-16) atoms were incubated with 0.5-5 mM of L-carnitine and its esters for 60 min at 37°C. A decrease in the order parameter, calculated from the ESR spectra of the 5-doxylstearic acid derivative was observed at all the concentrations of propionyl-L-carnitine. The extent of the decrease depends on the dose and temperature. An increase of the chain length between the doxyl ring and the carboxylic group of the spin label, results in a much lower efficacy of propionyl-L-carnitine in decreasing the order parameter. Acetyl-L-carnitine also shows a significant effect on the molecular order but only at lower temperatures and higher concentration of the drug. L-Carnitine does not modify the molecular dynamics at the temperatures and concentrations used. [20].

The EPR technique using a spin label (16PC) was employed to investigate the effect of gramicidin A aggregation on the phase structure of dipalmitoylphosphatidylcholine (DPPC) lamillar vesicles [21]. An analogous data are obtained by 2D EPR spectroscopy [22].

The EPR spin trapping technique data show that, during the autoxidation of tetracycline antibiotics, cytotoxic oxygen radicals such as $HO^{\bullet}$, $O_2^{-\bullet}$, H_2O_2 and $^1O_2$ are formed [23].

The reduction of the colony-forming ability, induced by DNA strand breaks is detected in human fibroblasts after treatment with tetracycline in the presence of light. The strand break formation is suppressed by adding catalase. This is due to the formation of EPR-detectable tetracycline-derived radicals [24].

Streptonigrin displays a clearly defined antitumoral activity. Its mechanism of action is based on DNA synthesis inhibition. The EPR spectra of zinc(II)- and cadmium(II)-streptonigrin semiquinone complexes have been examined. The spectral data show that metal binding significantly affects density of unpaired electron located in the quinolinesemiquinone fragment of the above complex. This is proved by superfine structure constants in the EPR spectra of the complexes [25].

An EPR monitoring of etoposide (VP-16) cytotoxicity against human leukemic cells K562 has been carried out. It is found that the toxicity can be decreased on addition of reduced glutathione [26].

Using EPR spectroscopy, the interaction of anti-inflammatory preparation, indomethacin with PGH synthase and hydroperoxides was studied [27].

EPR, NMR and UV spectroscopic techniques were used in investigation of the Cu^{2+}-diisopropylsalicilate complex, which attracts much attention due to its wide spectrum of pharmacological activity. The data obtained provide evidence for the presence of more than one Cu-binding sites on human serum albumine [28].

The EPR technique was used to study an antimicrobial preparation, sulfacetamide (N-[4-(amino(phenyl)sulfonyl]acetamide) and Co^{2+}, Ni^{2+}, and Cu^{2+} complexes with sulfacetamide. Their structure and antimicrobial activity have been evaluated [29].

The interaction between heptacaine (local anesthetic agent) homologs and phosphatidyl was studied by the EPR method using spin labels [30].

EPR spectroscopy was used to study pharmacokinetics of recombinant human superoxide dismutase, NK341, upon intravenous administration to healthy male volunteers at doses of 1750-35000 U/kg, and to prove the SOD safety (good tolerance). The values of maximal plasma SOD concentration linearly increases with the doses given. The superoxide dismutase concentration in the urine also correlates with its plasma concentration. It is suggested that the recombinant human superoxide dismutase NK341 is present in blood as the active form of superoxide dismutase [31].

Using EPR spectroscopy the cardioprotective effects of human recombinant extracellular superoxide dismutase type C (SOD-C) were compared with those of bovine Cu,Zn-superoxide dismutase in isolated working rat heart subjected to 35-min global normothermic ischemia followed by 55-min reperfusion. Human superoxide dismutase C or bovine Cu,Zn-superoxide dismutase (3×10^4 and 6×10^4 IU/L, respectively) was added to a cardioplegic solution infused 5 min before and 10 min after the ischemic period. Intact hearts were treated with St. Thomas' Hospital cardioplegic solution without superoxide dismutase. By the end of reperfusion, superoxide dismutase C-treated hearts recovered the left ventricular systolic pressure, aortic flow and cardiac output to (95 ± 4) %, (60 ± 4) %, (69 ± 6)% of preischemic value, respectively, and as compared with (86 ± 3) %, (44 ± 5) %, and (52 ± 6)% in the control. Cardioplegia with superoxide dismutase C significantly reduces lactate dehydrogenase release into myocardial effluent during reperfusion and increases ATP, AMP, and total creatine contents in reperfused hearts by (21 ± 3) %, (42 ± 4) %, and (34 ± 3)%, respectively. The effects of bovine superoxide dismutase are similar but not statistically proven. Treatment with human recombinant superoxide dismutase C, but not with bovine superoxide dismutase, reduces the EPR-assessed hydroxyl radical formation in coronary effluent at early reperfusion. The results suggest that myocardial protection against ischemia/reperfusion injury by superoxide dismutase C is related to scavenging of oxygen-derived free radicals [32].

As shown by EPR spectroscopy, the antipsychotic drug JL-13 (closapine analog) is less sensitive to oxidation phenomenon compared to closapine abdloxarine. JL-16 is scavenges oxygen radicals and efficiently inhibits lipid peroxidation [33].

The EPR technique was used to investigate the effect of neuroleptics (trifluoperazine, alimemazine tartrate, chloropromazine), tricyclic antidepressants (imipramine, desipramine), an anticholinergic agent (trihexylphenidyl hydrochloride), sedative agents and lithium carbonate on the rotational correlation frequency (V+) of the spin label 16NS (fatty acid nitroxide probe) in whole human erythrocytes. It turned out, that V+ was about 40% increased by the three psychotropic agents (0.2 mM). By contrast, lithium carbonate did not induce any change in V+ at the same concentrations. It is suggested that the increase in "membrane fluidity", observed with these variety of drugs, is a non specific effect. It is unrelated to their psychotropic action, and can be ascribed to their amphiphilic properties [34].

The effect of vinblastin on HeLa K and HeLa CA membrane fluidity has been studied by EPR spectroscopy. At concentrations below the cytotoxic level, vinblastin enhances HeLa K membrane fluidity and does not affect much that of HeLa CA cells. [35].

Using radiolysis and EPR spectroscopy the oxidative denitration of an antitumoral drug hydroxyguanidine (HG) has been investigated. Azide radical ($N_3^\bullet$) reacts with HG to form a carbon-centered guanidine radical, which rapidly decomposes to NO and urea. Further HG reacts with molecular oxygen to form peroxyl radical ($HGOO^\bullet$) [36].

The interaction between 5,6-benzyliden-L-sodium ascorbate (BSA) and dopamine was studied taking into account three different parameters: radical intensity, prooxidant and cytotoxic activities. In the EPR spectra under alkaline conditions BSA and dopamine give rise to a doublet and a quartet, respectively. When taken separately, these two compounds stimulate methionine oxidation in the culture medium. In combination, however, their joint stimulating effect is much lower. The two compounds dose dependently decrease the viability of human promyelocytic leukemic HL-60 cells, whereas combined cytotoxic effect is lowered [37].

During the last time drug sterilization by exposure to γ-irradiation has been widely accepted. In the solution of this problem much attention is given to EPR spectroscopy, distinguished among other analytical approaches as a highly sensitive, non-invasive, simple in use and precision method. In this way the sterilization data on dopamine and norepinephrine [38], cefotaxime [39], salbutamol [40], cefoperazone [41] and vitamins (riboflavin, biotin, thiamine mononitrate, thiamine hydrochloride, folic acid and pyridoxine) [42] have been examined.

4.3 FORENSIC MEDICAL EXAMINATION

EPR spectroscopy has been used in forensic medical examination. The constant of spin probe recovery by the myocardium determined by EPR is used to identify the time of death. These studies were performed 1, 3, 6, 9, 12, 15, 18, and 21 days postmortem in subjects of both sexes who died from craniocerebral injury. The rate constant of spin probe recovery by the myocardium did not change during the first six days postmortem. On the seventh day certain changes of the myocardial EPR spectra were observed, which may be used in the diagnosis of the time of death up to day 21 postmortem [43].

The rate of the spin probe reduction by tongue homogenates was measured on postmortem days (15, 30, 45, 60, 75, 105, 135, 150, 165, 180, 195, 210, 225, and 240) in order to determine much earlier time of death. The rate constant of this reaction was used as the diagnostic sign of the time of death within the period from 4 to 8 months postmortem [44].

4.4 ARCHAEOLOGY

EPR find use in determination of the archaeological age of human bone remnants [45]. These remnants are able to trap electronic charges as a results of bombardment by radioactive radiation from the surrounding sediment.

Free radicals produced in this case can be detected by EPR. The intensity of the EPR signal is a measure of the accumulated radiation dose and, thus, of the age. Tooth remnants are frequently found in archaeological sites and tooth enamel is well suited for EPR dating, with a precision of about 10-20%. The EPR data for sites in Israel and Africa show the existence of anatomically modern humans more than 100 ka ago [46]. Even earlier materials were found in China [47].

Tooth enamel EPR examination has found much use for dating the biological and cultural evolution of modern man [48-57].

The EPR spectra of fossil tooth enamel found in human and animal remnants show a singlet radiation sensitive signal at g = 2.0018, absent in the spectrum of anatomically modern teeth [58, 59]. That is why this signal is used for dating the human remnants [59-62].

REFERENCES

1. E. N. Frankel, J. Kanner, J. B. German, E. Parks and J. E. Kinsella, Inhibition of oxidation of lowdensity lipoprotein by phenolic substances in red wine, Lancet. **341**:454 (1993).
2. G. J. Troup, D. R. Hutton, D. G. Hewitt and C. R. Hunter, Free radicals in red wine, but not in white?, Free Radic. Res. **20**:63 (1994).
3. J. Gonis, D. G. Hewitt, G. Troup, D. R. Hutton and C. R. Hunter , The chemical origin of free radicals in coffee and other beverages, Free Radic. Res. **23**:393 (1995).
4. Y. Y. Bilto and S. S. Abdalla, Effects of selected flavonoids on deformability, osmotic fragility and aggregation of human erythrocytes, Clin. Hemorheol. Microcirc. **18**:165 (1998).
5. T. Matsuzaki and Y. Hara, Antioxidative activity of tea leaf catechins, J. Agr. Chem. Soc. Jap. **59**:129 (1985).
6. S. Uchida, R. Edamatsu, M. Hiramatsu, A. Mori, G. Nonaka, I. Nishioka, M. Niwa and M. Ozaki, Condensend tannins scavenge active oxygen free radicals, Med. Sci. Res. Biochem. **15**:831 (1987).
7. Y. Noda, K. Anzai, A. Mori, M. Kohno, M. Shinmei and L. Packer, Hydroxy and superoxide anion radical scaverning activities of natural source antioxidants using the computerized JES-FR30 ESR spectrometer system, Biochem. Mol. Biol. Int. **42**:35 (1997).
8. T. Yoneda, M. Hiramatsu, M. Sakamoto, K. Togasaki, M. Komatsu and K. Yamaguchi, Antioxidant effects of "beta catechin", Biochem. Mol. Biol. Int. **35**:995 (1995).
9. Q. Guo, B. Zhao, M. Li, S. Shen and W. Xin, Studies on protective mechanisms of four components of green tea polyphenols against lipid peroxidation in synaptosomes, Biochim. Biophys. Acta. **1304**:210 (1996).
10. M. V. Kumari, T. Yoneda and M. Hiramatsu, Scavenging activity of "beta catechin" on reactive oxygen species generated by photosensitization of riboflavin, Biochem. Mol. Biol. Int. **38**:1163 (1996).

11. Y. Ueda, M. Komatsu and M. Hiramatsu, Free radical scavenging activity of the Japanese herbal medicine toki-shakuyaku-san (TJ-23) and its effect on superoxide dismutase activity, lipid peroxides, glutamate, and monoamine metabolites in aged rat brain, Neurochem. Res. **21**:909 (1996).
12. L. Pronai and S. Arimori, BG-104 enhances the decreased plasma superoxide scavenging activity in patients with Behcet's disease, Sjogren's syndrome or hematological malignancy, Biotherapy. **3**:365 (1991).
13. L. Douifi, J. Raffi, P. Stocker and F. Dole, A point about electron paramagnetic resonance detection of irradiated foodstuffs, Spectrochim. Acta **A**, Mol. Biomol. Spectrosc. **54A**:2403 (1998).
14. C. Watala and K. Gwozdzinski, Melittin-induced alterations in dynamic properties of human red blood cell membranes, Chem. Biol. Interact. **82**:135 (1992).
15. H. S. Ju, X. J. Li, B. L. Zhao, Z. W. Han and W. J. Xin, Scavenging effect of berbamine on active oxygen radicals in phorbol ester-stimulated human polymorphonuclear leukocytes, Biochem. Pharmacol. **39**:1673 (1990).
16. V. Misik and D. Gergel, Kampo medicines possess OH radical scavenging activity, Pharmazie. **46**:545 (1991).
17. T. Hino, S. Kawanishi, H. Yasui, S. Oka and H. Sakurai, HTHQ (1-O-hexyl-2,3,5-trimethylhydroquinone), an anti-lipid-peroxidative compound: its chemical and biochemical characterizations, Biochim. Biophys. Acta **1425**:47 (1998).
18. H. Shi, B. Zhao and W. Xin, Scavenging effects of baicalin on free radicals and its protection on erythrocyte membrane from free radical injury, Biochem. Mol. Biol. Int. **35**:981 (1995).
19. E. Cossins, R. Lee and L. Packer, ESR studies of vitamin C regeneration, order of reactivity of natural source phytochemical preparations, Biochem. Mol. Biol. Int. **45**:583 (1998).
20. A. Arduini, N. Gorbunov, E. Arrigoni-Martelli, S. Dottori, F. Molajoni, F. Russo and G. Federici, Effects of L-carnitine and its acetate and propionate esters on the molecular dynamics of human erythrocyte membrane, Biochim. Biophys. Acta. **1146**:229 (1993).
21. M. Ge and J. H. Freed, Electron-spin resonance study of aggregation of gramicidin in dipalmitoylphosphatidylcholine bilayers and hydrophobic mismatch, Biophys. J. **76**(Pt 1):264 (1999).
22. B. R. Patyal, R. H. Crepeau and J. H. Freed, Lipid-gramicidin interactions using two-dimensional Fourier-transform electron spin resonance, Biophys. J. **73**:2201 (1997).
23. I. Kruk, K. Lichszteld, T. Michalska, K. Nizinkiewicz and J. Wronska, The extra-weak chemiluminescence generated during oxidation of some tetracycline antibiotics. 1. Autoxidation, J. Photochem. Photobiol. **14B**:329 (1992).
24. I. Witte, G. Oetken, C. Buschfort and A. Hartmann, A comparison of the DNA-damaging, the cytotoxic and genotoxic properties of tetracycline in human fibroblasts in the presence and absence of light, Mutat. Res. **315**:33 (1994).
25. H. S. Soedjak, R. E. Cano, L. Tran, B. L. Bales and J. Hajdu, Preparation and ESR spectroscopic characterization of the zinc(II) and cadmium(II) complexes of streptonigrin semiquinone, Biochim. Biophys. Acta. **1335**:305 (1997).
26. T. G. Gantchev and D. J. Hunting, Enhancement of etoposide (VP-16) cytotoxicity by enzymatic and photodynamically induced oxidative stress, Anticancer Drugs. **8**:164 (1997).

27. R. J. Kulmacz, Y. Ren, A. L. Tsai and G. Palmer, PGH synthase: interaction with hydroperoxides and indomethacin, Adv. Prostaglandin Thromboxane Leukot. Res. **21A**:137 (1991).
28. F. T. Greenaway, J. J. Hahn, N. Xi and J. R. Sorenson, Interaction of Cu(II) 3,5-diisopropylsalicylate with human serum albumin-an evaluation of spectroscopic data, Biometals. **11**:21 (1998).
29. F. Blasco, L. Perello, J. Latorre, J. Borras and S. Garcia-Granda, Cobalt(II), nickel(II), and copper(II) complexes of sulfanilamide derivatives: synthesis, spectroscopic studies, and antibacterial activity. Crystal structure of [Co(sulfacetamide)2(NCS)2], J. Inorg. Biochem. **61**:143 (1996).
30. J. Gallov, F. Andriamainty, D. Uhrnikova and P. Balgave, Interaction of local anesthetic heptacaine homologs with phosphatidylcholine bilayers: spin label ESR study, Biochim. Biophys. Acta. **1325**:189 (1997).
31. T. Uematsu, S. Nagashima, K. Umemura, M. Kanamaru and M. Nakashima, Pharmacokinetics and safety of intravenous recombinant human superoxide dismutase (NK341) in healthy subjects, Int. J. Clin. Pharmacol. Ther. **32**:638 (1994).
32. O. I. Pisarenko, I. M. Studneva, V. L. Lakomkin, A. A. Timoshin and V. I. Kapelko, Human recombinant extracellular-superoxide dismutase type C improves cardioplegic protection against ischemia/reperfusion injury in isolated rat heart, J. Cardiovasc. Pharmacol. **24**:655 (1994).
33. J. F Liugeois, A. Mouithys Mickalad, J. Bruhwyler, J. Delarge, C. Petit, J. M. Kauffmann and M. Lamy, JL-13, a potential successor to clozapine, is less sensitive to oxidative phenomena, Biochem. Biophys. Res. Commun. **238**:252 (1997).
34. M. Lejoyeux, D. Daveloose, J. C. Maziere, J. Ades and J. Viret, A spin label study of the membrane effect of various psychoactive drugs in human erythrocytes, Life Sci. **52**:L7 (1993).
35. M. Sentjurc, M. Zorec, M. Cemazar, M. Auersperg and G. Sersa, Effect of vinblastine on cell membrane fluidity in vinblastine-sensitive and -resistant HeLa cells, Cancer Lett. **130**:183 (1998).
36. S. A. Everett, K. B. Patel, M. F. Dennis, K. A. Smith, M. R Stratford and P. Wardman, Oxidative denitrification of the antitumour drug hydroxyguanidine, Free Radic. Biol. Med. **24**:1 (1998).
37. K. Satoh, Y. Ida, M. Kochi and H. Sakagami, Interaction between sodium 5,6-benzylidene-L-ascorbate and dopamine, Anticancer Res. **18A**:3565 (1998).
38. J. P. Basly, I. Longy and M. Bernard, Radiation effects on dopamine and norepinephrine, Pharm. Res. **14**:1192 (1997).
39. J. P. Basly, J. L. Duroux and M. Bernard, The effect of gamma radiation on the degradation of salbutamol, J. Pharm. Biomed. Anal. **15**:1137 (1997).
40 J. P. Basly, I. Basly and M. Bernard, Radiation-induced effects on cefotaxime: ESR study, Free Radic. Res. **29**:67 (1998).
41. J. P. Basly, I. Basly and M. Bernard, ESR spectroscopy applied to the study of pharmaceuticals radiosterilization: cefoperazone, J. Pharm. Biomed. Anal. **17**:871 (1998).
42. J. P. Basly, I. Basly and M. Bernard, Radiosterilization dosimetry of vitamins: an ESR study, Int. J. Radiat. Biol. **74**:521 (1998).
43. V. V. Zharov, The dynamics of the EPR spectra of the myocardium at different times following the onset of death, Sud. Med. Ekspert. **39**:3 (1996) (Russ.).

44. G. A. Pashinian, V. V. Zharov, I. I. Reznikov, M. A. Korsunskaya, Establishing the time of death by the constant of the reaction rate of spin-label reduction, Sud. Med. Ekspert. **39**:4 (1996) (Russ.).
45. A. D. Oduwole and K. D. Sales, ESR signals in bones: Interference from Fe3+ ions and a new method of dating, Nucl. Tracks. Radiat. Meas. **18**:213 (1991).
46. H. P. Schwarcz and R. Grun, Electron spin resonance (ESR) dating of the origin of modern man, Philos. Trans. R. Soc. Lond. B. Biol. Sci. **337**:145 (1992).
47. H. Pei-Hua, J. Si-Zhao, P. Zi-cheng, et al., ESR dating of tooth enamel: comparison with U-series, FT and TL dating ROR Peking man site, China, Third International Symposium on ESR Dosimetry and Applications, October 14-18, Japan (1991), p. 61.
48. R. Grun, P. B. Beaumont and C. B. Stringer, ESR dating evidence for early modern humans at Border Cave in South Africa, Nature. **344**(N6266):537 (1990).
49. F. H. Smith, Models and realities in modern human origins: the African fossil evidence, Philos. Trans. R. Soc. Lond. B. Biol. Sci. **337**:243 (1992).
50. O. Bar-Yosef, The role of western Asia in modern human origins, Philos. Trans. R. Soc. Lond. B. Biol. Sci. **337**:193 (1992).
51. J. J. Hublin, Recent human evolution in northwestern Africa, Philos. Trans. R. Soc. Lond. B. Biol. Sci. **337**:185 (1992).
52. J. J. Bahain, M. N. Sarcia, C. Falgueres and Y. Yokoyama, Attempt at ESR dating of tooth enamel of French middle Pleistocene sites, Appl. Radiat. Isot. **44**:267 (1993).
53. C. Tiemei, Y. Quan and W. En, Antiquity of Homo sapiens in China, Nature. **368**(N6466):55 (1994).
54. W. Huang, R. Ciochon, Y. Gu, R. Larick, F. Qiren, H. Schwarcz, C. Yonge, J. de Vos and W. Rink, Early Homo and associated artefacts from Asia [see comments], Nature. **378**(N6554):275 (1995).
55. C. C. Swisher, 3rd, W. J. Rink, S. C. Anton, H. P. Schwarcz, G. H. Curtis, A. Suprijo and Widiasmoro, Latest Homo erectus of Java: potential contemporaneity with Homo sapiens in southeast Asia [see comments], Science. **274**(N5294):1870 (1996).
56. R. Ciochon, V. T. Long, R. Larick, L. Gonzalez, R. Grun, J. de Vos, C. Yonge, L. Taylor, H. Yoshida and M. Reagan, Dated co-occurrence of Homo erectus and Gigantopithecus from Tham Khuyen Cave, Vietnam, Proc. Natl. Acad. Sci. USA. **93**:3016 (1996).
57. R. Grun, P. H. Huang, W. Huang, F. McDermott, A. Thorne, C. B. Stringer and G. Yan, ESR and U-series analyses of teeth from the palaeoanthropological site of Hexian, Anhui Province, China, J. Hum. Evol. **34**:555 (1998).
58. B. A. Blackwell and H. P. Schwarcz, ESR isochron dating for teeth: a brief demonstration in solving the external dose calculation problem, Appl. Radiat. Isot. **44**:243 (1993).
59. B. A. Blackwell, H. P. Schwarcz, K. Schick and N. Toth, ESR dating tooth enamel from the Paleolithic site at Longola, Zambia, Appl. Radiat. Isot. **44**:253 (1993).
60. E. J. Rhodes and R. Grun, ESR behavior of the paramagnetic centre at g = 2.0018 in tooth enamel, Third International Symposium on ESR Dosimetry and Applications, October 14-18, Japan (1991), p. 62.
61. E. J. Rhodes and R. Grun, ESR behavior of the paramagnetic centre at g = 2.0018 in tooth enamel, Ancient. **9**:14 (1991).
62. B. A. Blackwell, H. P. Schwarcz, K. Schick and N. Toth, ESR dating of tooth enamel from the paleolithic site, Longola spiking mound, Zambia, Third International Symposium on ESR Dosimetry and Applications, October 14-18, Japan (1991), p. 63.

CONCLUSION

The influence of hyperthermia on human cells was investigated by means of EPR spectroscopy using nitroxyl radicals. The heat effect enhances the intracellular microviscosity thus damaging the erythrocytes. A noticeable increase in the Over 50 years have gone since the discovery of electron paramagnetic resonance by E.K. Zavoisky. During this time the EPR spectroscopy has provided a widely accepted method in both clinical practice and laboratory investigations. The above discussed literature data and the present authors' results indicate much promise for EPR spectroscopy in biochemistry, biology and medicine. The use of this method has not only provided valuable information concerning the presence and origin of free radicals in the organism of human and animals, but allowed deep understanding of the role of transition metal microelements in enzymatic systems (cytochrome P_{450}, cytochrome-c oxidase, etc.), changes of their valence in the process of enzymatic catalysis (superoxidedismutase. ceruloplasmin, *etc*), the character of their bonds with biological substrates, and the structure of close surrounding of metabolically active paramagnetic ion (nitrosyl complexes of hem- and non-hem iron).

In spite of the fact that EPR spectroscopy allows a clearer understanding of the nature and role of paramagnetic centers and, consequently, of transition elements in biological objects, the potential of this method increases greatly when used in enzymology in combination with spectrophotometry, multinuclear NMR, chemiluminescence and other physico-chemical methods.

The use of spin labels and probes extended considerably the capabilities of EPR spectroscopy in studying membranes and the structure of macro- and micromolecules. In this way, in particular, pH-sensitive spin-labeled proteins were examined. Spin pH-probes can be of help for measuring local pH values within a range of 1.8 ± 6.2. In clinical practice EPR spectroscopy is of special value in events of oral administration of pH-sensitive nitroxides. This enables intragastric acidity monitoring. The results thus obtained suggest that EPR spectroscopy provides a valuable non-invasive tool to determine *in vivo* pH of biological fluids.

EPR spectroscopy should occupy an important place in clinical practice. It will enable refinement of some aspects of pathogenesis of many diseases. Thus, for example, according to EPR data, treatment of heart diseases with prolonged nitrates often gives rise to side effects. As a result erythrocyte MetHb concentration increases which adds to hypoxia of tissues. Stenocardia attack results in an increase in the level of blood plasma free radicals due to ejection of catecholamines, which provides a differential criterion in cardialgia. An important mechanism of ischemic tissue injury involves generation of endothelial oxidant, subsequent hemotaxis and leukocyte activation.

EPR spectroscopy is of help in determining the effect of oxidant molecules and free oxygen radicals such as $O_2^{-\bullet}$, $HO^{\bullet}$ and H_2O_2 on integration rate as well as the effect of both selected adhesion of molecules on the surface of human polymorphonuclear leukocytes and changes in adhesion on endothelium. From EPR data in patients with progressive stenocardia the erythrocyte MetHb level and blood plasma free radical content increase with the severity of state, whereas in patients with hypertensive disease the blood lipid α-tfq concentration decreases.

Using EPR spectroscopy a Hem-NO signal characterizing the relationship between the organism and microflora in the gastric contents was revealed. Besides, this method provides a valuable information concerning the Hem-NO role in functioning the gastric-intestinal tract.

EPR spectroscopy using spin-trapping allowed elucidation of the role of hydroxy radicals in coal pneumoconiosis pathogenesis. The catalytic effect of coal dust on the generation of radicals from hydrogen peroxide has been studied. By this method it was shown that the activity of kidney copper-zinc superoxidedismutase in patients with Berge's disease and glomerulonephritis decreased with deteriorating the kidneys function. As a result the level of scrambling the superoxide anion radicals decreased and this made the kidney tissue more sensitive to oxidation stress.

EPR spectroscopy has found the proper place in clinical and experimental neurology. It was established, for example, that free radicals played an essential role in nervous root inflammation, whereas the cerebrospinal fluid superoxidedismutase activity is reduced in patients with unilateral radiculopathy caused by discal hernia. FR formation in cerebral lobe homogenates in Alzheimer patients has been studied. It turned out that free radicals of NO type participated actively in Parkinson's disease pathogenesis.

In ophthalmology the EPR method found much use for the examination of alterations in human lenses. Investigations in this field are mainly related with the problem of cataract development and the ways of its treatment. Special attention is given to free radicals responsible for the formation of cataract. Hypoxanthine-xanthine oxidase producing anion-radicals $O_2^{-\bullet}$ and the Fenton reaction generating $HO^•$ radicals and their influence on cataract formation were studied using EPR. It is also used in stomatology for elucidating the mechanism of development of diseases of teeth and mouth cavity. The EPR method provides a tool for operative diagnosis of iron-deficient states in human. This method was of help in establishing this pathology in many diseases of internal organs. According to EPR data iron is localized not only in transferrin, but in albumin as well. This changes the existing concepts concerning Fe^{3+} ions transportation and localization in blood plasma.

EPR signal intensity of blood free radicals was observed in studying the effect of hyperbaric oxygenation of human organism.

Narcotic-induced biochemical changes in human were studied by EPR. The results show a decrease in plasma superoxidedismutase-like activity under surgery independently of the type of anesthesia, the anesthetic agent used and the degree of surgical invasion.

The spin-labeled method used for examination of the effect of heavy metals on human erythrocyte membranes showed the copper and mercury ions to increase the membrane lipid bilayer rigidity.

EPR spectroscopy turned out to be a valuable method for assessment of the severity of clinical course of many diseases. Thus, it was possible to establish a decrease in the level of Fe^{3+} ions in blood plasma and of Mn^{2+} ions in saliva as well as a decrease in the fecal concentration of Fe^{3+} ions in patients with iron-deficient anemia as the disease was developing.

The EPR method was used in the pharmakinetic and pharmadynamic studies of medicines of both natural (bee venom - melittin, berbamine, baikalin, L.carnitin, etc.) and synthetic origin (captopryl - an angyotension-transforming enzyme inhibitor, nifidipid, anti tumoral drugs, etc.). The role of β-catechol, a "scavenger" of oxygen radicals, consisting of green tea extract, sun flower seeds, vitamins A, C and E, was studied by EPR spectroscopy. A solution of this substance produces a strong scavenging action on DPPG radicals, superoxide radical-anions generated by hypoxanthine-xanthine oxidase and system, and hydroxyl radicals generated in the Fenton reaction. A compound of Chinese herbs, BG-104 also inactivates superoxide ion-radical in cell free systems.

Using EPR spectroscopy cytotoxic oxygen radicals $HO^•$, $O_2^{-•}$ and etc. formed during auto-oxidation of antibiotic tetracycline were studied.

EPR spectroscopy enabled investigation of the pharmakinetics of recombinant human superoxidedismutase and showed its safety and tolerance upon intravenous administration. Using this method the mechanism of effect of omeprosol, a medicine reducing the gastric acid-forming function, a "proton" pump blocker, on the $K/[H^+]$ value of ascorbic acid was studied. With the use of EPR method it was found that rebamipide, a new antiulcerous drug, exhibits an antioxidant effect. It acts as an $HO^•$ radical scanvenger and inhibits the formation of superoxide anion radicals in neurophiles. Rebamipide also inhibits the generation of active oxygen with neuriphiles stimulated with extracts from *Helicobacter pylori*. This indicates rebamipide to exert anti-inflammatory protection of gastric mucosa. EPR spectroscopy is also of help in revealing side effects of treatment with some drugs such as organic nitrates, vitamins C and E, symptomimetics, etc.

The use of this method also showed that in chronic diseases of upper respiratory tracts the nasal secretion Fe^{3+} content is noticeably reduced. In cases of sub- and atrophic alteration in the nasal cavity the Cu^{2+} content in the nasal mucous membrane is increased. Vasomotor rhinitis is accompanied with a reduced Fe^{3+} level in the nasal secretion.

The increased content of desquamative cells in the nasal secretion in events of pathological processes in respiratory tracts is accompanied by accumulation of intracellular enzymes and agrees well with an increased level of free radicals.

The EPR data extend the present knowledge of absorptive and excretory functions of the nasal mucous membrane to the notion of rhino-hematic barrier. The changes in Fe^{3+} and Cu^{2+} concentration and, consequently, in Cu^{2+}/Fe^{3+} ratio which is a factor of nasal secretion antioxidant properties, are in some way related to the rhino-hematic barrier functional state and to the dynamics of metallo-proteid supply from the blood flow.

Examination of paramagnetic centers in the blood, duodenal contents and bile of patients with cholecystitis suggests that the EPR method will find much use in diagnosis of chronic cholecystitis and will be of help in elucidating some aspects of its pathogenesis.

Investigation of methemoglobin and Hem-NO in biological fluids of patients with peritonitis reveals wide possibilities of this method in clearing up pathogenic aspects in the course of the disease at a molecular level.

This forms the basis for pathophysiological analysis of the reaction of nitrates and nitrites with hemoglobin and for assessment of the role of the resultant products in pathogenesis of toxemia and polyorganic deficiency in events of peritonitis.

These results suggest the study of paramagnetic centers using EPR spectroscopy to be of value as supplementary diagnostic tests as well as criteria in assessment of the severity of clinical course of many diseases and the efficiency of treatment.

By the present time, EPR investigation has been used with relatively readily accessible substrates such as blood, saliva, nasal secretion, gastric and duodenal contents, feces, some tissues taken in the process of surgical invasion (teeth, bones, skin, uterus, placenta, synovial fluid).

This method is of help for the determination of occasional radioactive irradiation basing on the signal of free radicals in bones, dentine and, especially, in tooth enamel. Therefore, the determination of doses accumulated in human as a result of γ- and X-ray irradiation, basing on teeth enamel EPR spectra is widely accepted now.

Along with other methods, EPR spectroscopy is employed in studying AIDS and in forensic medical examination (to identify the time of death).

The EPR method has found much use in determination of archeological age of human osseous remains. From EPR data it was established the existence of anatomically contemporary man over 100 ka years ago in Israel and Africa. Remains of even earlier age were revealed in China. The method has also found widespread application in dating the biological and cultural evolution of man.

Summing up the above said the EPR method is suggested to extend our knowledge of the role of transition microelements in vital activities of man, to provide an understanding of thin mechanisms of pathogenic processes of various diseases and to offers a great promise for new therapeutic and prophylactic measures. This provides evidence on the limitless possibilities of EPR spectroscopy for use in biochemistry and medicine.

We hope that the goal of this monograph will be accomplished by convincing biochemists and clinicians of the necessity of using spectroscopy in clinical practice and research.

The authors would consider their task fulfilled if the present book enhances the interest in clinical aspects of the use of EPR spectroscopy and initiates new research and developments supplementing and illustrating the role of paramagnetic centers in the organism.

ABBREVIATIONS

AIDS	acquired immunodeficiency syndrome
ATP	adenosinetriphosphate
CO	carbon monoxide
CP	ceruloplasmin
CRF	chronic renal failure
CSF	cerebrospinal fluid
DBNBS	3,5-dibromo-4-nitrosobenzenesulfonate
DEDTC	diethyldithiocarbamate
DETAPAC	diethylenetriaminepentaacetic acid
DMPO	2,2-dimethyl-1-pyrroline *N*-oxide
DMSO	dimethylsulfoxides
DOA	deforming osteoarthrosis
DPPH	1,1-diphenyl-2-picrylhydrazyl
EDRF	endothelial-derived relaxing factor
EPR	electron paramagnetic resonance
FALS	familial amyotrophic lateral sclerosis
FR	free radicals
GR	glutathione reductase
GRS	glutathione reductase system
GSSG	oxidized glutathione
GSH	reduced form of glutathione
Hb	hemoglobin
HbM	mutant hemoglobin
Hb-NO	hemoglobin-NO
HbO_2	oxyhemoglobin
HDL	high-density lipoproteides
Hem-NO	complex heme with NO
HF	high frequency
HFS	hyperfine structure
HMP-S	hexosomonophosphate shunt
$HO^•$	hydroxyl radicals
H_2O_2	peroxide hydrogen

HSA	human serum albumin
IDA	iron-deficient iron-deficient anemia
IHD	ischemic heart disease
I-1-β	interluekin-1-β
LDL	low-density lipoproteides
LOOH	lipid hydroperoxides
LPO	lipid peroxidation
LPS	lipopolysaccharide
Mb-NO	myoglobin-NO
MDA	malonic dialdehyde
MetHb	methemoglobin
MI	myocardial infarction
NMMA	N-monomethyl-L-arginine
MNP	2-methyl-2-nitrosopropane
NADP.H	reduced nicatinamidedinucleotidephosphate
NO	nitric oxide
NOS	NO-synthtase
$O_2^{-\bullet}$	superoxide anion radical
PBN	α-phenyl-*tert*-butylnitrone
PEG	polyethyleneglycol
PMA	4 β-phorbol 12 β-myristate 13 α-acetate
PMC	paramagnetic centers
PMT	phenylmethylthiolate
Q_{10}	ubiquinone-10
r	correlation coefficient
RA	rheumatoid arthritis
SDS	sodium dodecyl sulfate
SF	synovial fluid
SFS	superfine structure
SHF	super high frequence
SOD	superoxide dismutase
TEMPO	2,2,6,6-tetramethyl-4-oxypiperidin-1-oxyl
TF	transferrin
α-T	α-tocopherol
α-Tq	α-tocopherylquinone
α-Tpr	α-tocopheroxyl radical
UE	unpaired electron
URT	upper respiratory tract
VLDL	very low density lipoproteins
ΔH	line half-width

Index

MIX
Papier aus verantwortungsvollen Quellen
Paper from responsible sources
FSC® C105338
FSC
www.fsc.org

If you have any concerns about our products,
you can contact us on
ProductSafety@springernature.com

In case Publisher is established outside the EU,
the EU authorized representative is:
Springer Nature Customer Service Center GmbH
Europaplatz 3, 69115 Heidelberg, Germany

Printed by Libri Plureos GmbH
in Hamburg, Germany